Birkhäuser

Trends in Mathematics

Trends in Mathematics is a series devoted to the publication of volumes arising from conferences and lecture series focusing on a particular topic from any area of mathematics. Its aim is to make current developments available to the community as rapidly as possible without compromise to quality and to archive these for reference.

Proposals for volumes can be submitted using the Online Book Project Submission Form at our website www.birkhauser-science.com.

Material submitted for publication must be screened and prepared as follows:

All contributions should undergo a reviewing process similar to that carried out by journals and be checked for correct use of language which, as a rule, is English. Articles without proofs, or which do not contain any significantly new results, should be rejected. High quality survey papers, however, are welcome.

We expect the organizers to deliver manuscripts in a form that is essentially ready for direct reproduction. Any version of TEX is acceptable, but the entire collection of files must be in one particular dialect of TEX and unified according to simple instructions available from Birkhäuser.

Furthermore, in order to guarantee the timely appearance of the proceedings it is essential that the final version of the entire material be submitted no later than one year after the conference.

Kaïs Ammari • Anna Doubova • Stéphane Gerbi •
Manuel González-Burgos
Editors

Advances in Partial Differential Equations and Control

The 2023 Conference in Seville, Spain

Birkhäuser

Editors
Kaïs Ammari
Department of Mathematics
University of Monastir
Monastir, Tunisia

Anna Doubova
Faculty of Mathematics
University of Seville
Seville, Spain

Stéphane Gerbi
Department of Mathematics
Université Savoie Mont Blanc
Chambery, France

Manuel González-Burgos
Faculty of Mathematics
University of Seville
Seville, Spain

ISSN 2297-0215 ISSN 2297-024X (electronic)
Trends in Mathematics
ISBN 978-3-031-62264-9 ISBN 978-3-031-62265-6 (eBook)
https://doi.org/10.1007/978-3-031-62265-6

Mathematics Subject Classification: 37L15, 35B40, 35B30, 65M32, 34K20, 65N21, 93B05, 93C27, 35Q84, 93D15, 35R30, 35P25, 35L04

This book is published under the imprint Birkhäuser, www.birkhauser-science.com by the registered company Springer Nature Switzerland AG
The registered company address is: Gewerbestrasse 11, 6330 Cham, Switzerland

Preface

This volume constitutes the conference "Control & related fields". This conference was held at the University of Seville in the period 27–29 March 2023 (see http://departamento.us.es/edan/CRF23/ for details). It was organized by Kaïs Ammari (University of Monastir, Tunisia), Anna Doubova and Manuel Gonzàlez-Burgos (University of Seville, Spain and Stéphane Gerbi, Université Savoie Mont Blanc, France).

This conference was devoted to promote, expose and develop some new results in the mathematical control theory of partial differential equations (PDE) and also in some related fields such as Calderón problem inverse problem, reconstruction problem from internal or boundary data, stability of PDE in presence of time delay, control theory. Twenty invited speakers presented their recent works and some of them agreed to write review papers related to their contributions.

This conference brought together internationally leading researchers among the community of control theory, inverse problem and generally analysis of PDE and young researchers who came from all around the world. The organizers' intention was to provide a wide-angle snapshot of this exciting and fast moving area and facilitate the exchange of ideas on recent advances in its various aspects. The numerous formal, informal and sometimes lively discussions that resulted from this interaction were for us a sign that we achieved something in the direction of fulfilling this aim.

We believe that this volume therefore provides an accessible summary of a wide range of active research topics, along with some exciting new results, and we hope that it will prove a useful resource for both graduate students new to the area and to more established researchers.

Our second aim was to ensure that the diffusion of these recent results was not limited to established researchers in the area and present at the conference, but also available to newcomers and more junior members of the research community. This was reflected by the presence of many unfamiliar and/or young faces in the audience. The present proceedings should hopefully complete the fulfillment of our second aim.

This summer school would not have materialized without the help and support of the following institutions.

We are very grateful to the CNRS (Centre National de la Recherche Scientifique), the University of Seville and IMUS, CNRS-IRN-GE2MI for their financial supports without whom this conference would not be accessible without fees.

Finally, we would like to thank all the participants of the conference who have made this event a success, the contributors to these proceedings and the reviewers for their hard work.

Monastir, Tunisia — Kaïs Ammari
Seville, Spain — Anna Doubova
Chambéry, France — Stéphane Gerbi
Seville, Spain — Manuel González-Burgos
March, 2024

Contents

Part I
Control of Partial Differential Equations

Energy Decay Estimate for a Wave-Plate Interface Transmission Problem with Only Two Dynamical Boundary Controls

Zahraa Abdallah, Stéphane Gerbi, and Ali Wehbe

2010 Mathematics Subject Classification 35L04, 37L15

1 Introduction

In this chapter, we investigate a wave-plate system coupled through transmission conditions along a steady interface between the domains in which the wave and plate equations evolve, respectively, with dynamical boundary controls. More precisely, let $\Omega \subset \mathbb{R}^2$ be an open bounded set with Lipschitz boundary $\partial\Omega$ such that $\overline{\Omega} = \overline{\Omega}_1 \cup \overline{\Omega}_2$, where Ω_i is a bounded set with Lipschitz boundary $\partial\Omega_i$, $i = 1, 2$, such that $\Omega_1 \cap \Omega_2 = \emptyset$. We denote by I the interior of $\overline{\Omega}_1 \cap \overline{\Omega}_2$, which is called the interface between Ω_1 and Ω_2, and $\Gamma_i = \partial\Omega_i \backslash \overline{I}$ represents the exterior boundary of Ω_i with a positive measure, $i = 1, 2$. We consider the following interface transmission wave-plate model:

Z. Abdallah (✉)
Lebanese University, Faculty of Sciences Khawarizmi Laboratory of Mathematics and Applications-KALMA Hadath-Beirut, Hadath, Lebanon

Laboratoire de Mathématiques-LAMA UMR 5127-CNRS and Université Savoie Mont Blanc, Le Bourget-du-Lac Cedex, France

S. Gerbi
Laboratoire de Mathématiques-LAMA UMR 5127-CNRS and Université Savoie Mont Blanc, Le Bourget-du-Lac Cedex, France
e-mail: stephane.gerbi@univ-smb.fr

A. Wehbe
Lebanese University, Faculty of Sciences Khawarizmi Laboratory of Mathematics and Applications-KALMA Hadath-Beirut, Hadath, Lebanon
e-mail: ali.wehbi@ul.edu.lb

K. Ammari et al. (eds.), *Advances in Partial Differential Equations and Control*, Trends in Mathematics, https://doi.org/10.1007/978-3-031-62265-6_1

$$\begin{cases} u_{tt} - \Delta u = 0, & \text{in } \Omega_1 \times \mathbb{R}_+^*, \\ w_{tt} + \Delta^2 w = 0, & \text{in } \Omega_2 \times \mathbb{R}_+^*, \\ u = w, \ \ \mathcal{B}_1 w = 0, \ \ \mathcal{B}_2 w = \partial_{\nu_1} u, & \text{on } I \times \mathbb{R}_+^*, \\ \partial_{\nu_1} u + \eta = 0, & \text{on } \Gamma_1 \times \mathbb{R}_+^*, \\ \mathcal{B}_1 w = 0, & \text{on } \Gamma_2 \times \mathbb{R}_+^*, \\ \mathcal{B}_2 w - \zeta = 0, & \text{on } \Gamma_2 \times \mathbb{R}_+^*, \end{cases} \tag{1}$$

with the initial conditions:

$$u(x,0) = u_0(x), \ \ u_t(x,0) = u_1(x), \ \ x \in \Omega_1, \ \ \eta(x,0) = \eta_0(x), \ \ x \in \Gamma_1, \tag{2}$$

$$w(x,0) = w_0(x), \ \ w_t(x,0) = w_1(x), \ \ x \in \Omega_2, \ \ \zeta(x,0) = \zeta_0(x), \ \ x \in \Gamma_2, \tag{3}$$

where η and ζ denote the dynamical boundary controls, $\nu_i = (\nu_{i1}, \nu_{i2})$ is the unit outward normal vector along $\partial\Omega_i$, and $\tau_i = (-\nu_{i2}, \nu_{i1})$ is the unit tangent vector along $\partial\Omega_i$, $i = 1, 2$. Now, let us proceed with defining the gradient of a function f as follows:

$$\nabla f = \left(f_{x_1}, f_{x_2}\right),$$

where f_{x_1} and f_{x_2} represent the partial derivatives of f with respect to x_1 and x_2, respectively. Next, the Laplacian operator is defined as the divergence of the gradient of f, given by

$$\Delta f = \operatorname{div}(\nabla f) = f_{x_1x_1} + f_{x_2x_2},$$

where $f_{x_1x_1}$ and $f_{x_2x_2}$ are the second partial derivatives of f with respect to x_1 and x_2. Additionally, in this problem's context, we introduce the constant $\mu \in \left(0, \frac{1}{2}\right)$ as the Poisson coefficient. Moreover, we define the boundary operators $\mathcal{B}_1$ and $\mathcal{B}_2$ on $\partial\Omega_2$ as follows:

$$\mathcal{B}_1 f = \Delta f + (1-\mu)\mathcal{C}_1 f$$

and

$$\mathcal{B}_2 f = \partial_{\nu_2}\Delta f + (1-\mu)\partial_{\tau_2}\mathcal{C}_2 f,$$

where

$$\mathcal{C}_1 f = 2\nu_{21}\nu_{22} f_{x_1x_2} - \nu_{21}^2 f_{x_2x_2} - \nu_{22}^2 f_{x_1x_1} \quad \text{and}$$
$$\mathcal{C}_2 f = (\nu_{21}^2 - \nu_{22}^2) f_{x_1x_2} - \nu_{21}\nu_{22}\left(f_{x_1x_1} - f_{x_2x_2}\right).$$

The damping of the system is made via the indirect damping mechanism that involves the following first-order differential equations:

$$\begin{cases} \eta_t - u_t + \eta = 0, & \text{on } \Gamma_1 \times \mathbb{R}_+^*, \\ \zeta_t - w_t + \zeta = 0, & \text{on } \Gamma_2 \times \mathbb{R}_+^*. \end{cases} \tag{4}$$

Dynamical controls are part of the indirect damping mechanisms proposed by Russell in [28], which do not arise from the insertion of damping terms into the original equations describing the mechanical motion, but by coupling those equations to further equations describing other processes in the structure. A significant number of papers dealing with the mathematical theory of dynamical boundary control in elastic structures have appeared. The emphasis is usually placed upon elastic beams modeled by the Euler–Bernoulli beam or Timoshenko beam, including [21, 22], and [23] in our list of references. Additionally, our study also aligns with investigations pertaining to Kirchhoff plates and wave equations, which are relevant to this study (see, for instance, [26, 27]).

On the other hand, transmission systems are mathematical models that arise most naturally in the description of structures that are partly composed of interactive or interconnected materials. Transmission systems can be found in many practical applications, such as spacecraft [7], satellite antennas [20], road traffic [16], and many other interactive physical processes. Such problems have also received attention, and there have been fruitful results concerning the control design and stability analysis of the solutions to different types of problems (see, for instance, [2, 3, 18, 30, 31], and [13]).

In [1], we considered the stabilization problem of a transmission model involving a wave equation and a Kirchhoff plate equation, employing a set of three dynamical boundary controls denoted as η, ξ, and ζ. We proved the strong stability of the system without imposing any specific geometric conditions. However, we proved that the energy of the system does not decay uniformly to zero. Thus, under certain geometric conditions, we showed the polynomial stability of the system with an energy decay rate $1/t$. We remark that the preceding results remain constructed with some slight modifications when reducing the number of dynamical controls to 2, specifically ξ and ζ. This is due to the useful information provided by the transmission conditions at the interface, which plays an essential role in this case. Yet, an intriguing issue that has remained unexplored is the stabilization problem when one of the dynamical controls, either ξ or ζ, is entirely removed. So, the problem of the energy decay rate is still an open problem when $\xi = 0$ or $\zeta = 0$. This chapters's primary objective is to estimate the decay rate of the energy under the condition where the dynamical control ξ is omitted, thereby subjecting the system to the influence of two dynamical controls, namely η and ζ. We establish, under specific geometric assumptions, the strong stability of the system, confirming that the removal of control ξ does not affect the polynomial decay rate of the system's energy. In conclusion, while the dynamical boundary control ξ plays a pivotal role in establishing the system's strong stability in the absence of geometric conditions, it does not affect the energy decay rate of the system.

Let u, η, w, and ζ be smooth solutions of system (1)–(4). We define their associated energy by

$$E(t)=\frac{1}{2}\left\{\int_{\Omega_1}\left(|\nabla u|^2+|u_t|^2\right)dx+\int_{\Gamma_1}|\eta|^2d\Gamma\right\}$$
$$+\frac{1}{2}\left\{a(w,\overline{w})+\int_{\Omega_2}|w_t|^2dx+\int_{\Gamma_2}|\zeta|^2d\Gamma\right\},$$

where the sesquilinear form $a : H^2(\Omega_2)\times H^2(\Omega_2)\longmapsto\mathbb{C}$ is defined by

$$a(f,\overline{g})=\int_{\Omega_2}\left[f_{x_1x_1}\overline{g}_{x_1x_1}+f_{x_2x_2}\overline{g}_{x_2x_2}+\mu\left(f_{x_1x_1}\overline{g}_{x_2x_2}+f_{x_2x_2}\overline{g}_{x_1x_1}\right)\right.$$
$$\left.+2(1-\mu)f_{x_1x_2}\overline{g}_{x_1x_2}\right]dx. \tag{5}$$

For further analysis, the following Green formula (see [15]) is used in the study:

$$a(f,\overline{g})=\int_{\Omega_2}\Delta^2 f\overline{g}dx+\int_{\partial\Omega_2}\left(\mathcal{B}_1 f\partial_{\nu_2}\overline{g}-\mathcal{B}_2 f\overline{g}\right)d\Gamma,\ \ \forall f\in H^4(\Omega_2),$$
$$g\in H^2(\Omega_2). \tag{6}$$

Lemma 1 *Let $U=(u,u_t,\eta,w,w_t,\zeta)$ be a regular solution of system* (1)–(4). *Then, the energy $E(t)$ satisfies the following estimation:*

$$E'(t)=-\int_{\Gamma_1}|\eta|^2d\Gamma-\int_{\Gamma_2}|\zeta|^2d\Gamma\leq 0.$$

Proof Firstly, multiplying the first and second equations of (1) by $\overline{u}_t$ and $\overline{w}_t$, respectively, then using Green's formula in Ω_1 and Ω_2, and taking the real part, we get

$$\frac{1}{2}\frac{d}{dt}\left\{\int_{\Omega_1}\left(|\nabla u|^2+|u_t|^2\right)dx+a(w,\overline{w})+\int_{\Omega_2}|w_t|^2dx\right\}$$
$$+\Re\left\{\int_{\Gamma_1}\eta\overline{u}_td\Gamma+\int_{\Gamma_2}\zeta\overline{w}_td\Gamma\right\}=0, \tag{7}$$

where $\Re$ stands for the real part of a complex number. Secondly, by inserting equation (4) into the second term of (7), we get

$$\Re\left\{\int_{\Gamma_1}\eta\overline{u}_td\Gamma+\int_{\Gamma_2}\zeta\overline{w}_td\Gamma\right\}=\frac{1}{2}\frac{d}{dt}\left\{\int_{\Gamma_1}|\eta|^2d\Gamma+\int_{\Gamma_2}|\zeta|^2d\Gamma\right\}$$
$$+\int_{\Gamma_1}|\eta|^2d\Gamma+\int_{\Gamma_2}|\zeta|^2d\Gamma. \tag{8}$$

Combining equations (7) and (8), we obtain

$$E'(t) = -\int_{\Gamma_1} |\eta|^2 d\Gamma - \int_{\Gamma_2} |\zeta|^2 d\Gamma.$$

Thus, from Lemma 1, we deduce that the system (1)–(4) is dissipative in the sense that the energy $E(t)$ is nonincreasing with respect to time variable t.

The rest of this chapter is organized as follows: In Sect. 2, we formulate system (1)–(4) as an evolution equation and discuss the well-posedness property by using semigroup theory. Some regularity results needed in the following sections are presented in Sect. 3. In Sect. 4, a general criterion of Arendt–Batty theorem is used to prove the strong stability of the system under certain geometric conditions. Finally, Sect. 5 is devoted to studying the energy decay rate under certain geometric conditions given by Theorem 4. We prove that the energy of our system has a polynomial decay rate of type $1/t$.

2 Well-Posedness of the Problem

This section is devoted to study the well-posedness property of system (1)–(4) using the semigroup approach. We first introduce the following spaces:

$$H^1_*(\Omega_1) = \left\{ u \in H^1(\Omega_1) \mid \int_{\Omega_1} u dx = 0 \right\},$$

$$H^2_*(\Omega_2) = \left\{ w \in H^2(\Omega_2) \mid \int_{\Omega_2} w dx = \int_{\Omega_2} \nabla w dx = 0 \right\}$$

and the energy space $\mathcal{H}$ by

$$\mathcal{H} = \Big\{ (u, v, \eta, w, z, \zeta) \in H^1_*(\Omega_1) \times L^2(\Omega_1) \times L^2(\Gamma_1) \times H^2_*(\Omega_2) \\ \times L^2(\Omega_2) \times L^2(\Gamma_2) \mid u = w \text{ on } I \Big\},$$

which is endowed with the following usual inner product:

$$\left(U, \tilde{U}\right)_{\mathcal{H}} = \int_{\Omega_1} \left(\nabla u \cdot \nabla \overline{\tilde{u}} + v \overline{\tilde{v}}\right) dx + \int_{\Gamma_1} \eta \overline{\tilde{\eta}} d\Gamma + a(w, \overline{\tilde{w}}) + \int_{\Omega_2} z \overline{\tilde{z}} dx \\ + \int_{\Gamma_2} \zeta \overline{\tilde{\zeta}} d\Gamma, \tag{9}$$

where $U = (u, v, \eta, w, z, \zeta)$ and $\tilde{U} = (\tilde{u}, \tilde{v}, \tilde{\eta}, \tilde{w}, \tilde{z}, \tilde{\zeta}) \in \mathcal{H}$.

Remark 1 The condition

$$\int_{\Omega_2} \nabla w \, dx = 0$$

signifies that both

$$\int_{\Omega_2} w_{x_1} \, dx = 0 \quad \text{and} \quad \int_{\Omega_2} w_{x_2} \, dx = 0.$$

Remark 2 By considering the conditions of the function spaces $H_*^1(\Omega_1)$ and $H_*^2(\Omega_2)$, it follows that $(\cdot, \cdot)_{\mathcal{H}}$ is a scalar product in the space $\mathcal{H}$. Consequently, this observation confirms that $\mathcal{H}$ is a Hilbert space, a fundamental prerequisite for the application of Theorem 3.

We next define the linear unbounded operator $\mathring{A} : D(\mathring{A}) \subset \boldsymbol{\mathcal{H}} \longmapsto \boldsymbol{\mathcal{H}}$ by

$$D(\mathring{A}) = \left\{ \begin{array}{l} U = (u, v, \eta, w, z, \zeta) \in \boldsymbol{\mathcal{H}} \mid \Delta u \in L^2(\Omega_1), \;\; v \in H_*^1(\Omega_1), \;\; \Delta^2 w \in L^2(\Omega_2), \;\; z \in H_*^2(\Omega_2), \\ \partial_{\nu_1} u + \eta = 0 \text{ on } \Gamma_1, \;\; \mathcal{B}_1 w = 0 \text{ on } \Gamma_2, \;\; \mathcal{B}_2 w - \zeta = 0 \text{ on } \Gamma_2, \\ v = z, \;\; \mathcal{B}_1 w = 0 \text{ and } \mathcal{B}_2 w = \partial_{\nu_1} u \text{ on } I \end{array} \right\} \tag{10}$$

and

$$\begin{aligned} \mathring{A} U &= \left(v, \Delta u, \gamma_1(v) - \eta, z, -\Delta^2 w, \gamma_{2,1}(z) - \zeta\right), \\ \forall U &= (u, v, \eta, w, z, \zeta) \in D(\mathring{A}), \end{aligned} \tag{11}$$

where $\gamma_1 : H^1(\Omega_1) \longmapsto L^2(\Gamma_1)$ and $\gamma_2 : H^2(\Omega_2) \longmapsto L^2(\Gamma_2) \times L^2(\Gamma_2)$ are the trace operators such that $\gamma_1(\varphi) = \varphi|_{\Gamma_1}$ and $\gamma_2(\psi) = \left(\gamma_{2,1}(\psi), \gamma_{2,2}(\psi)\right) = \left(\psi|_{\Gamma_2}, \partial_{\nu_2} \psi|_{\Gamma_2}\right)$. Now, setting $U = (u, u_t, \eta, w, w_t, \zeta)$ as the state of system (1)–(4), we rewrite the problem into a first-order evolution equation

$$\begin{cases} U_t = \mathring{A} U, \\ U(0) = U_0, \end{cases} \tag{12}$$

where the initial value $U_0 = (u_0, u_1, \eta_0, w_0, w_1, \zeta_0) \in \boldsymbol{\mathcal{H}}$ is given by Eqs. (2) and (3).

Remark 3 From (5) and the fact that $2\Re\left(w_{x_1x_1}\overline{w}_{x_2x_2}\right) = |w_{x_1x_1} + w_{x_2x_2}|^2 - |w_{x_1x_1}|^2 - |w_{x_2x_2}|^2$, we remark that

$$
\begin{aligned}
a(w, \overline{w}) &= \int_{\Omega_2} \Big[|w_{x_1x_1}|^2 + |w_{x_2x_2}|^2 + \mu \left(w_{x_1x_1} \overline{w}_{x_2x_2} + w_{x_2x_2} \overline{w}_{x_1x_1} \right) \\
&\quad +2(1-\mu)|w_{x_1x_2}|^2 \Big] dx \\
&= \int_{\Omega_2} \Big[|w_{x_1x_1}|^2 + |w_{x_2x_2}|^2 + 2\mu \Re \left(w_{x_1x_1} \overline{w}_{x_2x_2} \right) + 2(1-\mu)|w_{x_1x_2}|^2 \Big] dx \\
&= \int_{\Omega_2} \Big[(1-\mu)|w_{x_1x_1}|^2 + (1-\mu)|w_{x_2x_2}|^2 + \mu |w_{x_1x_1} + w_{x_2x_2}|^2 \\
&\quad +2(1-\mu)|w_{x_1x_2}|^2 \Big] dx \geq 0.
\end{aligned}
$$

For further purposes, we set the Hilbert space H by

$$
\mathsf{H} = \left\{ (f, g) \in H_*^1(\Omega_1) \times H_*^2(\Omega_2) \mid f = g \quad \text{on} \quad I \right\},
$$

equipped with the norm

$$
\|(f, g)\|_{\mathsf{H}}^2 = \|\nabla f\|_{L^2(\Omega_1)}^2 + a(g, \overline{g}). \tag{13}
$$

Note that using the compactness of the injection from $H^1(\Omega_1)$ into $L^2(\Omega_1)$ and from $H^2(\Omega_2)$ into $H^1(\Omega_2)$, we can easily show that $\|\nabla f\|_{L^2(\Omega_1)}$ is equivalent to the usual norm of $H^1(\Omega_1)$ on $H_*^1(\Omega_1)$ and $\sqrt{a(g, \overline{g})}$ is equivalent to the usual norm of $H^2(\Omega_2)$ on $H_*^2(\Omega_2)$.

Proposition 1 *The unbounded linear operator $\mathring{A}$ is m-dissipative in the energy space $\mathcal{H}$.*

Proof First, let $U = (u, v, \eta, w, z, \zeta) \in D(\mathring{A})$. Using (9) and (11), we have

$$
\begin{aligned}
\Re \left(\mathring{A}U, U \right)_{\mathcal{H}} &= \Re \Big\{ \int_{\Omega_1} \nabla v \cdot \nabla \overline{u} dx + \int_{\Omega_1} \Delta u \overline{v} dx + \int_{\Gamma_1} (v - \eta) \overline{\eta} d\Gamma + a(z, \overline{w}) \\
&\quad - \int_{\Omega_2} \Delta^2 w \overline{z} dx + \int_{\Gamma_2} (z - \zeta) \overline{\zeta} d\Gamma \Big\}.
\end{aligned}
$$

We can proceed by integrating by parts and utilizing Green's formula (6). Given that $U \in D(\mathcal{A})$, we have

$$
\Re \left(\mathring{A}U, U \right)_{\mathcal{H}} = - \int_{\Gamma_1} |\eta|^2 d\Gamma - \int_{\Gamma_2} |\zeta|^2 d\Gamma \leq 0, \tag{14}
$$

which implies that $\mathring{A}$ is dissipative. Now, let us prove that $\mathring{A}$ is maximal. For this aim, let $F = (f^1, g^1, h^1, f^2, g^2, h^2) \in \mathcal{H}$, and we look for $U = (u, v, \eta, w, z, \zeta) \in D(\mathring{A})$ unique solution of

$$
U - \mathring{A}U = F. \tag{15}
$$

Equivalently, we have the following system:

$$u - v = f^1, \quad \text{in } \Omega_1, \tag{16}$$

$$v - \Delta u = g^1, \quad \text{in } \Omega_1, \tag{17}$$

$$\eta - v + \eta = h^1, \quad \text{on } \Gamma_1, \tag{18}$$

$$w - z = f^2, \quad \text{in } \Omega_2, \tag{19}$$

$$z + \Delta^2 w = g^2, \quad \text{in } \Omega_2, \tag{20}$$

$$\zeta - z + \zeta = h^2, \quad \text{on } \Gamma_2, \tag{21}$$

with the following transmission and boundary conditions:

$$\begin{cases} u = w, \;\; \mathcal{B}_1 w = 0, \;\; \mathcal{B}_2 w = \partial_{\nu_1} u, & \text{on } I, \\ \partial_{\nu_1} u + \eta = 0, & \text{on } \Gamma_1, \\ \mathcal{B}_1 w = 0, & \text{on } \Gamma_2, \\ \mathcal{B}_2 w - \zeta = 0, & \text{on } \Gamma_2. \end{cases} \tag{22}$$

It follows from (16), (18), (19), and (21) that

$$v = u - f^1 \text{ in } \Omega_1, \;\; 2\eta = v + h^1 \text{ on } \Gamma_1, \;\; z = w - f^2 \text{ in } \Omega_2 \text{ and}$$
$$2\zeta = z + h^2 \text{ on } \Gamma_2. \tag{23}$$

By elimination of v, z, η, and ζ in (17), (20), and (22), we find that u and w satisfy the following system:

$$\begin{cases} u - \Delta u = f^1 + g^1, & \text{in } \Omega_1, \\ w + \Delta^2 w = f^2 + g^2, & \text{in } \Omega_2, \\ u = w, \;\; \mathcal{B}_1 w = 0, \;\; \mathcal{B}_2 w = \partial_{\nu_1} u, & \text{on } I, \\ 2\partial_{\nu_1} u + u = f^1 - h^1, & \text{on } \Gamma_1, \\ \mathcal{B}_1 w = 0 & \text{on } \Gamma_2, \\ 2\mathcal{B}_2 w - w = h^2 - f^2, & \text{on } \Gamma_2. \end{cases} \tag{24}$$

Let $\phi = (\varphi, \psi) \in \mathsf{H}$. Multiplying the first equation of (24) by $\overline{\varphi}$ and integrating over Ω_1, multiplying the second equation of (24) by $\overline{\psi}$ and integrating over Ω_2, and then using Green's formula, we get

$$\int_{\Omega_1} u\overline{\varphi} dx + \int_{\Omega_1} \nabla u \cdot \nabla \overline{\varphi} dx + \frac{1}{2}\int_{\Gamma_1} u\overline{\varphi} d\Gamma - \int_I \partial_{\nu_1} u \overline{\varphi} d\Gamma$$
$$= \int_{\Omega_1} F^1 \overline{\varphi} dx + \frac{1}{2}\int_{\Gamma_1} F^2 \overline{\varphi} d\Gamma, \tag{25}$$

$$\int_{\Omega_2} w\overline{\psi}dx + a(w,\overline{\psi}) + \frac{1}{2}\int_{\Gamma_2} w\overline{\psi}d\Gamma - \int_I \left(\mathcal{B}_1 w \partial_{\nu_2}\overline{\psi} - \mathcal{B}_2 w\overline{\psi}\right) d\Gamma$$
$$= \int_{\Omega_2} F^3\overline{\psi}dx + \frac{1}{2}\int_{\Gamma_2} F^4\overline{\psi}d\Gamma, \tag{26}$$

where $F^1 = f^1 + g^1$, $F^2 = f^1 - h^1$, $F^3 = f^2 + g^2$, and $F^4 = f^2 - h^2$. Adding the resulting equations and using the transmission conditions, we obtain a variational formulation of (24): Find $(u, w) \in \mathsf{H}$ such that

$$b((u, w), (\varphi, \psi)) = l(\varphi, \psi), \quad \forall (\varphi, \psi) \in \mathsf{H}, \tag{27}$$

where

$$b((u, w), (\varphi, \psi)) = \int_{\Omega_1} u\overline{\varphi}dx + \int_{\Omega_1} \nabla u \cdot \nabla\overline{\varphi}dx + \frac{1}{2}\int_{\Gamma_1} u\overline{\varphi}d\Gamma$$
$$+ \int_{\Omega_2} w\overline{\psi}dx + a(w,\overline{\psi}) + \frac{1}{2}\int_{\Gamma_2} w\overline{\psi}d\Gamma$$

and

$$l(\varphi, \psi) = \int_{\Omega_1} F^1\overline{\varphi}dx + \frac{1}{2}\int_{\Gamma_1} F^2\overline{\varphi}d\Gamma + \int_{\Omega_2} F^3\overline{\psi}dx + \frac{1}{2}\int_{\Gamma_2} F^4\overline{\psi}d\Gamma.$$

It is easy to see that b is a sesquilinear, continuous, and coercive form on the space $\mathsf{H}\times\mathsf{H}$ and l is a linear and continuous form on H. Then, it follows by Lax-Milgram's theorem that (27) admits a unique solution $(u, w) \in \mathsf{H}$. By choosing $\varphi \in C_0^\infty(\Omega_1)$ and $\psi = 0$ in (27) and applying Green's formula, we have

$$\int_{\Omega_1} (u - \Delta u)\overline{\varphi}dx = \int_{\Omega_1} \left(f^1 + g^1\right)\overline{\varphi}dx, \quad \forall \varphi \in C_0^\infty(\Omega_1),$$

which implies that the first equation of (24) holds in the sense of distributions in Ω_1 and hence is satisfied in $L^2(\Omega_1)$. As $u - f^1 - g^1$ belongs to $L^2(\Omega_1)$, the same holds for Δu, i.e., $\Delta u \in L^2(\Omega_1)$. In the same way, choosing $\varphi = 0$ and $\psi \in C_0^\infty(\Omega_2)$ in (27), we see that the second equation of (24) holds as equality in $L^2(\Omega_2)$, and therefore $\Delta^2 w \in L^2(\Omega_2)$. Now, let us define the spaces

$$H^1_{*,I}(\Omega_1) = \left\{f \in H^1_*(\Omega_1) \mid f = 0 \text{ on } I\right\},$$

and

$$H^2_{*,I}(\Omega_2) = \left\{f \in H^2_*(\Omega_2) \mid f = \partial_{\nu_2} f = 0 \text{ on } I\right\}.$$

By taking $\varphi \in H^1_{*,I}(\Omega_1)$ and $\psi = 0$ and applying Green's formula in (27), we calculate

$$\int_{\Gamma_1} (2\partial_{\nu_1} u + u - f^1 + h^1)\overline{\varphi} d\Gamma = 0, \quad \forall \varphi \in H^{\frac{1}{2}}_{*,I}(\Gamma_1),$$

where $H^{\frac{1}{2}}_{*,I}(\Gamma_1)$ is the corresponding trace space of $H^1_{*,I}(\Omega_1)$ through the operator γ_1. This implies that

$2\partial_{\nu_1} u + u - f^1 + h^1 \in (H^{\frac{1}{2}}_{*,I}(\Gamma_1))^\perp$ in $L^2(\Gamma_1)$, and since $H^{\frac{1}{2}}_{*,I}(\Gamma_1)$ is dense in $L^2(\Gamma_1)$, we deduce that u satisfies

$$2\partial_{\nu_1} u + u = f^1 - h^1, \quad \text{on } \Gamma_1.$$

Similarly, by taking $\varphi = 0$ and $\psi \in H^2_{*,I}(\Omega_2)$ and applying Green's formula in (27), we deduce that w satisfies

$$\mathcal{B}_1 w = 0, \quad \text{on } \Gamma_2,$$

as well as

$$2\mathcal{B}_2 w - w = h^2 - f^2, \quad \text{on } \Gamma_2.$$

Coming back to (27) and again applying Green's formula, we calculate

$$\int_I \mathcal{B}_1 w \partial_{\nu_2} \overline{\psi} d\Gamma - \int_I \left(\mathcal{B}_2 w - \partial_{\nu_1} u\right) \overline{\psi} d\Gamma = 0, \quad \forall \psi \in H^{\frac{3}{2}}_*(I),$$

where $H^{\frac{3}{2}}_*(I)$ is the corresponding trace space of $H^2_*(\Omega_2)$ through the operator $\psi \longmapsto \psi|_I$. Due to the density of $H^{\frac{3}{2}}_*(I)$ into $L^2(I)$, we can easily check that u and w satisfy the transmission conditions of (24). Therefore, we deduce that system (24) has a unique solution $(u, w) \in \mathsf{H}$ such that $\Delta u \in L^2(\Omega_1)$ and $\Delta^2 w \in L^2(\Omega_2)$. Finally, by setting

$$v := u - f^1, \ \eta := -\partial_{\nu_1} u, \ z := w - f^2 \text{ and } \zeta := \mathcal{B}_2 w,$$

we conclude that there exists a unique $U = (u, v, \eta, w, z, \zeta) \in D(\mathring{A})$ solution of Eq. (15), and thus the operator $\mathcal{A}$ is m-dissipative on $\mathcal{H}$. The proof is thus complete.

According to the Lumer–Philips theorem (see [24]), Proposition 1 implies that the operator $\mathring{A}$ generates a C_0-semigroup of contractions $\left(e^{t\mathring{A}}\right)_{t\geq 0}$ in $\mathcal{H}$ which gives the well-posedness of (12). Then, we have the following result:

Theorem 1 *For all $U_0 \in \mathcal{H}$, system* (12) *admits a unique weak solution*

$$U(t) \in C^0(\mathbb{R}^+, \mathcal{H}).$$

Moreover, if $U_0 \in D(\mathring{A})$*, then the system* (12) *admits a unique strong solution*

$$U(t) \in C^0(\mathbb{R}^+, D(\mathring{A})) \cap C^1(\mathbb{R}^+, \mathcal{H}).$$

3 Some Regularity Results

In the following sections, our goal is to establish energy estimates of the system (12). Our stability results are based on a frequency domain approach combined with a multiplier technique and hence require some Green's formulas in Ω_1 and Ω_2. In order to justify these formulas, we will need some regularity results proved in [8] and [11]. Hence, from now on, we assume that Ω_1 and Ω_2 have a polygonal boundary in the sense that their boundary is piecewise smooth. For $i = 1, 2$, we denote by $\omega_{i,j}$, $j = 1, \ldots, N_i$, the interior angles at the corners of Ω_i enclosed between two consecutive curves. These angles may vary within the range $0 < \omega_{i,j} \leq 2\pi$. To obtain the needed regularity, we require the following assumptions concerning the angles at the corners of Ω_1 and Ω_2:

Assumption 1 The inner angles $\omega_{1,j} < \pi$, for all $j = 1, \ldots, N_1$.

Assumption 2 There exists a minimal angle ω_0 where ω_0 depends on the Poisson coefficient μ (for instance, $\omega_0 \simeq 77.753311\ldots^\circ$ when $\mu = 0.3$) such that $\omega_{2,j} < \omega_0$, for all $j = 1, \ldots, N_2$.

The regularity results are summarized by the following proposition:

Proposition 2 *Let* $U = (u, v, \eta, w, z, \zeta) \in D(\mathring{A})$*. Assume that* Ω_1 *and* Ω_2 *are polygonal domains as described above, satisfying Assumptions 1 and 2, respectively. Then, there exists* $\varepsilon \in (0, \frac{1}{2})$ *such that* u *belongs to* $H^{\frac{3}{2}-\varepsilon}(\Omega_1)$*, and there exists a sequence* $\{w_k\}_{k\geq 0} \subset H^4(\Omega_2)$ *such that* w_k *converges to* w *in* $H^2(\Omega_2)$ *and* $\Delta^2 w_k$ *converges to* $\Delta^2 w$ *in* $L^2(\Omega_2)$.

Proof We use arguments inspired by Lemma 3.1 and Lemma 3.2 of [1].

4 Strong Stability with Non-compact Resolvent

In this section, we will prove the strong stability of the system (1)–(4) in the sense that its energy $E(t)$ converges to zero as t goes to infinity for all initial data in $\mathcal{H}$. As the resolvent of $\mathring{A}$ is not compact, classical methods such as Lasalle's invariance principle [29] or the spectrum decomposition theory of Benchimol [6] are not applicable in this case. Instead, we will prove the strong stability using a more general criterion of Arendt–Batty [4], which states that in a reflexive Banach space, a C_0-semigroup of contractions $(e^{t\mathcal{A}})_{t\geq 0}$ is strongly stable if $\mathcal{A}$ has no eigenvalues on the imaginary axis, and $\sigma(\mathcal{A}) \cap i\mathbb{R}$ is countable, where $\sigma(\mathcal{A})$ denotes the spectrum of $\mathcal{A}$. We require the following geometric assumption on the plate, which is due to the requirements of integration in Ω_2:

Assumption 3 Assume that there exists a fixed point $x_0 \in \mathbb{R}^2$ such that, putting $m(x) = x - x_0$, we have

$$(m \cdot \nu_2) = 0, \quad \forall x \in I, \tag{28}$$

and

$$(m \cdot \nu_2) \geq 0, \quad \forall x \in \Gamma_2, \tag{29}$$

where $(\,\cdot\,)$ designates the scalar product in $\mathbb{R}^2$.

Remark 4 We remark that if condition (28) is satisfied, then the interface I is straight. Hence, from now on, we assume that the interface is straight, and the interior angles $\omega_{i,1}$ and ω_{i,N_i}, for $i = 1, 2$, are the angles at the extremities of I.

Our main result in this section is summarized by the following theorem:

Theorem 2 *Assume that Ω_1 and Ω_2 are polygonal domains as described above. Assume also that Assumptions 1, 2, and 3 hold. Then, the semigroup of contractions $e^{t\mathring{A}}$ is strongly stable in the energy space $\mathcal{H}$ in the sense that*

$$\lim_{t\to\infty} \|e^{t\mathring{A}} U_0\|_{\mathcal{H}} = 0, \quad \forall U_0 \in \mathcal{H}.$$

In the proof of Theorem 2, we shall use the following lemmas:

Lemma 2 *Assume that Assumptions 1, 2, and 3 hold. Then, the operator $\mathring{A}$ has no pure imaginary eigenvalues.*

Proof Let $\lambda \in \mathbb{R}$, and let $U = (u, v, \eta, w, z, \zeta) \in D(\mathring{A})$ such that

$$\mathring{A}U = i\lambda U. \tag{30}$$

Equivalently, we have the following system of equations:

$$v = i\lambda u, \quad \text{in } \Omega_1, \tag{31}$$

$$\Delta u = i\lambda v, \quad \text{in } \Omega_1, \tag{32}$$

$$v - \eta = i\lambda \eta, \quad \text{on } \Gamma_1, \tag{33}$$

$$z = i\lambda w, \quad \text{in } \Omega_2, \tag{34}$$

$$-\Delta^2 w = i\lambda z, \quad \text{in } \Omega_2, \tag{35}$$

$$z - \zeta = i\lambda\zeta, \quad \text{on } \Gamma_2. \tag{36}$$

By using Eqs. (30) and (14), a direct computation leads to the following:

$$0 = \Re\left\{i\lambda \|U\|_{\mathcal{H}}^2\right\} = \Re\{\left(\mathring{A}U, U\right)_{\mathcal{H}}\} = -\int_{\Gamma_1} |\eta|^2 d\Gamma - \int_{\Gamma_2} |\zeta|^2 d\Gamma,$$

which implies that

$$\eta = 0 \text{ on } \Gamma_1 \quad \text{and} \quad \zeta = 0 \text{ on } \Gamma_2. \tag{37}$$

It follows from (10) that

$$\partial_{\nu_1} u = 0 \text{ on } \Gamma_1 \quad \text{and} \quad \mathcal{B}_2 w = 0 \text{ on } \Gamma_2 \tag{38}$$

and from (33) and (36) that

$$v = 0 \text{ on } \Gamma_1 \quad \text{and} \quad z = 0 \text{ on } \Gamma_2. \tag{39}$$

Now, we need to consider two distinct cases:
Case 1 If $\lambda = 0$, then $v = 0$ in Ω_1 and $z = 0$ in Ω_2, and we obtain the following system:

$$-\Delta u = 0, \quad \text{in } \Omega_1, \tag{40}$$

$$\Delta^2 w = 0, \quad \text{in } \Omega_2, \tag{41}$$

$$\partial_{\nu_1} u = 0, \quad \text{on } \Gamma_1, \tag{42}$$

$$\mathcal{B}_1 w = \mathcal{B}_2 w = 0, \quad \text{on } \Gamma_2, \tag{43}$$

with the following transmission conditions:

$$u = w, \quad \mathcal{B}_1 w = 0, \quad \mathcal{B}_2 w = \partial_{\nu_1} u \text{ on } I. \tag{44}$$

Multiplying equations (40) and (41) by $\overline{u}$ and $\overline{w}$, respectively, integrating over Ω_1 and Ω_2, and then using Green's formula and Eqs. (42) and (43), we get

$$\int_{\Omega_1} |\nabla u|^2 dx - \int_I \partial_{\nu_1} u\overline{u} d\Gamma = 0, \tag{45}$$

$$a(w, \overline{w}) - \int_I \left(\mathcal{B}_1 w \partial_{\nu_2}\overline{w} - \mathcal{B}_2 w\overline{w}\right) d\Gamma = 0. \tag{46}$$

Adding the resulting equations and taking (44) into consideration, we obtain

$$\int_{\Omega_1} |\nabla u|^2 dx + a(w, \overline{w}) = 0, \tag{47}$$

which leads to

$$u = 0 \ \text{ in } \ \Omega_1 \quad \text{and} \quad w = 0 \ \text{ in } \ \Omega_2. \tag{48}$$

Hence, we get

$$U = 0.$$

Case 2 If $\lambda \neq 0$, then using Eqs. (31), (34), and (39), we get

$$u = 0 \ \text{ on } \ \Gamma_1 \quad \text{and} \quad w = 0 \ \text{ on } \ \Gamma_2. \tag{49}$$

Inserting (31) in (32), we obtain

$$\begin{cases} \lambda^2 u + \Delta u = 0, \text{ in } \ \Omega_1, \\ u = \partial_{\nu_1} u = 0, \quad \text{on } \ \Gamma_1. \end{cases} \tag{50}$$

Thus, from the above system and by using a unique continuation theorem (see [17]) , we obtain

$$u = 0 \ \text{ in } \ \Omega_1. \tag{51}$$

Inserting (34) into (35) and using the definition of $D(\mathcal{A})$, we notice that w satisfies

$$\begin{cases} \lambda^2 w - \Delta^2 w = 0, & \text{in } \ \Omega_2, \\ w = \mathcal{B}_1 w = \mathcal{B}_2 w = 0, & \text{on } \ \partial\Omega_2. \end{cases} \tag{52}$$

Multiplying the first equation of (52) by $-m \cdot \nabla\overline{w}$ and integrating over Ω_2, we obtain

$$-\int_{\Omega_2} \lambda^2 w \,(m \cdot \nabla\overline{w})\, dx + \int_{\Omega_2} \Delta^2 w \,(m \cdot \nabla\overline{w})\, dx = 0. \tag{53}$$

Applying Green's formula and identity 3.4 of [27] (see also [15] and [25]), we have

$$\int_{\Omega_2} \Delta^2 w (m \cdot \nabla\overline{w}) dx = a(w, \overline{w}) - \int_{\partial\Omega_2} \left[\mathcal{B}_1 w \partial_{\nu_2}(m \cdot \nabla\overline{w}) - \mathcal{B}_2 w (m \cdot \nabla\overline{w})\right] d\Gamma$$
$$+ \frac{1}{2}\int_{\partial\Omega_2} (m \cdot \nu_2) b(w) d\Gamma,$$

where

$$b(w) = \left(\frac{\partial^2 w}{\partial x_1^2}\right)^2 + \left(\frac{\partial^2 w}{\partial x_2^2}\right)^2 + 2\mu \frac{\partial^2 w}{\partial x_1^2}\frac{\partial^2 w}{\partial x_2^2} + 2(1-\mu)\left(\frac{\partial^2 w}{\partial x_1 \partial x_2}\right)^2.$$

It follows from the geometric condition (28) and the boundary condition $\mathcal{B}_1 w = \mathcal{B}_2 w = 0$ on $\partial\Omega_2$ that

$$\int_{\Omega_2} \Delta^2 w(m \cdot \nabla \overline{w})dx = a(w, \overline{w}) + \frac{1}{2}\int_{\Gamma_2} (m \cdot \nu_2) b(w) d\Gamma. \tag{54}$$

A straightforward computation shows that

$$\begin{aligned} b(w) &= \left(\frac{\partial^2 w}{\partial x_1^2}\right)^2 + \left(\frac{\partial^2 w}{\partial x_2^2}\right)^2 + 2\mu \frac{\partial^2 w}{\partial x_1^2}\frac{\partial^2 w}{\partial x_2^2} + 2(1-\mu)\left(\frac{\partial^2 w}{\partial x_1 \partial x_2}\right)^2 \\ &= \mu \left\{\left(\frac{\partial^2 w}{\partial x_1^2}\right)^2 + \left(\frac{\partial^2 w}{\partial x_2^2}\right)^2 + 2\frac{\partial^2 w}{\partial x_1^2}\frac{\partial^2 w}{\partial x_2^2}\right\} \\ &\quad + (1-\mu)\left\{\left(\frac{\partial^2 w}{\partial x_1^2}\right)^2 + \left(\frac{\partial^2 w}{\partial x_2^2}\right)^2 + 2\left(\frac{\partial^2 w}{\partial x_1 \partial x_2}\right)^2\right\} \\ &= \mu\,(\Delta w)^2 + (1-\mu)\left\{\left(\frac{\partial^2 w}{\partial x_1^2}\right)^2 + \left(\frac{\partial^2 w}{\partial x_2^2}\right)^2 + 2\left(\frac{\partial^2 w}{\partial x_1 \partial x_2}\right)^2\right\} \geq 0. \end{aligned} \tag{55}$$

Taking the geometric condition (29) into consideration, it follows from (54) and (55) that

$$\int_{\Omega_2} \Delta^2 w(m \cdot \nabla \overline{w})dx \geq a(w, \overline{w}). \tag{56}$$

On the other hand, applying Green's formula to the first term of (53) yields

$$-\int_{\Omega_2} \lambda^2 w\,(m \cdot \nabla \overline{w})\,dx = -\frac{1}{2}\int_{\partial\Omega_2} (m \cdot \nu_2)\,|\lambda w|^2\,d\Gamma + \int_{\Omega_2} |\lambda w|^2\,dx. \tag{57}$$

Inserting (56) and (57) into (53) and using the boundary condition $w = 0$ on $\partial\Omega_2$, we obtain

$$a(w, \overline{w}) + \int_{\Omega_2} |\lambda w|^2\,dx \leq 0,$$

from which one concludes that

$$w = 0 \text{ in } \Omega_2. \tag{58}$$

Consequently, from Eqs. (31), (34), (37), (51), and (58), we get

$$U = 0.$$

The proof is thus complete.

Lemma 3 *Assume that Assumptions 1, 2, and 3 hold. Then, the operator* $i\lambda\mathcal{I} - \mathring{\mathcal{A}}$ *is surjective for all real numbers* $\lambda \in \mathbb{R}$.

Proof For any given $F = (f^1, g^1, h^1, f^2, g^2, h^2) \in \mathcal{H}$, we are looking for $U = (u, v, \eta, w, z, \zeta) \in D(\mathcal{A})$ that solves the following system:

$$(i\lambda\mathcal{I} - \mathring{\mathcal{A}})U = F, \tag{59}$$

which is equivalent to the following equations:

$$i\lambda u - v = f^1, \text{ in } \Omega_1, \tag{60}$$

$$i\lambda v - \Delta u = g^1, \text{ in } \Omega_1, \tag{61}$$

$$i\lambda\eta - v + \eta = h^1, \text{ on } \Gamma_1, \tag{62}$$

$$i\lambda w - z = f^2, \text{ in } \Omega_2, \tag{63}$$

$$i\lambda z + \Delta^2 w = g^2, \text{ in } \Omega_2, \tag{64}$$

$$i\lambda\zeta - z + \zeta = h^2, \text{ on } \Gamma_2. \tag{65}$$

By eliminating v, z, η, and ζ from the above equations and using (10), we obtain the following system:

$$\begin{cases} \lambda^2 u + \Delta u = -(i\lambda f^1 + g^1), & \text{in } \Omega_1, \\ \lambda^2 w - \Delta^2 w = -(i\lambda f^2 + g^2), & \text{in } \Omega_2, \\ u = w, \ \mathcal{B}_1 w = 0, \ \mathcal{B}_2 w = \partial_{\nu_1} u, & \text{on } I, \\ \partial_{\nu_1} u + \dfrac{i\lambda}{i\lambda + 1} u = \dfrac{1}{i\lambda + 1}(f^1 - h^1), & \text{on } \Gamma_1, \\ \mathcal{B}_1 w = 0, & \text{on } \Gamma_2, \\ \mathcal{B}_2 w - \dfrac{i\lambda}{i\lambda + 1} w = \dfrac{1}{i\lambda + 1}(h^2 - f^2), & \text{on } \Gamma_2. \end{cases} \tag{66}$$

Let $\phi = (\varphi, \psi) \in \mathsf{H}$. Multiplying the first equation of (66) by $\overline{\varphi}$ and integrating over Ω_1 and multiplying the second equation of (66) by $\overline{\psi}$ and integrating over Ω_2, we use Green's formula to get the following equations:

$$
\begin{aligned}
&-\lambda^2\int_{\Omega_1} u\overline{\varphi}dx+\int_{\Omega_1}\nabla u\cdot\nabla\overline{\varphi}dx+\frac{i\lambda}{i\lambda+1}\int_{\Gamma_1}u\overline{\varphi}d\Gamma-\int_I\partial_{\nu_1}u\overline{\varphi}d\Gamma\\
&=\int_{\Omega_1}F^1\overline{\varphi}dx+\frac{1}{i\lambda+1}\int_{\Gamma_1}F^2\overline{\varphi}d\Gamma, \qquad (67)
\end{aligned}
$$

and

$$
\begin{aligned}
&-\lambda^2\int_{\Omega_2} w\overline{\psi}dx+a(w,\overline{\psi})+\frac{i\lambda}{i\lambda+1}\int_{\Gamma_2}w\overline{\psi}d\Gamma-\int_I\left(\mathcal{B}_1 w\partial_{\nu_2}\overline{\psi}-\mathcal{B}_2 w\overline{\psi}\right)d\Gamma\\
&=\int_{\Omega_2}F^3\overline{\psi}dx+\frac{1}{i\lambda+1}\int_{\Gamma_2}F^4\overline{\psi}d\Gamma, \qquad (68)
\end{aligned}
$$

where

$$
F^1=i\lambda f^1+g^1,\quad F^2=f^1-h^1,\quad F^3=i\lambda f^2+g^2\quad\text{and}\quad F^4=f^2-h^2.
$$

Adding equations (67) and (68) and using the transmission conditions, we obtain

$$
\begin{aligned}
&-\lambda^2\left(\int_{\Omega_1}u\overline{\varphi}dx+\int_{\Omega_2}w\overline{\psi}dx\right)+\int_{\Omega_1}\nabla u\cdot\nabla\overline{\varphi}dx+i\lambda\lambda\int_{\Gamma_1}u\overline{\varphi}d\Gamma+a(w,\overline{\psi})\\
&+i\lambda\lambda\int_{\Gamma_2}w\overline{\psi}d\Gamma\\
&=\int_{\Omega_1}F^1\overline{\varphi}dx+\lambda\int_{\Gamma_1}F^2\overline{\varphi}d\Gamma+\int_{\Omega_2}F^3\overline{\psi}dx\\
&+\lambda\int_{\Gamma_2}F^4\overline{\psi}d\Gamma, \qquad (69)
\end{aligned}
$$

where

$$
\lambda=\frac{1-i\lambda}{\lambda^2+1}. \qquad (70)
$$

Here we note that Lax–Milgram's theorem cannot be applied because coercivity is not available. Therefore, we use a compact perturbation argument. For that purpose, let us introduce the sesquilinear form

$$
\begin{aligned}
b_\lambda((u,w),(\varphi,\psi))=&\int_{\Omega_1}\nabla u\cdot\nabla\overline{\varphi}dx+i\lambda\lambda\int_{\Gamma_1}u\overline{\varphi}d\Gamma+a(w,\overline{\psi})\\
&+i\lambda\lambda\int_{\Gamma_2}w\overline{\psi}d\Gamma,\quad\forall(\varphi,\psi)\in\mathsf{H}. \qquad (71)
\end{aligned}
$$

This sesquilinear form b_λ is continuous and coercive on $\mathsf{H} \times \mathsf{H}$. Then, by Lax–Milgram's theorem, the operator

$$B_\lambda : \mathsf{H} \to \mathsf{H}' : (\boldsymbol{u}, \boldsymbol{w}) \to B_\lambda(\boldsymbol{u}, \boldsymbol{w}),$$

with $B_\lambda(\boldsymbol{u}, \boldsymbol{w})((\boldsymbol{\varphi}, \boldsymbol{\psi})) := b_\lambda((\boldsymbol{u}, \boldsymbol{w}), (\boldsymbol{\varphi}, \boldsymbol{\psi}))$ $\forall(\varphi, \psi) \in \mathsf{H}$, is an isomorphism, where H' is the dual space of H. Now, let us set

$$R_\lambda : \mathsf{H} \to \mathsf{H}' : (\boldsymbol{u}, \boldsymbol{w}) \to R_\lambda(\boldsymbol{u}, \boldsymbol{w}),$$

with

$$R_\lambda(\boldsymbol{u}, \boldsymbol{w})((\boldsymbol{\varphi}, \boldsymbol{\psi})) = -\lambda^2 \left(\int_{\Omega_1} u\overline{\varphi} dx + \int_{\Omega_2} w\overline{\psi} dx \right), \quad \forall (\varphi, \psi) \in \mathsf{H}.$$

As R_λ is a compact operator, we deduce that $B_\lambda + R_\lambda$ is a Fredholm operator of index zero from H to H'. Now by setting

$$L_\lambda(\varphi, \psi) = \int_{\Omega_1} F^1\overline{\varphi} dx + \lambda \int_{\Gamma_1} F^2\overline{\varphi} d\Gamma + \int_{\Omega_2} F^3\overline{\psi} dx + \lambda \int_{\Gamma_2} F^4\overline{\psi} d\Gamma,$$

we notice that (69) is equivalent to

$$(B_\lambda + R_\lambda)(\boldsymbol{u}, \boldsymbol{w}) = L_\lambda \quad \text{in } \mathsf{H}'. \tag{72}$$

Hence, problem (69) admits a unique solution $(\boldsymbol{u}, \boldsymbol{w}) \in \mathsf{H}$ if and only if $B_\lambda + R_\lambda$ is invertible. Since $B_\lambda + R_\lambda$ is a Fredholm operator, it is enough to prove that $B_\lambda + R_\lambda$ is injective, i.e.,

$$\ker(B_\lambda + R_\lambda) = \{0\}.$$

Let us now fix $(\boldsymbol{u}, \boldsymbol{w}) \in \ker(B_\lambda + R_\lambda)$, which means it satisfies

$$\begin{aligned} -\lambda^2 \left(\int_{\Omega_1} \boldsymbol{u}\overline{\boldsymbol{\varphi}} dx + \int_{\Omega_2} \boldsymbol{w}\overline{\boldsymbol{\psi}} dx \right) + \int_{\Omega_1} \nabla \boldsymbol{u} \cdot \nabla \overline{\boldsymbol{\varphi}} dx + i\lambda\lambda \int_{\Gamma_1} \boldsymbol{u}\overline{\boldsymbol{\varphi}} d\Gamma \\ + a(\boldsymbol{w}, \overline{\boldsymbol{\psi}}) + i\lambda\lambda \int_{\Gamma_2} \boldsymbol{w}\overline{\boldsymbol{\psi}} d\Gamma = 0, \quad \forall (\boldsymbol{\varphi}, \boldsymbol{\psi}) \in \mathsf{H}. \end{aligned} \tag{73}$$

Thus, if we set

$$\boldsymbol{v} = i\lambda \boldsymbol{u}, \quad \boldsymbol{\eta} = i\lambda\lambda \boldsymbol{u}, \quad \boldsymbol{z} = i\lambda \boldsymbol{w}, \quad \text{and} \quad \boldsymbol{\zeta} = i\lambda\lambda \boldsymbol{w},$$

we can conclude that $\boldsymbol{U} = (\boldsymbol{u}, \boldsymbol{v}, \boldsymbol{\eta}, \boldsymbol{w}, \boldsymbol{z}, \boldsymbol{\zeta}) \in D(\mathcal{A})$ is a solution of

$$(i\lambda - \mathcal{A})\boldsymbol{U} = 0.$$

Using Lemma 2, we deduce that $U = 0$. This shows that $B_\lambda + R_\lambda$ is invertible, and therefore a unique solution $(u, w) \in \mathsf{H}$ of (72) exists. At this stage, by setting

$$v = i\lambda u - f^1, \quad \eta = i\lambda\lambda u + \lambda(h^1 - f^1), \quad z = i\lambda w - f^2, \quad \text{and}$$
$$\zeta = i\lambda\lambda w + \lambda(h^2 - f^2),$$

we conclude that $(u, v, \eta, w, z, \zeta) \in D(\mathcal{A})$ is a solution of (59), and the proof is thus complete.

Proof of Theorem 2 Using Lemma 2, we conclude that the operator $\mathcal{A}$ has no pure imaginary eigenvalues. Additionally, by Lemma 3, we have $Im(i\lambda\mathcal{I} - \mathring{A}) = \mathcal{H}$ for all real numbers $\lambda \in \mathbb{R}$. Therefore, the closed graph theorem of Banach implies that $\sigma(\mathcal{A}) \cap i\mathbb{R} = \emptyset$. Following Arendt–Batty (see [4]), the C_0-semigroup of contractions $(e^{t\mathcal{A}})_{t\geq 0}$ is strongly stable, and the proof is complete.

5 Polynomial Stability

Following Theorem 5.2 in [1], it is evident that exponential decay fails when three dynamical controls are employed. Consequently, the lack of exponential stability persists within the system (1)–(4) even when only two dynamical controls are applied. Hence, in this section, it is natural to hope for polynomial stability of the system (1)–(4) for smooth solutions using a multiplier method. One of the main ingredients is to use the frequency domain approach, specifically the key ingredient for the proof of polynomial stability is Theorem 2.4 of [9] (see also [5] and [19]), which we will partially recall:

Theorem 3 (Borichev–Tomilov (See [9])) *Let $\left(e^{tA}\right)_{t\geq 0}$ be a bounded C_0-semigroup of contractions on a Hilbert space H generated by A such that $i\mathbb{R} \subset \rho(A)$. Then, for a fixed $\ell > 0$, the following conditions are equivalent:*

(i) $\sup_{\beta\in\mathbb{R}} \frac{1}{|\beta|^\ell} \left\|(i\beta\mathcal{I} - A)^{-1}\right\|_{\mathcal{L}(H)} < +\infty$

(ii) there exists a constant $C > 0$ such that for all $U_0 \in D(A)$ we have $\|e^{tA}U_0\|_H^2 \leq \frac{C}{t^{\frac{2}{\ell}}}\|U_0\|_{D(A)}^2, \quad \forall t > 0.$

To proceed with our polynomial energy decay result, we require the following additional geometric assumption on the wave, which is due to the requirements of integration in Ω_1:

Assumption 4 Assume that there exists a fixed point $x_0 \in \mathbb{R}^2$ such that, putting $m(x) = x - x_0$, we have

$$(m \cdot \nu_1) \geq \delta, \quad \forall x \in \Gamma_1, \tag{74}$$

where δ is a positive real number and $(\,\cdot\,)$ designates the scalar product in $\mathbb{R}^2$.

Now, we can state our energy decay rate result.

Theorem 4 *Assume that Ω_1 and Ω_2 are polygonal domains as described above. Also, assume that Assumptions 1, 2, 3, and 4 hold. Then, there exists a constant $\mathcal{C} > 0$ such that for all initial data $U_0 = (u_0, u_1, \eta_0, w_0, w_1, \zeta_0) \in D(\mathring{A})$, the energy of system* (1)–(4) *satisfies the following estimate:*

$$E(t) \leq \frac{\mathcal{C}}{t}\|U_0\|^2_{D(\mathcal{A})}, \quad \forall t > 0. \tag{75}$$

Before stating the proof of the above theorem, we shall preliminarily present the following two technical lemmas to be used in the proof of our polynomial-type decay estimate:

Lemma 4 *Assume that Ω_1 is a polygonal domain as described above. Assume also that Assumptions 1, 3, and 4 hold. Then, there exists $\varepsilon \in (0, \frac{1}{2})$ such that the solution $y \in H^1(\Omega_1)$ of system*

$$\begin{cases} \Delta y \in L^2(\Omega_1), \\ \partial_{\nu_1} y = v_1 \in L^2(\Gamma_1), & \text{on } \Gamma_1, \\ y = v_2 \in H^1(I), & \text{on } I \end{cases} \tag{76}$$

satisfies the following inequality:

$$-\Re\left\{\int_{\Omega_1} \Delta y(m\cdot\nabla\overline{y})dx\right\} \geq -\frac{R_1^2}{\delta}\int_{\Gamma_1}|\partial_{\nu_1} y|^2 d\Gamma - \Re\left\{\left\langle\partial_{\nu_1} y, m\cdot\nabla\overline{y}\right\rangle_{H^{-\varepsilon}(I)\times H^{\varepsilon}(I)}\right\}, \tag{77}$$

where $R_1 = \max_{x\in\Gamma_1}|m(x)|$.

Proof Before starting the proof, we assume that $y \in H^2(\Omega_1)$. Using Green's formula, we have the following Rellich identity (see [14], and also see Identity 3.17 of [26]):

$$-\int_{\Omega_1}\Delta y(m\cdot\nabla\overline{y})dx = \frac{1}{2}\int_{\partial\Omega_1}(m\cdot\nu_1)|\nabla y|^2 d\Gamma - \int_{\partial\Omega_1}\partial_{\nu_1} y(m\cdot\nabla\overline{y})d\Gamma. \tag{78}$$

According to the geometric condition (28), we have from (78) that

$$-\int_{\Omega_1}\Delta y(m\cdot\nabla\overline{y})dx = \frac{1}{2}\int_{\Gamma_1}(m\cdot\nu_1)|\nabla y|^2 d\Gamma - \int_{\Gamma_1}\partial_{\nu_1} y(m\cdot\nabla\overline{y})d\Gamma - \int_I \partial_{\nu_1} y(m\cdot\nabla\overline{y})d\Gamma. \tag{79}$$

Next, for any $\varepsilon_1 > 0$, using Young's inequality, we have

$$\Re\left\{\int_{\Gamma_1} \partial_{\nu_1} y(m \cdot \nabla \overline{y}) d\Gamma\right\} \leq \frac{1}{2\varepsilon_1}\int_{\Gamma_1} |\partial_{\nu_1} y|^2 d\Gamma + \frac{R_1^2 \varepsilon_1}{2}\int_{\Gamma_1} |\nabla y|^2 d\Gamma. \tag{80}$$

Using the geometric condition (74), we deduce from (80) that

$$\begin{aligned}
&\frac{1}{2}\int_{\Gamma_1} (m \cdot \nu_1)|\nabla y|^2 d\Gamma - \Re\left\{\int_{\Gamma_1} \partial_{\nu_1} y(m \cdot \nabla \overline{y}) d\Gamma\right\} \\
&\quad \geq \left(\frac{\delta}{2} - \frac{R_1^2 \varepsilon_1}{2}\right)\int_{\Gamma_1} |\nabla y|^2 d\Gamma - \frac{1}{2\varepsilon_1}\int_{\Gamma_1} |\partial_{\nu_1} y|^2 d\Gamma.
\end{aligned} \tag{81}$$

Choosing $\varepsilon_1 = \frac{\delta}{2R_1^2} > 0$ in (81), we get

$$\begin{aligned}
&\frac{1}{2}\int_{\Gamma_1} (m \cdot \nu_1)|\nabla y|^2 d\Gamma - \Re\left\{\int_{\Gamma_1} \partial_{\nu_1} y(m \cdot \nabla \overline{y}) d\Gamma\right\} \\
&\quad \geq \frac{\delta}{4}\int_{\Gamma_1} |\nabla y|^2 d\Gamma - \frac{R_1^2}{\delta}\int_{\Gamma_1} |\partial_{\nu_1} y|^2 d\Gamma \\
&\quad \geq -\frac{R_1^2}{\delta}\int_{\Gamma_1} |\partial_{\nu_1} y|^2 d\Gamma.
\end{aligned} \tag{82}$$

Taking the real part of (79) and then inserting (82), we obtain

$$-\Re\left\{\int_{\Omega_1} \Delta y(m \cdot \nabla \overline{y}) dx\right\} \geq -\frac{R_1^2}{\delta}\int_{\Gamma_1} |\partial_{\nu_1} y|^2 d\Gamma - \Re\left\{\int_I \partial_{\nu_1} y(m \cdot \nabla \overline{y}) d\Gamma\right\}. \tag{83}$$

Now, for $y \in H^1(\Omega_1)$, the solution of (76), we deduce from Lemma 3.1 of [1] and Assumption 1 that $y \in H^{\frac{3}{2}-\varepsilon}(\Omega_1)$ for some $\varepsilon \in (0, \frac{1}{2})$. By density, there exists a sequence $\{y_k\}_{k \geq 0} \subset H^2(\Omega_1)$ such that

$$y_k \underset{k \to \infty}{\longrightarrow} y \text{ in } H^{\frac{3}{2}-\varepsilon}(\Omega_1).$$

Applying (83) to y_k and then passing to the limit in k, we obtain

$$\begin{aligned}
&-\Re\left\{\langle \Delta y, m \cdot \nabla \overline{y}\rangle_{H^{-\frac{1}{2}-\varepsilon}(\Omega_1) \times H^{\frac{1}{2}+\varepsilon}(\Omega_1)}\right\} \\
&\geq -\frac{R_1^2}{\delta}\|\partial_{\nu_1} y\|^2_{H^{-\varepsilon}(\Gamma_1)} - \Re\left\{\langle \partial_{\nu_1} y, m \cdot \nabla \overline{y}\rangle_{H^{-\varepsilon}(I) \times H^{\varepsilon}(I)}\right\}.
\end{aligned}$$

Since y satisfies (76), we see that the duality pairings become integrals in the first two terms of the above identity, which gives (77), as desired.

Lemma 5 *Assume that Ω_2 is a polygonal domain as described above. Assume also that Assumptions 2 and 3 hold. Then, there exists $\varepsilon \in (0, \frac{1}{2})$ such that the solution $y \in H^2(\Omega_2)$ of system*

$$\begin{cases} \Delta^2 y \in L^2(\Omega_2), \\ \mathcal{B}_1 y = 0, & on\ I, \\ \mathcal{B}_2 y = v_3 \in H^{-\varepsilon}(I), & on\ I, \\ \mathcal{B}_1 y = v_4 \in L^2(\Gamma_2), & on\ \Gamma_2, \\ \mathcal{B}_2 y = v_5 \in L^2(\Gamma_2), & on\ \Gamma_2 \end{cases} \tag{84}$$

satisfies the following inequality:

$$\Re\left\{\int_{\Omega_2} \Delta^2 y(m \cdot \nabla \overline{y})dx\right\} \geq \frac{1}{2}a(y, \overline{y}) - \frac{R_2^2 M}{2}\int_{\Gamma_2} |\mathcal{B}_2 y|^2 d\Gamma + \Re\left\{\langle \mathcal{B}_2 y, m \cdot \nabla \overline{y}\rangle_{H^{-\varepsilon}(I)\times H^{\varepsilon}(I)}\right\}, \tag{85}$$

where $R_2 = \max_{x\in\Gamma_2} |m(x)|$ and M is a positive constant.

Proof Before starting, we assume that $\mathcal{B}_2 y$ is smooth enough so that $y \in H^4(\Omega_2)$ (see [12]). Using Green's formula and Identity 3.4 of [27] (see also [15] and [25]), we have

$$\int_{\Omega_2} \Delta^2 y(m \cdot \nabla \overline{y})dx = a(y, \overline{y}) - \int_{\partial\Omega_2} \left[\mathcal{B}_1 y \partial_{\nu_2}(m \cdot \nabla \overline{y}) - \mathcal{B}_2 y(m \cdot \nabla \overline{y})\right] d\Gamma + \frac{1}{2}\int_{\partial\Omega_2} (m \cdot \nu_2) b(y) d\Gamma, \tag{86}$$

where

$$b(y) = \left(\frac{\partial^2 y}{\partial x_1^2}\right)^2 + \left(\frac{\partial^2 y}{\partial x_2^2}\right)^2 + 2\mu\frac{\partial^2 y}{\partial x_1^2}\frac{\partial^2 y}{\partial x_2^2} + 2(1-\mu)\left(\frac{\partial^2 y}{\partial x_1 \partial x_2}\right)^2.$$

It follows from the geometric condition (28) and the boundary conditions $\mathcal{B}_1 y = 0$ on I and $\mathcal{B}_1 y = 0$ on Γ_2 that

$$\int_{\Omega_2} \Delta^2 y(m \cdot \nabla \overline{y})dx = a(y, \overline{y}) + \int_{\Gamma_2} \mathcal{B}_2 y(m \cdot \nabla \overline{y}) d\Gamma + \frac{1}{2}\int_{\Gamma_2} (m \cdot \nu_2) b(y) d\Gamma + \int_I \mathcal{B}_2 y(m \cdot \nabla \overline{y}) d\Gamma. \tag{87}$$

Applying Young's inequality to the second term on the right-hand side of (87) gives

$$\Re\left\{\int_{\Gamma_2} \mathcal{B}_2 y(m\cdot\nabla\overline{y})d\Gamma\right\} \geq -\frac{1}{2\varepsilon_3}\int_{\Gamma_2} |\mathcal{B}_2 y|^2 d\Gamma - \frac{R_2^2\varepsilon_3}{2}\int_{\Gamma_2} |\nabla y|^2\, d\Gamma, \tag{88}$$

for any $\varepsilon_3 > 0$. Thanks to the trace theorem (see Theorem 1.6.6 of [10]), we know that there exists a constant $C_1 > 0$ such that

$$\|\nabla y\|^2_{L^2(\Gamma_2)} \leq C_1\|\nabla y\|_{L^2(\Omega_2)}\|\nabla y\|_{H^1(\Omega_2)} \leq C_1\|y\|^2_{H^2(\Omega_2)}. \tag{89}$$

Moreover, according to the fact that $a(y, \overline{y})$ is equivalent to the usual norm of $H^2(\Omega_2)$ on $H^2_*(\Omega_2)$, we know that there exists a constant $C_2 > 0$ such that

$$\|y\|^2_{H^2(\Omega_2)} \leq C_2 a(y, \overline{y}). \tag{90}$$

The combination of (89) and (90) yields

$$\|\nabla y\|^2_{L^2(\Gamma_2)} \leq M a(y, \overline{y}), \tag{91}$$

where the constant $M = C_1C_2 > 0$. This, along with (88), implies that

$$\Re\left\{\int_{\Gamma_2} \mathcal{B}_2 y(m\cdot\nabla\overline{y})d\Gamma\right\} \geq -\frac{1}{2\varepsilon_3}\int_{\Gamma_2} |\mathcal{B}_2 y|^2 d\Gamma - \frac{R_2^2 M\varepsilon_3}{2} a(y, \overline{y}). \tag{92}$$

By taking the real part of (87) and inserting (92), we obtain

$$\begin{aligned}\Re\left\{\int_{\Omega_2} \Delta^2 y(m\cdot\nabla\overline{y})dx\right\} \geq \left(1 - \frac{R_2^2 M\varepsilon_3}{2}\right) a(y, \overline{y}) - \frac{1}{2\varepsilon_3}\int_{\Gamma_2} |\mathcal{B}_2 y|^2 d\Gamma\\ + \frac{1}{2}\int_{\Gamma_2} (m\cdot\nu_2)b(y)d\Gamma\\ + \Re\left\{\int_{I} \mathcal{B}_2 y(m\cdot\nabla\overline{y})d\Gamma\right\}.\end{aligned} \tag{93}$$

By letting $\varepsilon_3 = \frac{1}{R_2^2 M} > 0$ in (93), we conclude that

$$\begin{aligned}\Re\left\{\int_{\Omega_2} \Delta^2 y(m\cdot\nabla\overline{y})dx\right\} \geq \frac{1}{2}a(y, \overline{y}) - \frac{R_2^2 M}{2}\int_{\Gamma_2} |\mathcal{B}_2 y|^2 d\Gamma\\ + \frac{1}{2}\int_{\Gamma_2} (m\cdot\nu_2)b(y)d\Gamma\\ + \Re\left\{\int_{I} \mathcal{B}_2 y(m\cdot\nabla\overline{y})d\Gamma\right\}.\end{aligned} \tag{94}$$

A straightforward computation shows that

$$\begin{aligned} b(y) &= \left(\frac{\partial^2 y}{\partial x_1^2}\right)^2 + \left(\frac{\partial^2 y}{\partial x_2^2}\right)^2 + 2\mu \frac{\partial^2 y}{\partial x_1^2}\frac{\partial^2 y}{\partial x_2^2} + 2(1-\mu)\left(\frac{\partial^2 y}{\partial x_1 \partial x_2}\right)^2 \\ &= \mu \left\{ \left(\frac{\partial^2 y}{\partial x_1^2}\right)^2 + \left(\frac{\partial^2 y}{\partial x_2^2}\right)^2 + 2\frac{\partial^2 y}{\partial x_1^2}\frac{\partial^2 y}{\partial x_2^2} \right\} \\ &\quad + (1-\mu)\left\{ \left(\frac{\partial^2 y}{\partial x_1^2}\right)^2 + \left(\frac{\partial^2 y}{\partial x_2^2}\right)^2 + 2\left(\frac{\partial^2 y}{\partial x_1 \partial x_2}\right)^2 \right\} \\ &= \mu\,(\Delta y)^2 + (1-\mu)\left\{ \left(\frac{\partial^2 y}{\partial x_1^2}\right)^2 + \left(\frac{\partial^2 y}{\partial x_2^2}\right)^2 + 2\left(\frac{\partial^2 y}{\partial x_1 \partial x_2}\right)^2 \right\}. \end{aligned} \tag{95}$$

Taking the geometric condition (29) into consideration, it follows from (95) that

$$\begin{aligned} \Re\left\{\int_{\Omega_2} \Delta^2 y (m\cdot\nabla\overline{y})dx\right\} \geq \frac{1}{2}a(y,\overline{y}) &- \frac{R_2^2 M}{2}\int_{\Gamma_2} |\mathcal{B}_2 y|^2 d\Gamma \\ &+ \frac{1}{2}\int_{\Gamma_2} (m\cdot\nu_2)\,(\Delta y)^2\, d\Gamma \\ &+ \frac{1}{2}\int_{\Gamma_2} (m\cdot\nu_2) \left\{ \left(\frac{\partial^2 y}{\partial x_1^2}\right)^2 \right. \\ &\left. + \left(\frac{\partial^2 y}{\partial x_2^2}\right)^2 + 2\left(\frac{\partial^2 y}{\partial x_1 \partial x_2}\right)^2 \right\} d\Gamma \\ &+ \Re\left\{\int_I \mathcal{B}_2 y (m\cdot\nabla\overline{y}) d\Gamma\right\}, \end{aligned} \tag{96}$$

from which we conclude

$$\begin{aligned} \Re\left\{\int_{\Omega_2} \Delta^2 y (m\cdot\nabla\overline{y})dx\right\} \geq \frac{1}{2}a(y,\overline{y}) &- \frac{R_2^2 M}{2}\int_{\Gamma_2} |\mathcal{B}_2 y|^2 d\Gamma \\ &+ \Re\left\{\int_I \mathcal{B}_2 y (m\cdot\nabla\overline{y}) d\Gamma\right\}. \end{aligned} \tag{97}$$

Now, for $y \in H^2(\Omega_2)$, which is a solution of (84), we deduce, based on Lemma 3.2 of [1] and Assumption 2, that there exists a sequence $\{y_k\}_{k\geq 0} \subset H^4(\Omega_2)$ such that

$$y_k \underset{k\to\infty}{\longrightarrow} y \text{ in } H^2(\Omega_2) \text{ and } \Delta^2 y_k \underset{k\to\infty}{\longrightarrow} \Delta^2 y \text{ in } L^2(\Omega_2).$$

By applying (97) to y_k and passing the limit in k, we obtain the desired estimation (85).

We are now ready to present the proof of Theorem 4.

Proof In accordance with Theorem 3, a C_0-semigroup of contractions $(e^{t\mathcal{A}})_{t\geq 0}$ on a Hilbert space $\mathcal{H}$ verifies (75) if the following conditions

$$i\mathbb{R} \subset \rho(\mathcal{A}) \qquad (P1)$$

and

$$\lim \sup_{|\lambda|\to\infty} \frac{1}{\lambda^2} \left\| (i\lambda\mathcal{I} - \mathcal{A})^{-1} \right\|_{\mathcal{L}(\mathcal{H})} < +\infty \qquad (P2)$$

are satisfied. Since the resolvent of the operator $\mathring{A}$ is not compact in the energy space $\mathcal{H}$ and $0 \in \rho(\mathcal{A})$, proving $i\mathbb{R} \subset \rho(\mathcal{A})$ is equivalent to showing that $i\lambda\mathcal{I} - \mathring{A}$ is bijective in the energy space $\mathcal{H}$ for all $\lambda \in \mathbb{R}^*$. This is established in Sect. 4 using a unique continuation theorem and Fredholm's alternative. Then, we still need to prove condition $(P2)$. This is checked by a contradiction argument. Indeed, suppose that there exist a sequence $\{\lambda_n\}_{n\geq 1} \subset \mathbb{R}_+^*$ and a sequence $\{U_n := (u_n, v_n, \eta_n, w_n, z_n, \zeta_n)\}_{n\geq 1} \subset D(\mathring{A})$ such that

$$\lambda_n \longrightarrow +\infty, \qquad \|U_n\|_{\mathcal{H}} = 1, \tag{98}$$

and

$$\lambda_n^2(i\lambda_n\mathcal{I} - \mathring{A})U_n = (f_n^1, g_n^1, h_n^1, f_n^2, g_n^2, h_n^2) \longrightarrow 0 \quad \text{in } \mathcal{H}. \tag{99}$$

Our aim is to show that $\|(u_n, v_n, \eta_n, w_n, z_n, \zeta_n)\|_{\mathcal{H}} \longrightarrow 0$. This condition provides a contradiction with (98). By detailing equation (99), we get the following system:

$$\lambda_n^2(i\lambda_n u_n - v_n) = f_n^1 \to 0 \quad \text{in} \quad H_*^1(\Omega_1), \tag{100}$$

$$\lambda_n^2(i\lambda_n v_n - \Delta u_n) = g_n^1 \to 0 \quad \text{in} \quad L^2(\Omega_1), \tag{101}$$

$$\lambda_n^2(i\lambda_n \eta_n - v_n + \eta_n) = h_n^1 \to 0 \quad \text{in} \quad L^2(\Gamma_1), \tag{102}$$

$$\lambda_n^2(i\lambda_n w_n - z_n) = f_n^2 \to 0 \quad \text{in} \quad H_*^2(\Omega_2), \tag{103}$$

$$\lambda_n^2(i\lambda_n z_n + \Delta^2 w_n) = g_n^2 \to 0 \quad \text{in} \quad L^2(\Omega_2), \tag{104}$$

$$\lambda_n^2(i\lambda_n \zeta_n - z_n + \zeta_n) = h_n^2 \to 0 \quad \text{in} \quad L^2(\Gamma_2). \tag{105}$$

We notice from (98) that v_n and z_n are uniformly bounded in $L^2(\Omega_1)$ and $L^2(\Omega_2)$, respectively. It follows from Eqs. (100) and (103) that

$$\|u_n\|_{L^2(\Omega_1)} = \frac{O(1)}{\lambda_n} \quad \text{and} \quad \|w_n\|_{L^2(\Omega_2)} = \frac{O(1)}{\lambda_n}. \tag{106}$$

On the other hand, inserting equation (100) (respectively, (103)) into Eq. (101) (respectively, (104)), we obtain the following system:

$$-\lambda_n^2 u_n - \Delta u_n = \frac{i f_n^1}{\lambda_n} + \frac{g_n^1}{\lambda_n^2}, \tag{107}$$

$$-\lambda_n^2 w_n + \Delta^2 w_n = \frac{i f_n^2}{\lambda_n} + \frac{g_n^2}{\lambda_n^2}. \tag{108}$$

For clarity, the proof is divided into several lemmas:

Lemma 6 *The solution* $(u_n, v_n, \eta_n, w_n, z_n, \zeta_n) \in D(\mathring{A})$ *of system* (100)–(105) *satisfies the following asymptotic behavior estimation:*

$$\int_{\Gamma_1} |\eta_n|^2 d\Gamma = \frac{o(1)}{\lambda_n^2} \quad \textit{and} \quad \int_{\Gamma_2} |\zeta_n|^2 d\Gamma = \frac{o(1)}{\lambda_n^2}. \tag{109}$$

Proof Taking the inner product of (99) with $U_n = (u_n, v_n, \eta_n, w_n, z_n, \zeta_n)$ in $\mathcal{H}$ and then using the fact that U_n is uniformly bounded in $\mathcal{H}$, we get

$$\int_{\Gamma_1} |\eta_n|^2 d\Gamma + \int_{\Gamma_2} |\zeta_n|^2 d\Gamma = \Re\{\left((i\lambda_n \mathcal{I} - \mathring{A})U_n, U_n\right)_{\mathcal{H}}\} = \frac{o(1)}{\lambda_n^2},$$

which implies that

$$\int_{\Gamma_1} |\eta_n|^2 d\Gamma = \frac{o(1)}{\lambda_n^2} \quad \text{and} \quad \int_{\Gamma_2} |\zeta_n|^2 d\Gamma = \frac{o(1)}{\lambda_n^2}.$$

Lemma 7 *The solution* $(u_n, v_n, \eta_n, w_n, z_n, \zeta_n) \in D(\mathring{A})$ *of system* (100)–(105) *satisfies the following asymptotic behavior estimation:*

$$\|u_n\|_{L^2(\Gamma_1)} = \frac{o(1)}{\lambda_n}, \tag{110}$$

$$\|w_n\|_{L^2(\Gamma_2)} = \frac{o(1)}{\lambda_n}. \tag{111}$$

Proof First, using Eq. (102) and the first estimation of (109), we get

$$\|v_n\|_{L^2(\Gamma_1)} = o(1). \tag{112}$$

From Eq. (100) and the fact that the trace operator $\varphi \longmapsto \varphi|_{\Gamma_1}$ is a linear and continuous mapping from $H^1_*(\Omega_1)$ into $L^2(\Gamma_1)$, we have

$$i\lambda_n u_n - v_n = \frac{f_n^1}{\lambda_n^2} \to 0 \quad \text{in} \quad L^2(\Gamma_1).$$

Multiplying the above equation by $-i\lambda_n \overline{u_n}$ and integrating over Γ_1, we obtain

$$\int_{\Gamma_1} |\lambda_n u_n|^2 d\Gamma + \int_{\Gamma_1} i\lambda_n \overline{u_n} v_n d\Gamma = -\int_{\Gamma_1} \frac{i f_n^1 \overline{u_n}}{\lambda_n} d\Gamma.$$

Using estimations (106) and (112) and the fact that f_n^1 converges to zero in $L^2(\Gamma_1)$, we deduce from the above equation that

$$\int_{\Gamma_1} |\lambda_n u_n|^2 d\Gamma = o(1),$$

which gives the desired estimation (110). Next, using Eq. (105) and the second estimation of (109), we obtain

$$\|z_n\|_{L^2(\Gamma_2)} = o(1). \tag{113}$$

A similar computation performed on Eq. (103), using estimation (113) and the fact that f_n^2 converges to zero in $L^2(\Gamma_2)$, yields, as well,

$$\int_{\Gamma_2} |\lambda_n w_n|^2 d\Gamma = o(1),$$

which gives the desired estimation (111).

Lemma 8 *The solution* $(u_n, v_n, \eta_n, w_n, z_n, \zeta_n) \in D(\mathring{A})$ *of system* (100)–(105) *satisfies the following asymptotic behavior estimation:*

$$\|\partial_{\nu_1} u_n\|_{L^2(\Gamma_1)} = \frac{o(1)}{\lambda_n}, \tag{114}$$

$$\|\mathcal{B}_2 w_n\|_{L^2(\Gamma_2)} = \frac{o(1)}{\lambda_n}. \tag{115}$$

Proof From the boundary conditions of (10), we have $\partial_{\nu_1} u_n = -\eta_n$ on Γ_1 and $\mathcal{B}_2 w_n = \zeta_n$ on Γ_2. Therefore, using estimation (109), the desired asymptotic estimations (114) and (115) follow.

Lemma 9 *Assume that Assumptions 1, 2, 3, and 4 hold. Then, the solution* $(u_n, v_n, \eta_n, w_n, z_n, \zeta_n) \in D(\mathring{A})$ *of system* (100)–(105) *satisfies the following estimation:*

$$\Re\left\{\int_{\Omega_2} \Delta^2 w_n(m \cdot \nabla\overline{w_n})dx - \int_{\Omega_1} \Delta u_n(m \cdot \nabla\overline{u_n})dx\right\}$$
$$\geq \frac{1}{2}a(w_n, \overline{w_n}) - \frac{R_1^2}{\delta}\int_{\Gamma_1} |\partial_{\nu_1} u_n|^2 d\Gamma$$
$$- \frac{R_2^2 M}{2}\int_{\Gamma_2} |\mathcal{B}_2 w_n|^2 d\Gamma, \tag{116}$$

where $R_1 = \max_{x\in\Gamma_1} |m(x)|$, $R_2 = \max_{x\in\Gamma_2} |m(x)|$, *and* M *is a positive constant independent of* n.

Proof By the definition of $D(\mathring{A})$, u_n may be seen as the unique solution $u_n \in H^1(\Omega_1)$ of system (76) with $v_1 = -\eta_n$ and $v_2 = w_n$. As Assumptions 1, 3, and 4 hold, it suffices to apply Lemma 4 for u_n to obtain

$$-\Re\left\{\int_{\Omega_1} \Delta u_n(m \cdot \nabla\overline{u_n})dx\right\} \geq -\frac{R_1^2}{\delta}\int_{\Gamma_1} |\partial_{\nu_1} u_n|^2 d\Gamma$$
$$- \Re\left\{\left\langle\partial_{\nu_1} u_n, m \cdot \nabla\overline{u_n}\right\rangle_{H^{-\varepsilon}(I)\times H^{\varepsilon}(I)}\right\}, \tag{117}$$

for some $\varepsilon \in (0, \frac{1}{2})$. Again, since $U_n \in D(\mathring{A})$, we see that $w_n \in H^2(\Omega_2)$ is a solution of system (84) with $v_3 = \partial_{\nu_1} u_n$ and $v_4 = \zeta_n$. It follows from Lemma 5 and Assumptions 2 and 3 that

$$\Re\left\{\int_{\Omega_2} \Delta^2 w_n(m \cdot \nabla\overline{w_n})dx\right\} \geq \frac{1}{2}a(w_n, \overline{w_n}) - \frac{R_2^2 M}{2}\int_{\Gamma_2} |\mathcal{B}_2 w_n|^2 d\Gamma$$
$$+ \Re\left\{\langle\mathcal{B}_2 w_n, m \cdot \nabla\overline{w_n}\rangle_{H^{-\varepsilon}(I)\times H^{\varepsilon}(I)}\right\}. \tag{118}$$

Now, by adding (117) and (118), we obtain

$$\Re\left\{\int_{\Omega_2} \Delta^2 w_n(m \cdot \nabla\overline{w_n})dx\right\} - \Re\left\{\int_{\Omega_1} \Delta u_n(m \cdot \nabla\overline{u_n})dx\right\}$$
$$\geq \frac{1}{2}a(w_n, \overline{w_n}) - \frac{R_1^2}{\delta}\int_{\Gamma_1} |\partial_{\nu_1} u_n|^2 d\Gamma$$
$$- \frac{R_2^2 M}{2}\int_{\Gamma_2} |\mathcal{B}_2 w_n|^2 d\Gamma + \Re\left\{\langle\mathcal{B}_2 w_n, m \cdot \nabla\overline{w_n}\rangle_{H^{-\varepsilon}(I)\times H^{\varepsilon}(I)}\right.$$
$$\left. - \left\langle\partial_{\nu_1} u_n, m \cdot \nabla\overline{u_n}\right\rangle_{H^{-\varepsilon}(I)\times H^{\varepsilon}(I)}\right\}. \tag{119}$$

The boundary conditions $u_n = w_n$ and $\mathcal{B}_2 w_n = \partial_{\nu_1} u_n$ on the interface I, along with the geometric condition (28), lead to

$$\langle \mathcal{B}_2 w_n, m \cdot \nabla \overline{w_n} \rangle_{H^{-\varepsilon}(I)\times H^{\varepsilon}(I)} - \left\langle \partial_{\nu_1} u_n, m \cdot \nabla \overline{u_n} \right\rangle_{H^{-\varepsilon}(I)\times H^{\varepsilon}(I)}$$

$$= \left\langle \mathcal{B}_2 w_n, m \cdot (\nu_2 \partial_{\nu_2} \overline{w_n} + \tau_2 \partial_{\tau_2} \overline{w_n}) \right\rangle_{H^{-\varepsilon}(I)\times H^{\varepsilon}(I)}$$
$$- \left\langle \partial_{\nu_1} u_n, m \cdot (\nu_1 \partial_{\nu_1} \overline{u_n} + \tau_1 \partial_{\tau_1} \overline{u_n}) \right\rangle_{H^{-\varepsilon}(I)\times H^{\varepsilon}(I)}$$

$$= \left\langle \mathcal{B}_2 w_n, m \cdot \tau_2 \partial_{\tau_2} \overline{w_n} \right\rangle_{H^{-\varepsilon}(I)\times H^{\varepsilon}(I)} - \left\langle \partial_{\nu_1} u_n, m \cdot \tau_1 \partial_{\tau_1} \overline{u_n} \right\rangle_{H^{-\varepsilon}(I)\times H^{\varepsilon}(I)}$$

$$= \left\langle \mathcal{B}_2 w_n, m \cdot \tau_2 \partial_{\tau_2} \overline{w_n} \right\rangle_{H^{-\varepsilon}(I)\times H^{\varepsilon}(I)} - \left\langle \partial_{\nu_1} u_n, m \cdot \tau_2 \partial_{\tau_2} \overline{w_n} \right\rangle_{H^{-\varepsilon}(I)\times H^{\varepsilon}(I)}$$

$$= \left\langle \mathcal{B}_2 w_n - \partial_{\nu_1} u_n, m \cdot \tau_2 \partial_{\tau_2} \overline{w_n} \right\rangle_{H^{-\varepsilon}(I)\times H^{\varepsilon}(I)} = 0.$$

This, together with (119), proves (116), as desired.

Lemma 10 *Assume that Assumptions 1 and 2 hold. Then, the solution* $(u_n, v_n, \eta_n, w_n, z_n, \zeta_n) \in D(\mathring{A})$ *of system* (100)–(105) *satisfies the following asymptotic behavior estimation:*

$$\int_{\Omega_1} |\lambda_n u_n|^2 \, dx + \int_{\Omega_2} |\lambda_n w_n|^2 \, dx = \int_{\Omega_1} |\nabla u_n|^2 \, dx + a(w_n, \overline{w_n}) + \frac{o(1)}{\lambda_n^2}. \tag{120}$$

Proof Before starting the proof, we assume that $u_n \in H^2(\Omega_1)$ and $w_n \in H^4(\Omega_2)$. Multiplying equation (107) (respectively (108)) by $-\overline{u_n}$ (resp. $-\overline{w_n}$), integrating over Ω_1 (respectively, Ω_2), and applying Green's formula, we obtain

$$\int_{\Omega_1} |\lambda_n u_n|^2 \, dx - \int_{\Omega_1} |\nabla u_n|^2 \, dx + \int_{\partial\Omega_1} \partial_{\nu_1} u_n \overline{u_n} d\Gamma = -\int_{\Omega_1} \left(\frac{i f_n^1}{\lambda_n} + \frac{g_n^1}{\lambda_n^2} \right) \overline{u_n} dx, \tag{121}$$

and

$$\int_{\Omega_2} |\lambda_n w_n|^2 \, dx - a(w_n, \overline{w_n}) + \int_{\partial\Omega_2} \left(\mathcal{B}_1 w_n \partial_{\nu_2} \overline{w_n} - \mathcal{B}_2 w_n \overline{w_n} \right) d\Gamma$$
$$= -\int_{\Omega_2} \left(\frac{i f_n^2}{\lambda_n} + \frac{g_n^2}{\lambda_n^2} \right) \overline{w_n} dx. \tag{122}$$

Taking into consideration that f_n^1, g_n^1 converge to zero in $L^2(\Omega_1)$ and f_n^2, g_n^2 converge to zero in $L^2(\Omega_2)$ and using (106), we get from (121) and (122) that

$$\int_{\Omega_1} |\lambda_n u_n|^2 \, dx - \int_{\Omega_1} |\nabla u_n|^2 \, dx + \int_{\partial\Omega_1} \partial_{\nu_1} u_n \overline{u_n} d\Gamma = \frac{o(1)}{\lambda_n^2},$$

and

$$\int_{\Omega_2} |\lambda_n w_n|^2 \, dx - a(w_n, \overline{w_n}) + \int_{\partial\Omega_2} \left(\mathcal{B}_1 w_n \partial_{\nu_2} \overline{w_n} - \mathcal{B}_2 w_n \overline{w_n}\right) d\Gamma = \frac{o(1)}{\lambda_n^2}.$$

Adding both equations and taking into account the boundary conditions on the interface and Γ_2, as well as the estimations (110), (111), (114), and (115), we obtain

$$\int_{\Omega_1} |\lambda_n u_n|^2 \, dx - \int_{\Omega_1} |\nabla u_n|^2 \, dx + \int_{\Omega_2} |\lambda_n w_n|^2 \, dx - a(w_n, \overline{w_n}) = \frac{o(1)}{\lambda_n^2}.$$

By following a similar argument to the end of the proof of Lemma 4 and Lemma 5, we obtain the desired estimation (120).

Lemma 11 *Assume that Assumptions 1, 2, 3, and 4 hold. Then, the solution* $(u_n, v_n, \eta_n, w_n, z_n, \zeta_n) \in D(\mathring{A})$ *of system* (100)–(105) *satisfies the following asymptotic behavior estimation:*

$$\int_{\Omega_1} |\lambda_n u_n|^2 \, dx = o(1), \quad \int_{\Omega_2} |\lambda_n w_n|^2 \, dx = o(1) \quad \text{and} \quad a(w_n, \overline{w_n}) = o(1). \tag{123}$$

Proof Multiplying equation (107) by $m \cdot \nabla \overline{u_n}$ and integrating over Ω_1, and similarly for Eq. (108) over Ω_2, we obtain

$$\begin{aligned} &-\int_{\Omega_1} \lambda_n^2 u_n \,(m \cdot \nabla \overline{u_n}) \, dx - \int_{\Omega_1} \Delta u_n \,(m \cdot \nabla \overline{u_n}) \, dx \\ &= \int_{\Omega_1} \left(\frac{i f_n^1}{\lambda_n} + \frac{g_n^1}{\lambda_n^2}\right) (m \cdot \nabla \overline{u_n}) \, dx, \end{aligned} \tag{124}$$

and

$$\begin{aligned} &-\int_{\Omega_2} \lambda_n^2 w_n \,(m \cdot \nabla \overline{w_n}) \, dx + \int_{\Omega_2} \Delta^2 w_n \,(m \cdot \nabla \overline{w_n}) \, dx \\ &= \int_{\Omega_2} \left(\frac{i f_n^2}{\lambda_n} + \frac{g_n^2}{\lambda_n^2}\right) (m \cdot \nabla \overline{w_n}) \, dx. \end{aligned} \tag{125}$$

Using the Cauchy–Schwartz inequality and the fact that $\sqrt{a(w_n, \overline{w_n})}$ is equivalent to the usual norm of $H^2(\Omega_2)$ on $H^2_*(\Omega_2)$, we can observe that ∇u_n is uniformly bounded in $L^2(\Omega_1)$ and $a(w_n, \overline{w_n})$ is uniformly bounded. Additionally, f_n^1 and g_n^1 converge to zero in $L^2(\Omega_1)$, while f_n^2 and g_n^2 converge to zero in $L^2(\Omega_2)$. Hence, we obtain

$$\begin{aligned}
&\left| \int_{\Omega_1} \left(\frac{i f_n^1}{\lambda_n} + \frac{g_n^1}{\lambda_n^2} \right) (m \cdot \nabla \overline{u_n})\, dx \right| \\
&\quad \leq \|m\|_\infty \left(\frac{\|f_n^1\|_{L^2(\Omega_1)}}{\lambda_n} + \frac{\|g_n^1\|_{L^2(\Omega_1)}}{\lambda_n^2} \right) \|\nabla u_n\|_{L^2(\Omega_1)} = \frac{o(1)}{\lambda_n},
\end{aligned} \tag{126}$$

and

$$\begin{aligned}
&\left| \int_{\Omega_2} \left(\frac{i f_n^2}{\lambda_n} + \frac{g_n^2}{\lambda_n^2} \right) (m \cdot \nabla \overline{w_n})\, dx \right| \\
&\quad \leq \|m\|_\infty \left(\frac{\|f_n^2\|_{L^2(\Omega_2)}}{\lambda_n} + \frac{\|g_n^2\|_{L^2(\Omega_2)}}{\lambda_n^2} \right) \|\nabla w_n\|_{L^2(\Omega_2)} = \frac{o(1)}{\lambda_n}.
\end{aligned} \tag{127}$$

On the other hand, applying Green's formula to the first terms of (124) and (125) yields

$$\int_{\Omega_1} \lambda_n^2 u_n\, (m \cdot \nabla \overline{u_n})\, dx = \frac{1}{2} \int_{\partial\Omega_1} (m \cdot \nu_1)\, |\lambda_n u_n|^2\, d\Gamma - \int_{\Omega_1} |\lambda_n u_n|^2\, dx, \tag{128}$$

and

$$\int_{\Omega_2} \lambda_n^2 w_n\, (m \cdot \nabla \overline{w_n})\, dx = \frac{1}{2} \int_{\partial\Omega_2} (m \cdot \nu_2)\, |\lambda_n w_n|^2\, d\Gamma - \int_{\Omega_2} |\lambda_n w_n|^2\, dx. \tag{129}$$

Then, substituting (128) (respectively, (129)) into (124) (respectively, (125)) and considering the estimates (126) and (127), we obtain

$$\int_{\Omega_1} |\lambda_n u_n|^2\, dx - \frac{1}{2} \int_{\partial\Omega_1} (m \cdot \nu_1)\, |\lambda_n u_n|^2\, d\Gamma - \int_{\Omega_1} \Delta u_n\, (m \cdot \nabla \overline{u_n})\, dx = \frac{o(1)}{\lambda_n},$$

and

$$\int_{\Omega_2} |\lambda_n w_n|^2\, dx - \frac{1}{2} \int_{\partial\Omega_2} (m \cdot \nu_2)\, |\lambda_n w_n|^2\, d\Gamma + \int_{\Omega_2} \Delta^2 w_n\, (m \cdot \nabla \overline{w_n})\, dx = \frac{o(1)}{\lambda_n}.$$

Adding both equations, using the boundary condition $u_n = w_n$ on the interface I and the estimates (110) and (111), and then taking the real part, we obtain

$$\int_{\Omega_1} |\lambda_n u_n|^2\, dx + \int_{\Omega_2} |\lambda_n w_n|^2\, dx + \Re\left\{\int_{\Omega_2} \Delta^2 w_n (m \cdot \nabla\overline{w_n}) dx \right.$$

$$\left. - \int_{\Omega_1} \Delta u_n (m \cdot \nabla\overline{u_n}) dx \right\} = o(1). \tag{130}$$

As Assumptions 1, 2, 3, and 4 hold, it suffices to apply Lemma 9. Hence, inserting (116) into (130), we get

$$\int_{\Omega_1} |\lambda_n u_n|^2\, dx + \int_{\Omega_2} |\lambda_n w_n|^2\, dx + \frac{1}{2} a(w_n, \overline{w_n}) - \frac{R_1^2}{\delta} \int_{\Gamma_1} |\partial_{\nu_1} u_n|^2 d\Gamma$$

$$- \frac{R_2^2 M}{2} \int_{\Gamma_2} |\mathcal{B}_2 w_n|^2 d\Gamma = o(1). \tag{131}$$

Finally, using the estimations (114) and (115), it follows from (131) that

$$\int_{\Omega_1} |\lambda_n u_n|^2\, dx + \int_{\Omega_2} |\lambda_n w_n|^2\, dx + \frac{1}{2} a(w_n, \overline{w_n}) = o(1),$$

which proves the estimation (123), as desired.

Proof of Theorem 4 Using (120) and (123), we deduce that

$$\int_{\Omega_1} |\nabla u_n|^2 dx = o(1). \tag{132}$$

Therefore, combining the estimates (109), (123), and (132), we obtain $\|U_n\|_{\mathcal{H}} \longrightarrow 0$, which leads to the desired contradiction with (98). Consequently, condition $(P2)$ holds, and this permits us to conclude that the energy of system (1)–(4) decays polynomially to zero as t goes to infinity. The proof is thus complete.

6 Conclusion and Open Problems

This chapter focuses on investigating the stabilization of a transmission wave-plate model, where the coupling occurs solely through the interface and where we employ only two dynamical boundary controls. Using a general criterion proposed by Arendt–Batty, we have established the system's strong stability upon specific geometric conditions. Our analysis reveals a polynomial decay rate for the system's energy, characterized by $1/t$.

An intriguing question arising from the problem studied in this chapter is the stabilization when the wave-plate system is damped by two dynamical boundary controls η and ξ ($\zeta = 0$). We conjecture that the removal of control ζ will not affect the system's polynomial stability, and the decay rate will persist at the same level.

Acknowledgments Mr. Ali Wehbe would like to thank the LAMA Laboratory of Mathematics at the Université Savoie Mont Blanc for their support in the research development project. Ms. Zahraa Abdallah is grateful to the Doctoral School of Science and Technology at the Lebanese University for their financial support through the doctoral scholarship.

References

1. Z. Abdallah, S. Gerbi, C. Kassem, A. Wehbe, Stability for an interface transmission problem of wave-plate equations with dynamical boundary controls. Acta Applicandae Mathematicae **188**(12) (2024). https://doi.org/10.1007/s10440-023-00611-4
2. K. Ammari, M. Mehrenberger, Study of the nodal feedback stabilization of a string-beams network. J. Appl. Math. Comput. **36**, 441–458 (2011)
3. K. Ammari, S. Nicaise, Stabilization of a transmission wave/plate equation. J. Differ. Equ. **249**, 707–727 (2010)
4. W. Arendt, C.J. Batty, Tauberian theorems and stability of one-parameter semigroups. Trans. Amer. Math. Soc. **306**, 837–852 (1988)
5. C.J. Batty, T. Duyckaerts, Non-uniform stability for bounded semi-groups on Banach spaces. J. Evolut. Equ. **8**, 765–780 (2008)
6. C.D. Benchimol, A note on weak stabilizability of contraction semigroups. SIAM J. Control Optim. **16**, 373–379 (1978)
7. S.K. Biswas, N. Ahmed, Optimal control of large space structures governed by a coupled system of ordinary and partial differential equations. Math. Control Signals Syst. **2**, 1–18 (1989)
8. H. Blum, R. Rannacher, R. Leis, On the boundary value problem of the biharmonic operator on domains with angular corners. Math. Methods Appl. Sci. **2**, 556–581 (1980)
9. A. Borichev, Y. Tomilov, Optimal polynomial decay of functions and operator semigroups. Math. Ann. **347**, 455–478 (2010)
10. S.C. Brenner, L.R. Scott, L.R. Scott, *The Mathematical Theory of Finite Element Methods*, vol. 3 (Springer, Berlin, 2008)
11. M. Dauge, *Elliptic Boundary Value Problems on Corner Domains: Smoothness and Asymptotics of Solutions*, vol. 1341 (Springer, Berlin, 2006)
12. P. Grisvard, *Elliptic Problems in Nonsmooth Domains*. Pitman Monographs and Surveys in Pure and Applied Mathematics, vol. 24 (SIAM, Philadelphia, 1985).
13. Y.-P. Guo, J.-M. Wang, D.-X. Zhao, Energy decay estimates for a two-dimensional coupled wave-plate system with localized frictional damping. ZAMM-J. Appl. Math. Mechan./Zeitschrift für Angewandte Mathematik und Mechanik **100**, e201900030 (2020)
14. V. Komornik, *Exact Controllability and Stabilization: The Multiplier Method*, vol. 36 (Elsevier Masson, Gainesville, 1994)
15. J.E. Lagnese, *Boundary Stabilization of Thin Plates* (SIAM, Philadelphia, 1989)
16. C. Lattanzio, A. Maurizi, B. Piccoli, Moving bottlenecks in car traffic flow: a PDE-ODE coupled model. SIAM J. Math. Analy. **43**, 50–67 (2011)
17. J. Le Rousseau, G. Lebeau, On Carleman estimates for elliptic and parabolic operators. Applications to unique continuation and control of parabolic equations. ESAIM: Control Optim. Calculus Variat. **18**, 712–747 (2012)
18. Y.-F. Li, Z.-J. Han, G.-Q. Xu, Explicit decay rate for coupled string-beam system with localized frictional damping. Appl. Math. Lett. **78**, 51–58 (2018)
19. Z. Liu, B. Rao, Characterization of polynomial decay rate for the solution of linear evolution equation. Zeitschrift für angewandte Mathematik und Physik ZAMP **56**, 630–644 (2005)
20. Z.-H. Luo, B.-Z. Guo, Shear force feedback control of a single-link flexible robot with a revolute joint. IEEE Trans. Autom. Control **42**, 53–65 (1997)

21. O. Morgul, Dynamic boundary control of a Euler-Bernoulli beam. IEEE Trans. Autom. Control **37**, 639–642 (1992)
22. Ö. Morgül, Dynamic boundary control of the Timoshenko beam. Automatica **28**, 1255–1260 (1992)
23. Ö. Morgül, Control and stabilization of a rotating flexible structure. Automatica **30**, 351–356 (1994)
24. A. Pazy, *Semigroups of Linear Operators and Applications to Partial Differential Equations*, vol. 44 (Springer Science & Business Media, Cham, 2012)
25. B. Rao, Stabilization of Kirchhoff plate equation in star-shaped domain by nonlinear boundary feedback. Nonlinear Analy. Theory Methods Appl. **20**, 605–626 (1993)
26. B. Rao, L. Toufayli, A. Wehbe, Stability and controllability of a wave equation with dynamical boundary control. Math. Control Related Fields **5**, 305 (2015)
27. B. Rao, A. Wehbe, Polynomial energy decay rate and strong stability of Kirchhoff plates with non-compact resolvent. J. Evolut. Equ. **5**, 137–152 (2005)
28. D.L. Russell, A general framework for the study of indirect damping mechanisms in elastic systems. J. Math. Analy. Appl. **173**, 339–358 (1993)
29. M. Slemrod, Feedback stabilization of a linear control system in hilbert space with an a priori bounded control. Math. Control Signals Syst. **2**, 265–285 (1989)
30. X. Zhang, E. Zuazua, Polynomial decay and control of a 1- d hyperbolic-parabolic coupled system. J. Differ. Equ. **204**, 380–438 (2004)
31. Q. Zhang, J.-M. Wang, B.-Z. Guo, Stabilization of the Euler-Bernoulli equation via boundary connection with heat equation. Math. Control Signals Syst. **26**, 77–118 (2014)

Uniform Stabilization of an Acoustic System

Kaïs Ammari, Fathi Hassine, and Luc Robbiano

2010 Mathematics Subject Classification 35L04, 37L15

1 Introduction

Sound waves propagate due to the compressibility of a medium. In fact, it is a balance between the compressibility and the inertia of fluid that governs the propagation of sound waves through it. The assumptions made for the acoustic wave equation are gravitational forces is neglected, dissipative effects are neglected, the medium (fluid) is homogeneous, isotropic and perfectly elastic. In addition to these assumptions we also assume that particle velocity is small, and there are only very small perturbations (fluctuations) to the equilibrium pressure and density.

We consider the following acoustic damping system of equations:

K. Ammari
LR Analysis and Control of PDEs, UR 22ES03, Department of Mathematics, Faculty of Sciences of Monastir, University of Monastir, Monastir, Tunisia

Université Paris-Saclay, UVSQ, CNRS, Laboratoire de Mathématiques de Versailles, Université de Paris-Sud, Versailles, France
e-mail: kais.ammari@fsm.rnu.tn

F. Hassine
LR Analysis and Control of PDEs, UR 22ES03, Department of Mathematics, Faculty of Sciences of Monastir, University of Monastir, Monastir, Tunisia
e-mail: fathi.hassine@fsm.rnu.tn

L. Robbiano (✉)
Université Paris-Saclay, UVSQ, CNRS, Laboratoire de Mathématiques de Versailles, Versailles, France
e-mail: luc.robbiano@uvsq.fr

K. Ammari et al. (eds.), *Advances in Partial Differential Equations and Control*, Trends in Mathematics, https://doi.org/10.1007/978-3-031-62265-6_2

$$\begin{cases} u_t + \nabla r + b\,u = 0, \ \text{in } \Omega \times \mathbb{R}^+, \\ r_t + \operatorname{div} u = 0, \ \text{in } \Omega \times \mathbb{R}^+, \\ u \cdot n = 0, \ \text{on } \Gamma \times \mathbb{R}^+, \\ u(0,x) = u^0(x),\ r(0,x) = r^0(x),\ x \in \Omega, \end{cases} \tag{1}$$

where Ω is a bounded domain in $\mathbb{R}^d$, $d \geq 2$, with a smooth boundary Γ, $\operatorname{div} = \nabla \cdot$ is the divergence operator and $b \in L^\infty(\Omega)$, with $b \geq 0$ on Ω. As usual n denotes the unit outward normal vector along Γ.

The system of Eqs. (1) is a linearization of the *acoustic equation* governing the propagation of acoustic waves in a compressible medium, see Lighthill [18–20], where $b\,u$ represents a damping term of Brinkman type. This kind of damping arises also in the process of homogenization (see Allaire [1]) and is frequently used as a suitable *penalization* in fluid mechanics models, see Angot, Bruneau, and Fabrie [5]. Our main goal is to prove the exponential decay of solutions of (1) with growing time.

Let $L^2(\Omega)$ denote the standard Hilbert space of square integrable functions in Ω and its closed subspace $L^2_m(\Omega) = \{f \in L^2(\Omega) : \int_\Omega f(x)\,dx = 0\}$. To avoid abuse of notation, we shall write $\|\cdot\|$ for the $L^2(\Omega)$-norm or the $L^2(\Omega)^d$-norm.

Denoting $H = (L^2(\Omega))^d \times L^2_m(\Omega)$, we introduce the operator

$$\mathcal{A} = \begin{pmatrix} 0 & \nabla \\ \operatorname{div} & 0 \end{pmatrix} : \mathcal{D}(\mathcal{A}) = \left\{(u,r) \in H,\ (\nabla r,\ \operatorname{div} u) \in H,\ u \cdot n_{|\Gamma} = 0\right\} \subset H \to H,$$

and

$$\mathcal{B} = \begin{pmatrix} \sqrt{b} \\ 0 \end{pmatrix} \in \mathcal{L}((L^2(\Omega))^d, H),\ \mathcal{B}^* = \begin{pmatrix} \sqrt{b} & 0 \end{pmatrix} \in \mathcal{L}(H, (L^2(\Omega))^d).$$

We recall that for $u \in (L^2(\Omega))^d$ with $\operatorname{div} u \in L^2(\Omega)$, $u \cdot n_{|\Gamma}$ make sense in $H^{-1/2}(\Gamma)$ (see Girault-Raviart [14, Chap 1, Theorem 2.5]).

Accordingly, the problem (1) can be recasted in an abstract form:

$$\begin{cases} Z_t(t) + \mathcal{A}Z(t) + \mathcal{B}\mathcal{B}^* Z(t) = 0,\ t > 0, \\ Z(0) = Z^0, \end{cases} \tag{2}$$

where $Z = (u,r)$, or, equivalently,

$$\begin{cases} Z_t(t) = \mathcal{A}_d Z(t),\ t > 0, \\ Z(0) = Z^0, \end{cases} \tag{3}$$

with $\mathcal{A}_d = -\mathcal{A} - \mathcal{B}\mathcal{B}^*$ with $\mathcal{D}(\mathcal{A}_d) = \mathcal{D}(\mathcal{A})$.

It can be shown (see [2]) that for any initial data $(u^0, r^0) \in \mathcal{D}(\mathcal{A})$ the problem (1) admits a unique solution

$$(u, r) \in C([0, \infty); \mathcal{D}(\mathcal{A})) \cap C^1([0, \infty); H).$$

Moreover, the solution (u, r) satisfies, the energy identity

$$E(0) - E(t) = \int_0^t \left\| \sqrt{b}\, u(s) \right\|^2_{(L^2(\Omega))^d} \mathrm{d}s, \text{ for all } t \geq 0 \tag{4}$$

with

$$E(t) = \frac{1}{2} \, \|(u(t), r(t))\|^2_H, \; \forall t \geq 0, \tag{5}$$

where we have denoted

$$\langle (u, r), (v, p) \rangle_H = \int_\Omega (u(x).v(x) + r(x)p(x)) \, \mathrm{d}x,$$

$$\|(u, r)\|_H = \sqrt{\int_\Omega \left(|u(x)|^2 + r^2(x) \right) \mathrm{d}x}.$$

Using (4) and a standard density argument, we can extend the solution operator for data $(u^0, r^0) \in H$. Consequently, we associate with the problem (1) (or to the abstract Cauchy problems (2) or (3)) a semigroup that is globally bounded in H.

As the energy E is nonincreasing along trajectories, we want to determine the set of initial data (u^0, r^0) for which

$$E(t) \to 0 \text{ as } t \to \infty. \tag{6}$$

Such a question is of course intimately related to the structural properties of the function b, notably to the geometry of the set ω on which the damping is effective. In fact, when the damping term is globally distributed Ammari, Feireisl and Nicaise [2] showed an exponential decay rate of the energy by the means of an observability inequality associated with the conservative problem of (1). Besides, it is also shown that if the damping coefficient is not uniformly positive definite (i.e $\inf\limits_{x \in \Omega} b(x) = 0$) then the system (1) may be no exponentially stable. In this paper we consider a damping which is locally distributed over the domain Ω with a geometrical control conditions. In this article we prove uniform decay as given in [2] but under more restrictive geometrical conditions on the damping coefficient. More precisely, we prove exponential decay rates of the energy.

The paper is organized as follows. Section 2 summarizes some well known facts concerning the acoustic system (1). In Sect. 3, we introduce some tools and notation as Riemannian geometry, pseudo-differential operators, and Gearhart-Huang-Prüss theorem allowing to prove exponential decay from estimates on the resolvent. In Sect. 4 we introduce a semiclassical measure and we prove some properties, first the measure is supported on characteristic set and is not identically null. In Sect. 5

we prove local properties in a neighborhood of boundary. In Sect. 6 we give some properties on semiclassical measure. The complete proof is in [4].

2 Preliminaries and Main Results

We start with a simple observation that the problem (1) can be viewed as a bounded (in H) perturbation of the conservative system

$$\begin{cases} u_t + \nabla r = 0, & \text{in } \Omega \times \mathbb{R}^+, \\ r_t + \operatorname{div} u = 0, & \text{in } \Omega \times \mathbb{R}^+, \end{cases} \tag{7}$$

which can be recast as the standard *wave equation*

$$r_{tt} - \Delta r = 0.$$

Consequently, the basic existence theory for (1) derives from that of (7). Hence $\mathcal{A}_d$ generates a C_0-semigroup $(S(t))_{t\geq 0}$ in H that is even of contraction because $\mathcal{A}_d$ is dissipative (see (4)).

The first main difficulty is that the operator $\mathcal{A}_d$ possesses a non-trivial (and large) kernel that is left invariant by the evolution. Indeed if (u, r) belongs to $\ker \mathcal{A}_d$, then it is solution of the "stationary" problem

$$\nabla r + bu = 0, \quad \operatorname{div} u = 0, \quad \text{in } \Omega. \tag{8}$$

Thus multiplying the first identity of (8) by $\overline{u}$ and integrating over Ω yields

$$\int_\Omega (\nabla r \cdot \overline{u} + b|u|^2)\, dx = 0.$$

By an integration by parts, using the fact that u is solenoidal and the boundary condition $u \cdot n = 0$ on Γ, we get

$$\int_\Omega \nabla r \cdot \overline{u}\, dx = \int_\Omega r \operatorname{div} \overline{u}\, dx = 0,$$

and therefore we obtain

$$\int_\Omega b|u|^2\, dx = 0.$$

In other words, we have

$$u = 0 \text{ on } \operatorname{supp} b,$$

and coming back to (8), we find

$$\nabla r = 0.$$

Accordingly, we have shown that

$$\ker \mathcal{A}_d = \{(u, 0) \in \mathcal{D}(\mathcal{A}) \mid \operatorname{div} u = 0,\ u|_{\text{supp } b} = 0,\ u \cdot n|_{\Gamma} = 0\}.$$

For shortness set $E = \ker \mathcal{A}_d$ and introduce also its orthogonal complement H_0 in H.

It is easy to check that

$$\langle \mathcal{A}_d(w, s), (u, r) \rangle_H = 0 \text{ for any } (w, s) \in \mathcal{D}(\mathcal{A}),\ (u, r) \in E;$$

in particular, the semigroup associated with (1) leaves both E and H_0 invariant. Consequently, the decay property (6) may only hold for initial data emanating from the set H_0.

The following observation can be shown by a simple density argument:

Lemma 1 *The solution* (u, r) *of* (1) *with initial datum in* $\mathcal{D}(\mathcal{A}_d)$ *satisfies*

$$E'(t) = -\int_{\Omega} b\,|u|^2\,dx \leq 0. \tag{9}$$

Therefore the energy is nonincreasing and (4) *holds for all initial datum in* H.

As already shown in the above, the strong stability result (6) may hold only if we take the initial data

$$(u^0, r^0) \in H_0 = \ker[\mathcal{A}_d]^{\perp}.$$

There are several ways how to show (6), here we make use of the following result due to Arendt and Batty [6]:

Theorem 1 *Let* $(T(t))_{t \geq 0}$ *be a bounded* C_0*-semigroup on a reflexive Banach space* X. *Denote by* A *the generator of* $(T(t))$ *and by* $\sigma(A)$ *the spectrum of* A. *If* $\sigma(A) \cap i\mathbb{R}$ *is countable and no eigenvalue of* A *lies on the imaginary axis, then* $\lim_{t \to +\infty} T(t)x = 0$ *for all* $x \in X$.

In view of this theorem we need to identify the spectrum of $\mathcal{A}_d$ lying on the imaginary axis, and we have according to [2]:

- Suppose that $|\omega| > 0$. If λ is a non-zero real number, then $i\lambda$ is not an eigenvalue of $\mathcal{A}_d$.
- Suppose that $|\omega| > 0$. If λ is a non-zero real number, then $i\lambda$ belongs to the resolvent set $\rho(\mathcal{A}_d)$ of $\mathcal{A}_d$.

Now, Theorem 1 leads to

Corollary 1 ([2]) *Let (u, r) be the unique semigroup solution of the problem (1) emanating from the initial data $(u^0, r^0) \in H$. Let P_E be the orthogonal projection onto the space $E = \ker[\mathcal{A}_d]$ in H, and let*

$$(w, s) = P_E(u^0, r^0).$$

Then

$$\|(u, r)(t, \cdot) - (w, s)\|_H \to 0 \text{ as } t \to \infty.$$

Next we give uniform decay result under the Geometric Control Condition (GCC).

Definition 1 Let Ω be a smooth open domain in $\mathbb{R}^d$. Let ω be an open subset of Ω. We assume the geodesic associated with Δ are not tangent to infinity order to the boundary. We say that ω satisfies (GCC) if every generalized bicharacteristics starting from Ω meet ω.

We give an example of a geodesic in Fig. 1. The assumption is all geodesic must hit ω. Precise definition is given in [4].

We now state the main results of this article.

Theorem 2 *Let (u, r) be the unique semigroup solution of the problem (1) emanating from the initial data $(u^0, r^0) \in H$. Let P_E be the orthogonal projection onto the space $E = \ker[\mathcal{A}_d]$ in H, and let $(w, s) = P_E(u^0, r^0)$. We assume that the*

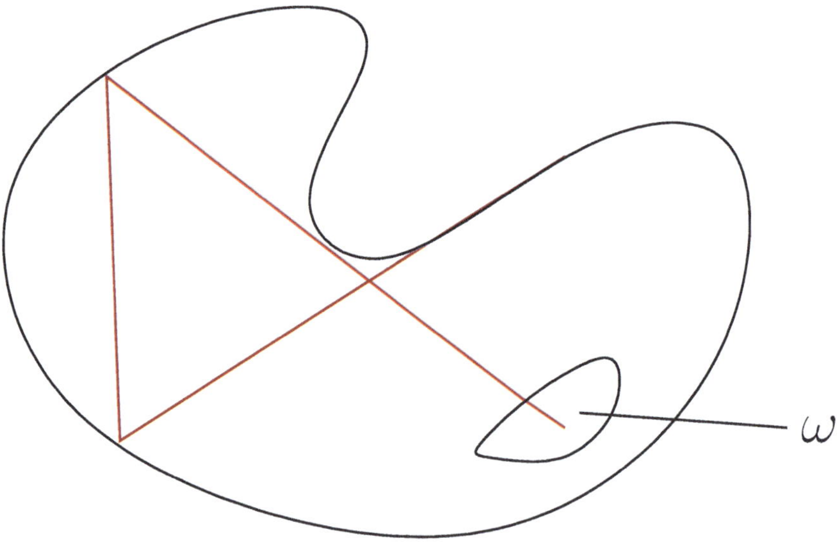

Fig. 1 The geodesic in red can glide on boundary, reflect on boundary following law of optic, or be glancing at a point of boundary, and hits ω

interior of the support of b, $\omega = \{x \in \Omega; b(x) > 0\}$, *satisfies the geometric control condition (GCC). Then,*

$$\|(u,r)(t,\cdot) - (w,s)\|_H \le C\, e^{-ct}\, \|(u^0,r^0)\|_H,\ \forall t \ge 0, \tag{10}$$

for some $C, c > 0$ *independent of* (u^0, r^0).

Before proving the previous theorem, we need some tools in Riemannian geometry, pseudo-differential calculus. Our goal is to apply characterization of exponential decay by an estimate of the resolvent on imaginary axis. To prove such an estimate we use semiclassical measure. In next section we give all the tools used in main part of the proof.

3 Tools

In this section we give some tools on pseudo-differential calculus and result on stabilization. If theses tools are classical, this introduction allows us to fix the notation used in this article.

3.1 Pseudo-Differential Tools

We recall in this section the notation and main definitions on pseudo-differential operators and semiclassical calculus. Let $a(x,\xi') \in \mathcal{C}(\mathbb{R}^d \times \mathbb{R}^{d-1})$, we say $a \in S^m$ if for all $\alpha \in \mathbb{N}^d$ and $\beta \in \mathbb{N}^{d-1}$ there exist $C_{\alpha,\beta}$ such that

$$|\partial_x^\alpha \partial_{\xi'}^\beta a(x,\xi')| \le C_{\alpha,\beta} \langle \xi' \rangle^{m-|\beta|},\ \text{for all } x \in \mathbb{R}^d, \xi' \in \mathbb{R}^{d-1}, \tag{11}$$

where $\alpha = (\alpha_1, \dots, \alpha_d)$, $|\alpha| = \alpha_1 + \cdots + \alpha_d$, $\partial_x^\alpha = \partial_{x_1}^{\alpha_1} \cdots \partial_{x_d}^{\alpha_d}$, and $\langle \xi' \rangle = (1 + |\xi'|^2)^{1/2}$.

An operator is associated with a symbol by the following formula

$$\mathrm{op}_{\mathrm{sc}}(a)u = \frac{1}{(2\pi)^{d-1}} \int_{\mathbb{R}^{d-1}} e^{ix'\xi'} a(x, h\xi') \hat{u}(\xi', x_d) d\xi', \tag{12}$$

where u is in Schwartz space, and

$$\hat{u}(\xi', x_d) = \int_{\mathbb{R}^{d-1}} e^{-ix'\xi'} u(x', x_d) dx'.$$

The symbolic calculus is given by the following results. Let $a \in S^m$ and $b \in S^q$.

Composition

There exists $c \in S^{m+q}$ such that $\mathrm{op}_{\mathrm{sc}}(a)\mathrm{op}_{\mathrm{sc}}(b) = \mathrm{op}_{\mathrm{sc}}(c)$ and there exists $c_{m+q-1} \in S^{m+q-1}$ such that $c = ab + hc_{m+q-1}$.

Commutator

There exists $d \in S^{m+q-1}$ such that $[\mathrm{op}_{\mathrm{sc}}(a), \mathrm{op}_{\mathrm{sc}}(b)] = h\mathrm{op}_{\mathrm{sc}}(d)$ and there exists $d_{m+q-2} \in S^{m+q-2}$ such that $d = -i\{a, b\} + hd_{m+q-2}$.

We also have an exact formula if a is a differential operator, and we shall use the following formula $[-ih\partial_d, \mathrm{op}_{\mathrm{sc}}(b)] = -ih\mathrm{op}_{\mathrm{sc}}(\{\xi_d, b\})$. Observe that $\{\xi_d, b\} = \partial_d b$ which is a tangential symbol.

Adjoint

There exists $\tilde{c} \in S^m$ such that $\mathrm{op}_{\mathrm{sc}}(a)^* = \mathrm{op}_{\mathrm{sc}}(\tilde{c})$ and there exists $\tilde{c}_{m-1} \in S^{m-1}$ such that $\tilde{c} = \bar{a} + hc_{m-1}$.

Operators Bounded on L^2

If $a \in S^0$ then there exists $C > 0$ such that $\|\mathrm{op}_{\mathrm{sc}}(a)u\|_{L^2} \leq C\|u\|_{L^2}$ for all u is in Schwartz space. Observe from this formula we can extend by density $\mathrm{op}_{\mathrm{sc}}(a)$ to an bounded operator on L^2. We keep the same notation if $\mathrm{op}_{\mathrm{sc}}(a)$ act on a function depending only on x'. In this case the estimate take the following form $|\mathrm{op}_{\mathrm{sc}}(a)u|_{L^2} \leq C|u|_{L^2}$. When no confusion is possible, we shall write $\|.\|$ instead of $\|.\|_{L^2} = \|.\|_{L^2(\mathbb{R}^d)}$ and $|.|$ instead of $|.|_{L^2} = |.|_{L^2(\mathbb{R}^{d-1})}$.

The constant C in the estimation of L^2-norm depends on a finite numbers of semi-norms of the symbol a. We have a precise result for h sufficiently small.

Let $a \in S^0$, there exists $C > 0$ depending on a finite numbers of semi-norms of the symbol a such that

$$\|\mathrm{op}_{\mathrm{sc}}(a)u\| \leq 2\|a\|_\infty\|u\| + C\|u\|h, \text{ for all } h \in (0, 1), \tag{13}$$

where $\|a\|_\infty = \sup_{x\in\mathbb{R}^d, \xi'\in\mathbb{R}^{d-1}} |a(x, \xi')|$. Observe that 2 in this estimate is arbitrary and may be replaced by $1 + \delta$ for all $\delta > 0$ but in this case C also depends on δ.

For $s \in \mathbb{R}$, let $\Lambda^s = \mathrm{op}_{\mathrm{sc}}(\langle\xi'\rangle^s)$. We deduce from L^2 bound an estimation for operators with symbol $a \in S^m$, we have

$$\|\mathrm{op}_{\mathrm{sc}}(a)u\| \lesssim \|\Lambda^m u\|.$$

We can define H^s_{sc}-norm by $\|u\|_{H^s_{sc}} = \|\Lambda^s u\|$.

Gårding Inequality

We state the Gårding inequality adapted to our purpose. We assume there exist $K \subset \mathbb{R}^d$ a compact, $W \subset \mathbb{R}^{d-1}$ satisfying for a fix $R_0 > 0$, if $|\xi| \geq R_0$ then for all $\lambda \geq 1$, $\lambda\xi \in W$. Let $m \in \mathbb{R}$ and $a \in S^m$, we assume there exists $C_0 > 0$ such that a satisfies $a(x, \xi') \geq C_0|\xi'|^m$ for all $x \in K$, $\xi' \in W \cap \{\xi',\ |\xi'| \geq R_0\}$. Let $\chi \in S^0$ supported in $K \times W$, there exist $C_1 > 0$ and $C_2 > 0$ such that

$$\Re(\mathrm{op}_{\mathrm{sc}}(a)\mathrm{op}_{\mathrm{sc}}(\chi)u, \mathrm{op}_{\mathrm{sc}}(\chi)u) \geq C\|\Lambda^{m/2}\mathrm{op}_{\mathrm{sc}}(\chi)u\|^2 - C_2h\|\Lambda^{(m-1)/2}u\|^2, \tag{14}$$

for all u in Schwartz space.

3.2 *Uniform Stabilization*

The proof of the first part of Theorem 2 is based on Gearhart-Huang-Prüss result we recall below.

Theorem 3 *Let A be a generator of a semigroup e^{tA} on H. We assume the following conditions*

1. $\exists M > 0,\ \|e^{tA}\|_{\mathcal{L}(H)} \leq M$ *for all* $t \geq 0$.
2. $A - i\mu I$ *is invertible on* H *for all* $\mu \in \mathbb{R}$.
3. $\exists M > 0,\ \|(A - i\mu I)^{-1}\|_{\mathcal{L}(H)} \leq M$.

Then there exist $\lambda > 0$ and $C > 0$, such that $\|e^{tA}\|_{\mathcal{L}(H)} \leq Ce^{-\lambda t}$ for all $t \geq 0$.

This result is classical and we refer to [11, 22], [15, Theorem 3], [10] for a proof. In our case, the first condition is satisfied as energy is nonincreasing. The first item is a consequence of the energy decaying. The second is not true for $\mu = 0$ as $\mathcal{A}_d$ as a non-trivial kernel. To avoid this difficulty we only have to consider $\mathcal{A}_d$ as an operator on $\tilde{H} = H/\ker \mathcal{A}_d$. On this space $A - i\mu I$ is invertible for all $\mu \in \mathbb{R}$, see [3] for a proof. In [3] we prove the following estimate $\|(\mathcal{A}_d - i\mu I)^{-1}\|_{\mathcal{L}(\tilde{H})} \leq Me^{C|\mu|}$ for some constants $M > 0$ and $C > 0$. In particular, for all $C > 0$, there exists $M > 0$ such that $\|(A - i\mu I)^{-1}\|_{\mathcal{L}(H)} \leq M$ for every $\mu \in [-C, C]$. This means, only large $|\mu|$ have to be considered to prove the third item of Theorem 3.

4 Introduction of a Semiclassical Measure

The goal is to prove the following estimate, there exists $C > 0$ such that

$$\|u\|_{(L^2(\Omega))^d} + \|r\|_{L^2(\Omega)} \leq C(\|f\|_{(L^2(\Omega))^d} + \|q\|_{L^2(\Omega)}), \tag{15}$$

for all $(f, q) \in (L^2(\Omega))^d \times L^2(\Omega)$ and $(u, r) \in (L^2(\Omega))^d \times L^2(\Omega)$ solution of system $\mathcal{A}_d(u, r) = (f, q)$.

We prove Estimate (15) by contradiction. If Estimate (15) is false, there exists a sequence $(h_j)_{j\in\mathbb{N}}$ and (f_{h_j}, q_{h_j}) and (u_{h_j}, r_{h_j}) satisfying Estimate (15) and $\mathcal{A}_d(u_{h_j}, r_{h_j}) = (f_{h_j}, q_{h_j})$.

$$h_j \to 0 \text{ as } j \to +\infty, \tag{16}$$
$$\|u_{h_j}\|_{(L^2(\Omega))^d} + \|r_{h_j}\|_{L^2(\Omega)} = 1,$$
$$\|f_{h_j}\|_{(L^2(\Omega))^d} + \|q_{h_j}\|_{L^2(\Omega)} \to 0 \text{ as } j \to +\infty.$$

As view above Estimate (15) is true for $|\mu|$ bounded, thus only $|\mu|$ goes to $+\infty$ can give such sequences. We can always normalized (u_{h_j}, r_{h_j}) by homogeneity. In what follows, we do not write the index j and we write h instead of h_j. The reader must keep in mind that h is in fact a sequence. Finally with these reductions and notation, we have sequences h, (u_h, r_h) and (f_h, q_h), such that

$$\begin{cases} -ih\nabla r_h + u_h - ihbu_h = ihf_h, \text{ in } \Omega, \\ -ih \operatorname{div} u_h + r_h = ihq_h, \text{ in } \Omega, \\ u_h.n = 0 \text{ on } \partial\Omega \end{cases} \tag{17}$$

and

$$\|u_h\|_{(L^2(\Omega))^d} + \|r_h\|_{L^2(\Omega)} = 1, \tag{18}$$
$$\|f_h\|_{(L^2(\Omega))^d} + \|q_h\|_{L^2(\Omega)} \to 0 \text{ as } h \to 0.$$

4.1 Semiclassical Measure

To define a semiclassical measure associated with $\gamma_h = (u_h, r_h)$ we have to prove that γ_h converges weakly to 0. We shall admit the following lemma.

Lemma 2 *The sequence* (u_h, r_h) *converges weakly to* 0.

As u_h and r_h are bounded functions in $L^2(\Omega)$ we extend these functions by 0 in exterior of Ω, and we denote respectively by $\underline{u}_h$ and $\underline{r}_h$ these extensions. Clearly $((\underline{u}_h, \underline{r}_h)$ converges weakly to 0 in $(L^2(\mathbb{R}^d))^d \times L^2(\mathbb{R}^d)$.

Let $a_j^k = a_k^j(x, \xi)$, $b^\ell(x, \xi)$, $c(x, \xi)$ be symbols of order 0. We assume a symmetric, i.e., $a_j^k = a_j^k$.

As $((\underline{u}_h, \underline{r}_h)$ converges weakly to 0, up to a subsequence of h, it is possible to define the semiclassical measure by the following relation,

$$((\mathrm{op}_{\mathrm{sc}}(a_j^k)\underline{u}_h^j), (\underline{u}_h^k) + ((\mathrm{op}_{\mathrm{sc}}(b^j)\underline{r}_h), (\underline{u}_h^j)) + (\mathrm{op}_{\mathrm{sc}}(b^k)\underline{u}_h^k, \underline{r}_h) + (\mathrm{op}_{\mathrm{sc}}(c)\underline{r}_h, \underline{r}_h)$$

converges to (19)

$$\langle \mu_j^k, a_k^j \rangle + 2\langle \nu_j, b^j \rangle + \langle m, c \rangle \text{ as } h \text{ goes to } 0,$$

where the measure $(\mu_j^k)_{1\le j,k\le d}$ is symmetric, i.e., $\mu_j^k = \mu_j^k$ which is related with the symmetry on $(a_j^k)_{1\le j,k\le d}$. From (17), the functions $\underline{u}_h$ and $\underline{r}_h$ satisfy the system

$$\begin{cases} \underline{u}_h - ih\nabla \underline{r}_h = hf_h - ih(r_h)_{|\partial\Omega}\nabla(\chi_\Omega), \\ -ih\,\mathrm{div}\underline{u}_h + \underline{r}_h = hq_h, \end{cases} \tag{20}$$

where $\nabla(\chi_\Omega)$ is a measure on $\partial\Omega$. As $u_h.n = 0$ at boundary, second equation does not have measure terms at the right hand side.

Proposition 1 *The measures μ_j^k, ν_k and m are supported on the characteristic set*

$$\{(x,\xi) \in \mathbb{R}^d \times \mathbb{R}^d,\ |\xi|^2 - 1 = 0 \text{ and } x \in \overline{\Omega}\}$$

and satisfy the following equations

$$\nu_k = -\xi_k m,$$

$$\mu_k^j = \xi_j \xi_k m \text{ for all } 1 \le j,k \le d.$$

Obviously, the measures are null in a neighborhood of (x,ξ) if $x \in \mathbb{R}^d \setminus \overline{\Omega}$ as $(\underline{u}_h, \underline{r}_h)$ is null in a neighborhood of x. We describe first the semiclassical measure in Ω. The description in a neighborhood of the boundary will be done in the following. For symbols a_j^k compactly supported in Ω, we consider

$$I_1(h) = \big((\mathrm{op_{sc}}(a_j^k)(\underline{u}_h^j - ih(\partial_{x_j}\underline{r}_h))), (\underline{u}_h^k)\big) + \big((\mathrm{op_{sc}}(a_j^k)\underline{u}_h^j), (\underline{u}_h^k - ih(\partial_{x_k}\underline{r}_h))\big). \tag{21}$$

From System (20), we have $I_1(h)$ goes to 0 as h. Observe that $-ih\partial_{x_j}\underline{r}_h = \mathrm{op_{sc}}(\xi_j)\underline{r}_h$ and $-ih\,\mathrm{div}w = \mathrm{op_{sc}}(\xi_k)w^k$. In what follows we denote by $O(h)$ all terms bounded by h in L^2-norm. From integration by parts formula and symbolic calculus, we have

$$\big(\mathrm{op_{sc}}(a_j^k)(-ih(\partial_{x_j}\underline{r}_h)), (\underline{u}_h^k)\big) = \big((\mathrm{op_{sc}}(a_j^k\xi_j)\underline{r}_h), \underline{u}_h^k\big) + O(h)$$

$$\begin{aligned}\big((\mathrm{op_{sc}}(a_j^k)\underline{u}_h^j), (-ih\partial_{x_k}\underline{r}_h)\big) &= \big(-ih\,\mathrm{div}(\mathrm{op_{sc}}(a_j^k)\underline{u}_h^j), \underline{r}_h\big) \\ &= (\mathrm{op_{sc}}(a_j^k\xi_k)\underline{u}_h^j, \underline{r}_h) + O(h).\end{aligned}$$

Then

$$I_1(h) = 2\big((\mathrm{op}_{\mathrm{sc}}(a_j^k)\underline{u}_h^j), (\underline{u}_h^k)\big) + \big((\mathrm{op}_{\mathrm{sc}}(a_j^k\xi_j)\underline{r}_h), (\underline{u}_h^k)\big)_g$$
$$+ \big(\mathrm{op}_{\mathrm{sc}}(a_j^k\xi_k)\underline{u}_h^j, \underline{r}_h\big)_g + O(h).$$

We can apply (19) as, by the symmetry on a, we have $a_j^k\xi_k = a_k^j\xi_k$. We obtain

$$2\langle\mu_k^j, a_j^k\rangle + 2\langle\nu_k, a_j^k\xi_j\rangle = 0,$$

for all symbols a_j^k compactly supported in Ω, then

$$\mu_k^j + \xi_j\nu_k = 0 \text{ for all } 1 \le j, k \le d. \tag{22}$$

For symbols b^k compactly supported in Ω, we consider

$$I_2(h) = \big((\mathrm{op}_{\mathrm{sc}}(b^k)(-ih\,\mathrm{div}_g\underline{u}_h + \underline{r}_h)), \underline{u}_h^k\big) + \big(\mathrm{op}_{\mathrm{sc}}(b^k)(\underline{u}_h^k - ih\partial_{x_k}\underline{r}_h), \underline{r}_h\big) \tag{23}$$
$$+ \big((\mathrm{op}_{\mathrm{sc}}(b^q)\underline{r}_h), (\underline{u}_h^k - ih\partial_{x_k}\underline{r}_h)\big) + \big(\mathrm{op}_{\mathrm{sc}}(b^j)\underline{u}_h^j, -ih\,\mathrm{div}_g\underline{u}_h + \underline{r}_h\big).$$

From System (20), we have $I_2(h)$ goes to 0 as h. From integration by parts formula and symbolic calculus, we have

$$\begin{aligned} I_2(h) &= \big((\mathrm{op}_{\mathrm{sc}}(b^k\xi_j)\underline{u}_h^j), (\underline{u}_h^k)\big) + \big((\mathrm{op}_{\mathrm{sc}}(b^k)\underline{r}_h), (\underline{u}_h^k)\big) + \big(\mathrm{op}_{\mathrm{sc}}(b^k)\underline{u}_h^k, \underline{r}_h\big) \\ &\quad + \big(\mathrm{op}_{\mathrm{sc}}(b^j\xi_j)\underline{r}_h, \underline{r}_h\big) + \big((\mathrm{op}_{\mathrm{sc}}(b^k)\underline{r}_h), (\underline{u}_h^k)\big) + \big(\mathrm{op}_{\mathrm{sc}}(b^q\xi_q)\underline{r}_h, \underline{r}_h\big) \\ &\quad + \big((\mathrm{op}_{\mathrm{sc}}(b^k\xi_j)\underline{u}_h^k), \underline{u}_h^j\big) + \big(\mathrm{op}_{\mathrm{sc}}(b^k)\underline{u}_h^k, \underline{r}_h\big) + O(h) \\ &= \big(\mathrm{op}_{\mathrm{sc}}(b^k\xi_j + b^j\xi_k)\underline{u}_h^k, \underline{u}_h^j\big) + 2\big(\mathrm{op}_{\mathrm{sc}}(b^k)\underline{r}_h, \underline{u}_h^k\big) + 2\big(\mathrm{op}_{\mathrm{sc}}(b^k)\underline{u}_h^k, \underline{r}_h\big) \\ &\quad + 2\big(\mathrm{op}_{\mathrm{sc}}(b^j\xi_j)\underline{r}_h, \underline{r}_h\big) + O(h). \end{aligned}$$

We then obtain, as

$$\begin{aligned} \langle\mu_k^j, b^k\xi_j + b^j\xi_k\rangle + 4\langle\nu_k, b^k\rangle + 2\langle m, b^k\xi_k\rangle &= \langle\xi_j\mu_k^j, b^k\rangle + \langle\xi_k\mu_k^q, b^q\rangle \\ &\quad + 4\langle\nu_k, b^k\rangle + 2\langle\xi_k m, b^k\rangle, \end{aligned}$$

the following equation

$$\xi_j\mu_k^j + 2\nu_k + \xi_k m = 0 \text{ for all } 1 \le k \le d. \tag{24}$$

For symbols c compactly supported in Ω, we consider

$$I_3(h) = \big(\mathrm{op}_{\mathrm{sc}}(c)(-ih\,\mathrm{div}\underline{u}_h + \underline{r}_h), \underline{r}_h\big) + \big(\mathrm{op}_{\mathrm{sc}}(c)\underline{r}_h, -ih\,\mathrm{div}\underline{u}_h + \underline{r}_h\big). \tag{25}$$

From System (20), we have $I_3(h)$ goes to 0 as h and integrating by parts we obtain

$$I_3(h) = 2\big(\mathrm{op_{sc}}(c)\underline{r}_h, \underline{r}_h\big) + \big(\mathrm{op_{sc}}(c\xi_j)\underline{u}_h^j, \underline{r}_h\big) + \big(\mathrm{op_{sc}}(c\xi_k)\underline{r}_h, \underline{u}_h^j\big) + O(h). \tag{26}$$

We apply (19) and we obtain

$$m + \xi_k \nu_k = 0. \tag{27}$$

From (22), (24), and (27) we obtain the following system on measure μ_k^j, ν_k and m.

$$\begin{cases} \mu_k^j + \xi_j \nu_k = 0 \text{ for all } 1 \le j, k \le d, \\ \xi_j \mu_k^j + 2\nu_k + \xi_k m = 0 \text{ for all } 1 \le k \le d, \\ m + \xi_k \nu_k = 0. \end{cases} \tag{28}$$

Replace μ_k^j in this equation from the first equation of (28) we have

$$-|\xi^2|\nu_k + 2\nu_k + \xi_k m = 0 \text{ for all } 1 \le k \le d. \tag{29}$$

Multiplying this equation by ξ_k and from the third equation of (28), this yields

$$|\xi|^2 m - 2m + |\xi|^2 m = 0.$$

Then

$$(|\xi|^2 - 1)m = 0, \tag{30}$$

which implies m supported on $(|\xi|^2 - 1)m = 0$. If $(|\xi|^2 - 1)m \neq 0$, from (28) and as $m = 0$ we have the following system

$$\begin{cases} \mu_k^j + \xi_j \nu_k = 0 \text{ for all } 1 \le j, k \le d, \\ \xi_j \mu_k^j + 2\nu_k = 0 \text{ for all } 1 \le k \le d, \\ \xi_k \nu_k = 0. \end{cases} \tag{31}$$

Multiplying the first equation by ξ_k we obtain

$$\xi_k \mu_k^j + \xi_k \xi_j \nu_k = 0 \text{ for all } 1 \le j \le d.$$

From the third equation of (31) and the symmetry on μ_k^j, we have

$$\xi_\ell \mu_j^\ell = 0 \text{ for all } 1 \le j \le d.$$

The second equation of (31) implies $\nu_k = 0$ for $1 \leq k \leq d$ and the first $\mu_k^j = 0$ for $1 \leq k, j \leq d$. Thus all the measures μ_k^j, ν_k and m are supported on $(|\xi|^2 - 1) = 0$.

Now from (29) as on the support of measures $|\xi|^2 = 1$, we have

$$\nu_k = -\xi_k m,$$

and from the first equation of (28) we have

$$\mu_k^j = \xi_j \xi_k m \text{ for all } 1 \leq j, k \leq d.$$

This achieves the proof of Proposition 1 in Ω.

Now we have to prove same result at the boundary. The proof is a consequence of the two following lemmas.

Lemma 3 *Let a be a symbol compactly supported, then there exists $C > 0$, such that, we have*

$$\|\mathrm{op}_{\mathrm{sc}}(a)(w \otimes \delta_{x_d=0})\| \leq Ch^{-1/2}|w|_{L^2(\mathbb{R}^{d-1})},$$

for all w in the Schwartz space of $\mathbb{R}^{d-1}_{x'}$.

Lemma 4 *Let $\underline{u}_h$ and $\underline{r}_h$ solution of System* (20) *then* $\lim_{h\to 0} h|\Lambda^{1/2}(r_h)_{|x_d=0}|^2 = 0$.

The proof of Lemma 4 is classical. We have to consider three kind of points, hyperbolic points, glancing points, and elliptic points. For each type we use a different method, for hyperbolic we use energy method and we do not use the boundary condition. For glancing points, we use the fact that we have a perturbation of equation $h^2 D_d^2 = hq_h$ for a q_h uniformly bounded on L^2. For elliptic points, we prove that we have a relation between the traces and the boundary conditions allow us to estimate all the traces. The proof is given in the preprint [4]

With these lemmas we can prove Proposition 1. In this part we assume that the boundary is given by $x_d = 0$. To prove the result in general we must make a change of variables and write the system in invariant form using the Riemannian geometry tools. The main difficulties are given here and the introduction of change of variables gives only few technical difficulties. We have

$$\begin{cases} \underline{u}_h - ih\nabla \underline{r}_h = hf_h - ih(r_h)_{|x_d=0} \otimes \delta_{x_d=0}, \\ -ih\, \mathrm{div}\underline{u}_h + \underline{r}_h = hq_h. \end{cases} \tag{32}$$

Following the proof given above, in $I_1(h)$ (see (21)), we have two extra terms to control,

$$(\mathrm{op}_{\mathrm{sc}}(a)(h(r_h)_{|x_d=0} \otimes \delta_{x_d=0}), \underline{u}_h^j) \text{ and}$$

$$(\mathrm{op}_{\mathrm{sc}}(a)(\underline{u}_h^j), h(r_h)_{|x_d=0} \otimes \delta_{x_d=0}) = (\underline{u}_h^j, \mathrm{op}_{\mathrm{sc}}(a)^*(h(r_h)_{|x_d=0} \otimes \delta_{x_d=0})).$$

For $I_2(h)$ (see (23)) we have to control same type of terms where $\underline{u}_h^j$ is replaced by $\underline{r}_h$. The term $I_3(h)$ (see (25)) does not have an extra boundary term. As all these terms write analogous, we only treat one term. We apply Lemma 3; we have

$$|(\mathrm{op_{sc}}(a)(h(r_h)_{|x_d=0} \otimes \delta_{x_d=0}), \underline{u}_h^j)| \lesssim h^{1/2} \|\underline{u}_h^j\| |(r_h)_{|x_d=0}| \to 0 \text{ as } h,$$

from Lemma 4.

Proof of Lemma 3 We can assume that $a(x, \xi) = a_1(x', \xi')\chi(x_d)\ell(\xi_d)$, where a_1, χ, and ℓ are compactly supported, as all symbols can be approximated by a sum of such symbols. From definition of operators associated with a symbol, we have

$$\mathrm{op_{sc}}(a)(w \otimes \delta_{x_d=0}) = \chi(\mathrm{op_{sc}}(a_1)w) \otimes \mathrm{op_{sc}}(\ell)\delta_{x_d=0}.$$

A direct computation in Fourier variables gives $\|\mathrm{op_{sc}}(\ell)\delta_{x_d=0}\|_{L^2(\mathbb{R})} \lesssim h^{-1/2}$, and as $a_1 \in S^{-N}$ for all $N \in \mathbb{N}$, $|\mathrm{op_{sc}}(a_1)w| \lesssim |w|_{L^2(\mathbb{R}^{d-1})}$, which implies the lemma.

In the following we prove that the measure defined is not null and it is null on the set $\{x,\ b(x) > 0\}$.

4.2 *The Measure Is Not Identically Null*

Proposition 2 *The measure m is not identically null.*

Proof From (17), multiplying the first equation by u_h, integrating by parts and taking account the boundary condition $(u_h^d)_{|x_d=0} = 0$, we have

$$O(h) = (-ih\nabla r_h + u_h, u_h) = \|u_h\|^2 + (r_h, -ih\ \mathrm{div} u_h).$$

From second equation of (17) we replace $-ih\ \mathrm{div}_g u_h$ by r_h and we obtain

$$O(h) = \|u_h\|^2 - \|r_h\|^2.$$

As from (16), we have $\|u_h\| + \|r_h\| = 1$, we have $\|r_h\| = 1/2 + O(h)$ and $\|u_h\| = 1/2 + O(h)$. From the first equation of (17), we obtain $\|h\nabla_g r_h\| \lesssim 1$.

We prove now that m is not null. Let ϕ be a smooth function such that $\phi(s) = 1$ for $|s| \leq 1$ and supported in $|s| \leq 2$. For $R > 0$, we set $\phi_R(\xi) = \phi(|\xi|/R)$. For $\alpha \in (0, 1/2)$ we have

$$\begin{aligned}\|\mathrm{op_{sc}}(1 - \phi_R)\underline{r}_h\| &\lesssim R^{-\alpha}\|\mathrm{op_{sc}}(\langle\xi\rangle^\alpha)\underline{r}_h\| \lesssim R^{-\alpha}(\|r_h\|_{L^2(\Omega)} + \|h\nabla_g r_h\|_{L^2(\Omega)}) \\ &\lesssim R^{-\alpha}. \qquad (33)\end{aligned}$$

(See for instance [9, Lemma 4.3].)

Then for h sufficiently small and R sufficiently large,

$$(\mathrm{op}_{\mathrm{sc}}(\phi_R)\underline{r}_h, \underline{r}_h) \geq 1/4,$$

and from the definition of m (see (19)), m is not null. This gives the result.

4.3 The Semiclassical Measure Null on Support of Damping

In this section we prove the following proposition.

Proposition 3 *Let m be the semiclassical measure defined in Proposition 1, we have $b(x)m = 0$.*

Proof From System (17) we have

$$(ihf_h, u_h) + (ihq_h, r_h) = (-ih\nabla r_h + u_h - ihbu_h, u_h) + (-ih\ \mathrm{div} u_h + r_h, r_h). \tag{34}$$

Integrating by parts and taking account the boundary condition $(u_h^d)_{|x_d=0} = 0$, we have

$$(-ih\nabla r_h, u_h) + (-ih\ \mathrm{div} u_h, r_h) = 2\Re(-ih\nabla r_h, u_h). \tag{35}$$

From the two previous formula, we obtain

$$\Im\big((ihf_h, u_h) + (ihq_h, r_h)\big) = ih(bu_h, u_h). \tag{36}$$

As the left hand side is a $o(h)$, we obtain $(bu_h, u_h) \to 0$ as $h \to 0$. Then from Proposition 1 and [4, Proposition 5.3] we have $\langle m|\xi|^2, b\rangle = 0$.

As b and m are nonnegative, $|\xi|^2 = 1$ on the support of m, we obtain the result.

Remark 1 Proposition 5.3 of [4] allows us to extend the result of convergence to term as (bu_h, u_h). As b is not compactly supported in ξ we cannot apply the definition but exploiting that m is supported on $|\xi|^2 = 1$, $\langle \mu_j^j, b\rangle$ has a sense and we can prove that $(bu_h, u_h) \to \langle \mu_j^j, b\rangle$.

5 Propagation of the Support

The goal is to prove the following proposition.

Proposition 4 *We assume that ω satisfy GCC. Let m the semiclassical measure constructed from $(u_h, r_h)_h$ and satisfying (17), we have $m = 0$.*

First in this section we give some formulas yielding to the propagation results. This is stated in the following proposition.

Proposition 5 *We denote by $p(\xi) = |\xi|^2$ and $R(\xi') = |\xi'|^2$.*
We have $\{p, m\} = 0$ in $\Omega \times \mathbb{R}^d$.
In a neighborhood of $\partial\Omega$, i.e., $x_d = 0$, let $a(x, \xi')$ be a tangential symbol compactly supported in a neighborhood of the boundary, we have,

$$\langle \{p, m\}, a\rangle = 0, \tag{37}$$

$$\langle \{p, m\}, a\xi_d\rangle = -\lim_{h\to 0} (\mathrm{op}_{\mathrm{sc}}(a(R-1))(r_h)_{|x_d=0}, (r_h)_{|x_d=0})_\partial. \tag{38}$$

Observe that we do not know if $(r_h)_{|x_d=0}$ is bounded in L^2, Lemma 4 is too weak. But the quantity $(\mathrm{op}_{\mathrm{sc}}(a(R-1))(r_h)_{|x_d=0}, (r_h)_{|x_d=0})_{\tilde{g}}$ has a limit. A priori the measure m is not defined on functions in (x, ξ') but as m is supported on $|\xi|^2 - 1 = 0$, which is a bounded set on ξ, it is easy to extend m to functions in (x, ξ'). We give only a proof in interior. The proof at boundary is given in [4].

Proof We introduce the following quantity, where a is a symbol compactly supported in $\Omega \times \mathbb{R}^d$. In what follows we shall choose b_k^j depending on a but the choice of b_k^j is more natural after the computations below. Note that b_k^j are also compactly supported in $\Omega \times \mathbb{R}^d$. In interior $\underline{u}_h$ and $\underline{r}_h$ are u_h and r_h. Rigorously we need to introduce a cutoff to do that but this introduces error terms of order $O(h^\infty)$.

$$\begin{aligned} A = {} & (\mathrm{op}_{\mathrm{sc}}(a)(u_h - ih\nabla r_h), u_h) + (\mathrm{op}_{\mathrm{sc}}(a)(r_h - ih\,\mathrm{div} u_h), r_h) \\ & - (\mathrm{op}_{\mathrm{sc}}(a)u_h, u_h - ih\nabla r_h) - (\mathrm{op}_{\mathrm{sc}}(a)r_h, r_h - ih\,\mathrm{div} u_h) \\ & + (ih\mathrm{op}_{\mathrm{sc}}(b_k^j)(u_h^k - ih\partial_{x_k} r_h), u_h^j). \end{aligned} \tag{39}$$

Here and in what follows, we use Einstein notation.

Obviously, the assumptions on System (17) imply that $h^{-1}A \to 0$ as $h \to 0$.

We can simplify A and we have,

$$\begin{aligned} A = {} & -(ih\mathrm{op}_{\mathrm{sc}}(a)\nabla r_h, u_h) - (ih\mathrm{op}_{\mathrm{sc}}(a)\,\mathrm{div} u_h, r_h) \\ & + (\mathrm{op}_{\mathrm{sc}}(a)u_h, ih\nabla r_h)_g + (\mathrm{op}_{\mathrm{sc}}(a)r_h, ih\,\mathrm{div} u_h) \\ & + (ih\mathrm{op}_{\mathrm{sc}}(b_k^j)u_h^k, u_h^\ell) + (h\mathrm{op}_{\mathrm{sc}}(b_k^j)(h\partial_{x_k} r_h, u_h^j). \end{aligned} \tag{40}$$

We integrate by parts and we write by R terms estimated by $h(\|u_h\|^2 + \|r_h\|^2)$ playing no role in the analysis. Observe that no boundary term appears as a is supported far away the boundary and we have

$$\begin{aligned} h^{-1}A & = (i([\nabla, \mathrm{op}_{\mathrm{sc}}(a)]r_h, u_h) + (i[\,\mathrm{div}, \mathrm{op}_{\mathrm{sc}}(a)]u_h, r_h) \\ & = (i\mathrm{op}_{\mathrm{sc}}\{\xi_j, a\}r_h, u_h^j) + (i\mathrm{op}_{\mathrm{sc}}\{\xi_k, a\}u_h^k, r_h) + R. \end{aligned} \tag{41}$$

Then $ih^{-1}A$ converges to $-2\langle \nu_j, \{\xi_j, a\}\rangle = 0$. From Proposition 1, $\nu_j = -m\xi_j$, we thus obtain $2\langle m\xi_j, \{\xi_j, a\}\rangle = \langle m, \{|\xi|^2, a\} = 0$. This proves Proposition 5 in $\Omega \times \mathbb{R}^d$.

Proposition 4 is a consequence of Proposition 5. To prove that we follows works [7, 8, 12, 13, 16, 17, 21]. The proof follows the one given in [9] and the proof is given in [4].

Below we give some lemmas based on Proposition 5. These lemmas give properties on the semiclassical measure m. There are not enough for the result.

6 Some Properties on Semiclassical Measure

6.1 Properties of Measure

In this section, we give several properties on measure m, in particular in neighborhood of boundary. This description is related to the property of point in $T^*(\partial\Omega) = \mathbb{R}^{d-1}_{x'} \times \mathbb{R}^{d-1}_{\xi'}$. We do not give all properties necessary to prove Proposition 5. Some of them are too technic to explain here. In particular one argument uses the tools introduced by Tataru [23] and also used in [9].

In next lemma, we decompose the measure into an interior measure and a measure supported at boundary. Moreover this last measure is supported on $\xi_d = 0$.

Lemma 5 *There exists a nonnegative Radon measure m^∂ on $x_d = \xi_d = 0$ such that $m = 1_{x_d>0}m + m^\partial \otimes \delta_{x_d=0} \otimes \delta_{\xi_d=0}$. Furthermore m^∂ is supported on $R(x', 0, \xi') - 1 = 0$.*

Proof We apply Formula (38). We observe that $H_p = 2\xi_d\partial_{x_d} - (\partial_{x_d}R)\partial_{\xi_d} + H'_R$ where $H'_R = \sum_{j=1}^{d-1} \big((\partial_{\xi_j}R)\partial_{x_j} - (\partial_{x_j}R)\partial_{\xi_j}\big)$. We have $H_p(a\xi_d) = 2\xi_d^2(\partial_{x_d}a) - (\partial_{x_d}R)a + \xi_d H'_R a$. Let $a = a^\varepsilon = \varepsilon\chi(x_d/\varepsilon)\ell(x, \xi')$, where $\chi \in C_0^\infty(\mathbb{R})$, such that $\chi(0) = 0$ and $\chi'(0) = 1$, $\ell \in C_0^\infty(\mathbb{R}^d \times \mathbb{R}^{d-1})$. We have

$$\begin{aligned} H_p(\xi_d a^\varepsilon) &= 2\xi_d^2\chi'(x_d/\varepsilon)\ell(x, \xi') + 2\varepsilon\chi(x_d/\varepsilon)\xi_d^2\partial_{x_d}\ell \\ &\quad - (\partial_{x_d}R)\varepsilon\chi(x_d/\varepsilon)\ell(x, \xi') + \xi_d\varepsilon\chi(x_d/\varepsilon)H'_R\ell. \end{aligned}$$

Clearly $H_p(\xi_d a^\varepsilon)$ is uniformly bounded on the support of m and $H_p(\xi_d a^\varepsilon) \to 2\xi_d^2\chi'(0)\ell 1_{x_d=0}$ everywhere as $\varepsilon \to 0$. Then by Lebesgue's dominated convergence theorem we have $\langle m, H_p(\xi_d a^\varepsilon)\rangle \to \langle m, 2\xi_d^2\ell 1_{x_d=0}\rangle$ as $\varepsilon \to 0$. As $\chi(0) = 0$, the right hand side of (38) is 0 for every h. Then $\langle m, 2\xi_d^2\ell 1_{x_d=0}\rangle = 0$. This means that $1_{x_d=0}m$ is supported on $\xi_d = 0$. We denote m^∂ the measure $1_{\xi_d=0}1_{x_d=0}m$. As $m = 1_{x_d>0}m + 1_{x_d=0}m$, we have $m = 1_{x_d>0}m + m^\partial \otimes \delta_{x_d=0} \otimes \delta_{\xi_d=0}$. We have by

Proposition 1, $(\xi_d^2 + R(x, \xi') - 1)1_{x_d=0}m = 0$, then $(R(x', 0, \xi') - 1)m^\partial = 0$. This gives the conclusion of Lemma.

The Hamiltonian of the interior measure is a priori a distribution of order one supported on $x_d = 0$. The following lemma says that this quantity is a measure if the Hamiltonian vector field is transverse to the boundary.

Lemma 6 *We have $H_p(m1_{x_d>0}) = \delta_{x_d=0} \otimes m_0$, where m_0 is a distribution of order 1 defined on $x_d = 0$. Moreover, in a neighborhood where $H_p x_d > 0$, m_0 is a nonnegative Radon measure, and in a neighborhood where $H_p x_d < 0$, m_0 is a non positive Radon measure.*

Proof The support of $H_p(m1_{x_d>0})$ as a distribution if $x_d \geq 0$ and $H_p(m) = 0$ on $x_d > 0$. Then $H_p(m1_{x_d>0})$ is supported on $x_d = 0$. This implies that there exist $n \geq 0$ and m_j distributions on $x_d = 0$ such that $H_p(m1_{x_d>0}) = \sum_{j=0}^{n} \delta^{(j)}_{x_d=0} \otimes m_k$.

Suppose that $n \geq 1$ and let $\chi \in C_0^\infty(\mathbb{R})$ be such that $\chi^{(k)}(0) = 0$ for $k = 0, \dots n-1$ and $\chi^{(n)}(0) = 1$. Let $b \in C_0^\infty(\mathbb{R}^d \times \mathbb{R}^{d-1})$, we have

$$\begin{aligned} -\langle m1_{x_d>0}, H_p\big(\varepsilon^n \chi(x_d/\varepsilon) b(x, \xi')\big)\rangle &= \langle H_p m, \varepsilon^n \chi(x_d/\varepsilon) b(x, \xi')\rangle \\ &= \langle \sum_{j=0}^{n} \delta^{(j)}_{x_d=0} \otimes m_k, \varepsilon^n \chi(x_d/\varepsilon) b(x, \xi')\rangle \\ &= (-1)^n \langle m_n, b(x', 0, \xi')\rangle. \end{aligned} \tag{42}$$

We also have

$$H_p\big(\varepsilon^n \chi(x_d/\varepsilon) b(x, \xi')\big) = 2\xi_d \varepsilon^{n-1} \chi'(x_d/\varepsilon) b(x, \xi') + \varepsilon^n \chi(x_d/\varepsilon) H_p b(x, \xi'),$$

which is bounded uniformly with respect ε and supported in a fixed compact set. Then if $n \geq 2$, $H_p\big(\varepsilon^n \chi(x_d/\varepsilon) b(x, \xi')\big) \to 0$ everywhere as $\varepsilon \to 0$ and (42), Lebesgue's dominated convergence theorem imply that $m_n = 0$ for $n \geq 2$. If $n = 1$, $H_p\big(\varepsilon \chi(x_d/\varepsilon) b(x, \xi')\big) \to 2\xi_d 1_{x_d=0} b(x, \xi')$ everywhere as $\varepsilon \to 0$. Lebesgue's dominated convergence theorem and (42) imply that

$$\langle m1_{x_d>0}, 2\xi_d 1_{x_d=0} b(x, \xi')\rangle = \langle m_1, b(x', 0, \xi')\rangle,$$

as $\langle m1_{x_d>0}, 2\xi_d 1_{x_d=0} b(x, \xi')\rangle = 0$ we find that $m_1 = 0$. Then we have $H_p(m1_{x_d>0}) = \delta_{x_d=0} \otimes m_0$, where m_0 is a distribution of order 1.

If $H_p x_d \neq 0$, let $\gamma(s; x, \xi) = (x(s; x', \xi), \xi(s; x', \xi))$ be the flow associated with H_p. We verify that the map $(s, x', \xi) \to (x(s; x', \xi), \xi(s; x', \xi))$ locally is one to one and transforms ∂_s in H_p. Moreover, $s = 0$ is transformed in $x_d = 0$ and if $H_p x_d > 0$, $s > 0$ is transformed in $x_d > 0$, if $H_p x_d < 0$, $s < 0$ is transformed in $x_d > 0$. In coordinates (s, x', ξ) the equation $H_p(m1_{x_d>0}) = \delta_{x_d=0} \otimes m_0$ is transformed in $\partial_s(m1_{s>0}) = \delta_{s=0} \otimes m_0$ if $H_p x_d > 0$ and $\partial_s(m1_{s<0}) = \delta_{s=0} \otimes m_0$ if $H_p x_d < 0$, where we keep the notations m, m_0 in variables (s, x', ξ) for the images of m, m_0. If $H_p x_d > 0$, we have $m1_{s>0} = (1_{s>0} ds) \otimes m_0$ and if $H_p x_d < 0$, we have $m1_{s<0} = -(1_{s<0} ds) \otimes m_0$. As m is nonnegative, we obtain that m_0 is a nonnegative measure in the first case and a nonpositive measure in the second case.

At the hyperbolic region the measure m_0 has a particular structure given by next lemma.

Lemma 7 *Let (x'_0, ξ'_0) be a hyperbolic point (i.e., $R(x'_0, 0, \xi'_0) < 1$). Locally in a neighborhood of (x'_0, ξ'_0), there exists m^+ a nonnegative measures on $\mathbb{R}^{d-1}_{x'} \times \mathbb{R}^{d-1}_{\xi'}$ such that*

$$m_0 = m^+ \otimes \delta_{\xi_d = \sqrt{1 - R(x', 0, \xi')}} - m^+ \otimes \delta_{\xi_d = -\sqrt{1 - R(x', 0, \xi')}}.$$

Proof As m is supported on $\xi_d^2 + R(x, \xi') - 1 = 0$, we have $m = m1_{x_d > 0}$ in a neighborhood of hyperbolic points and $\delta_{x_d = 0} \otimes m_0$ is supported on $\xi_d^2 + R(x, \xi') - 1 = 0$. Then m_0 is supported on $\xi_d = \pm\sqrt{1 - R(x', 0, \xi')}$. Moreover $H_p = 2\xi_d \partial_{x_d} + X$ where X is a vector field tangent to $x_d = 0$, in particular $H_p x_d = 2\xi_d$ on $x_d = 0$. This implies that $H_p x_d > 0$ if $\xi_d = \sqrt{1 - R(x', 0, \xi')}$ and $H_p x_d < 0$ if $\xi_d = -\sqrt{1 - R(x', 0, \xi')}$. From Lemma 6, we obtain $m_0 = m^+ \otimes \delta_{\xi_d = \sqrt{1 - R(x', 0, \xi')}} - m^- \otimes \delta_{\xi_d = -\sqrt{1 - R(x', 0, \xi')}}$. From (37) we have $\langle H_p m, a \rangle = 0$ which implies $\langle m^+ - m^-, a_{|x_d = 0} \rangle = 0$ as $a_{|x_d = 0}$ describes every C_0^∞ functions supported in a neighborhood of (x'_0, ξ'_0); this implies that $m^+ = m^-$ in a neighborhood of (x'_0, ξ'_0).

References

1. G. Allaire, Homogenization of the Navier-Stokes equations and derivation of Brinkman's law, in *Mathématiques appliquées aux sciences de l'ingénieur (Santiago, 1989), Cépaduès, Toulouse* (1991), pp. 7–20
2. K. Ammari, E. Feireisl, S. Nicaise, Polynomial stabilization of some dissipative hyperbolic systems. DCDS-A **34**, 4371–4388 (2014)
3. K. Ammari, F. Hassine, L. Robbiano, Logarithmic stabilization of an acoustic system with a damping term of Brinkman type. SIAM J. Control **61**, 170–197 (2023)
4. K. Ammari, F. Hassine, L. Robbiano, Uniform stabilization of an acoustic system. arXiv:2302.02726
5. P. Angot, C.H. Bruneau, P. Fabrie, A penalization method to take into account obstacles in incompressible viscous flows. Numer. Math. **81**, 497–520 (1999)
6. W. Arendt, C.J.K. Batty, Tauberian theorems and stability of one-parameter semigroups. Trans. Amer. Math. Soc. **305**, 837–852 (1988)
7. N. Burq, Mesures semi-classiques et mesures de défaut. Séminaire Bourbaki, Vol. 1996/97. Astérisque **245**, 167–195 (1997)
8. N. Burq, G. Lebeau, Mesures de défaut de compacité, application au système de Lamé. Ann. Sci. École Norm. Sup. **34**, 817–870 (2001)

9. P. Cornilleau, L. Robbiano, Exponential stabilization of waves for the Zaremba boundary condition. Arxiv:2110.05986
10. K.J. Engel, R. Nagel, *One-Parameter Semigroups for Linear Evolution Equations*. Graduate Texts in Mathematics, vol. 194 (Springer, New York, 2000)
11. L. Gearhart, Spectral theory for contraction semigroups on Hilbert space. Trans. Amer. Math. Soc. **236**, 385–394 (1978)
12. P. Gérard, Microlocal defect measures. Comm. Partial Differ. Equ. **16**, 1761–1794 (1991)
13. P. Gérard, E. Leichtnam, Ergodic properties of eigenfunctions for the Dirichlet problem. Duke Math. J. **71**, 559–607 (1993)
14. V. Girault, P.-A. Raviart, *Finite Element Methods for Navier-Stokes Equations*. Springer Series in Computational Mathematics (Springer, Berlin, 1986)
15. F.L. Huang, Characteristic conditions for exponential stability of linear dynamical systems in Hilbert spaces. Ann. Differ. Equ. **1**, 43–56 (1985)
16. G. Lebeau, Contrôle et stabilisation hyperboliques. *Séminaire Équations aux dérivées partielles (École Polytechnique)*. Exposé **16**, 1–16 (1990)
17. G. Lebeau, Équations des ondes amorties, in *Algebraic and Geometric Methods in Mathematical Physics (Kaciveli, 1993)*. Mathematical Physics Studies, vol. 19 (Kluwer Academic Publishers, Dordrecht, 1996), pp. 73–109
18. M.J. Lighthill, On sound generated aerodynamically. I. General theory. Proc. Roy. Soc. London. Ser. A. **211**, 564–587 (1952)
19. M.J. Lighthill, On sound generated aerodynamically. II. Turbulence as a source of sound. Proc. Roy. Soc. London. Ser. A. **222**, 1–32 (1954)
20. M.J. Lighthill, *Waves in Fluids* (Cambridge University Press, Cambridge, 1978)
21. L. Miller, Refraction of high-frequency waves density by sharp interfaces and semiclassical measures at the boundary. J. Math. Pures Appl. **79** (9), 227–269 (2000)
22. J. Prüss, On the spectrum of C_0-semigroups. Trans. Amer. Math. Soc. **284**, 847–857 (1984)
23. D. Tataru, On the regularity of boundary traces for the wave equation. Ann. Scuola Norm. Sup. Pisa Cl. Sci. **26**, 185–206 (1998)

Observability and Control of Parabolic Equations on Networks

Jone Apraiz and Jon Asier Bárcena-Petisco

2010 Mathematics Subject Classification 93B05, 35K15

1 Introduction

Network theory can be useful for studying complex systems such as those that arise, for example, in physical sciences, engineering, economics, and sociology. These systems can be modeled as networks, also known as metric graphs, and their elements and interactions or links are identified respectively by vertices and edges. During the last decades, the use of networks has been helpful and effective, among others, in the study of pipes, neural systems (the brain can be thought of as a network of neurons), the flow of traffic on roads, the global economy, and the human circulatory system (see, for example, [4, Chapter 9], [5, 10, 18, 21]).

In this work, we considered the propagation of heat on a network with loops. We sought to control these networks by acting in its interior with a source term, and to estimate the solutions with an observation domain located in the interior of the network. Indeed, the main purpose of this research was to extend the results of [15] to networks with loops. This is relevant considering that loops arise naturally in pipe systems, transport systems, etc.

Recent important works involving the control of parabolic equations on networks are the followings: [13], where the controllability of the discretized heat equation is studied, [6], where bilinear controls are analyzed on networks, [20], where the optimal control is studied in time-fractional diffusion equations and, [2], where the controllability is analyzed with vanishing viscosity. Note that the literature related to the controllability of hyperbolic equations on networks is more extensive, on which we may highlight the book [9] and the paper [14]. Particularly, [9] mainly analyzes the problem of propagation, observation, and control of waves on planar one-

J. Apraiz (✉) · J. A. Bárcena-Petisco
Department of Mathematics, University of the Basque Country UPV/EHU, Leioa, Spain
e-mail: jone.apraiz@ehu.eus; jonasier.barcena@ehu.eus

K. Ammari et al. (eds.), *Advances in Partial Differential Equations and Control*, Trends in Mathematics, https://doi.org/10.1007/978-3-031-62265-6_3

dimensional networks, using groundbreaking developments related to non-harmonic Fourier series, Diophantine approximation, graph theory, and wave propagation techniques (d'Alembert formula, for example).

1.1 Basic Definitions

We first define some concepts related to graph theory that we use in this work. Let $\mathcal{G} = (\mathcal{V}, \mathcal{E})$ be a graph.

- An edge $e \in \mathcal{E}$ that is incident to the vertices v and $\tilde{v} \in \mathcal{V}$ is expressed as $e = v\tilde{v}$, where v and $\tilde{v}$ are the *ends* of e. The set of ends of e is denoted by $\mathcal{V}(e)$. Similarly, for every vertex $v \in \mathcal{V}$, $\mathcal{E}(v)$ denotes the *set of edges incident to* v. The *degree of a vertex* $v \in \mathcal{V}$, denoted by $d(v)$, is $|\mathcal{E}(v)|$.
- $\mathcal{V}_0 = \{v : |\mathcal{E}(v)| \geq 2\}$ denotes the *set of inner vertices*, and $\mathcal{V}_\partial = \mathcal{V} \setminus \mathcal{V}_0$ denotes the *set of boundary vertices* of the graph.
- A sequence of vertices $v_0 v_1 ... v_{n-1} v_n$ such that $v_i \in \mathcal{V}$ for all $i = 0, ..., n$ and such that $v_{i-1} v_i \in \mathcal{E}$ for all $i = 1, ..., n$ is called a *walk*. If all the vertices are distinct, it is called a *path*, and if all the vertices are distinct except $v_0 = v_n$ it is called a *cycle*.
- A graph is *connected* if there is a walk joining each pair of vertices.
- A graph isomorphic to $(\{v_0, ..., v_n\}, \{v_0 v_1, ..., v_{n-1} v_n\})$ for some $n \in \mathbb{N}$ is called *path graph*.

We define a *network* as a tuple $\mathcal{G} = (\mathcal{V}, \mathcal{E}, \mathcal{I})$, where $\mathcal{V}$ is a finite *set of vertices*, $\mathcal{E} \subset \mathcal{V} \times \mathcal{V}$ is the *set of edges*, and $\mathcal{I}$ is the *identification* of each edge $e = v\tilde{v}$ and its ends v and $\tilde{v}$ with a closed interval $[0, L^e]$ and the ends 0 and L^e respectively. Formally, the identification can be viewed as a function from $\mathcal{E}$ to $X \times \mathcal{V} \times \mathcal{V}$, where X is the set of compact intervals. Notably, $I(e) = ([0, L^e], v, \tilde{v})$, where v is the end of e identified with 0, and $\tilde{v}$ is the other end of e, which we identify with L^e. This definition is also referred in the literature as *metric graph*. With the identification $\mathcal{I}$, for every edge $e \in \mathcal{E}$, we define the following numbers for the vertex v identified with 0 and for the vertex $\tilde{v}$ identified with L^e:

$$n^e(v) = -1 \quad \text{and} \quad n^e(\tilde{v}) = 1.$$

This allows to define the operator $\partial_{n^e(v)} y = n^e(v) \partial_x y^e(v)$, which can be shorten to $\partial_{n^e} y$ when the vertex is clear. Usually, we make a small abuse of notation and do not write the identification $\mathcal{I}$ explicitly when we denote the network $\mathcal{G}$.

In this article we work in the functional spaces $W^{k,p}_{pw}(\mathcal{E})$, which denotes the set of functions that belong to $W^{k,p}(e)$ for all $e \in \mathcal{E}$, $k \in \mathbb{N}$ and $1 \leq p \leq \infty$, and $W^{1,p}(\mathcal{E}) := W^{k,p}_{pw}(\mathcal{E}) \cap C^0(\overline{\mathcal{E}})$. Here "pw" stands for piecewise. Similar definitions apply to $H^k_{pw}(\mathcal{E})$ and $H^k(\mathcal{E})$. In this context, given a function $f \in W^{k,p}_{pw}(\mathcal{E})$, we define by $\partial_x f$ the derivative in each of the edges. Clearly, if $f \in W^{k,p}_{pw}(\mathcal{E})$, then $\partial_x f \in W^{k-1,p}_{pw}(\mathcal{E})$.

1.2 Controllability Results

The problem that we studied in this work was the dynamics of the flux and the control, which can be modeled by the following parabolic system:

$$\begin{cases} a\partial_t y - \mu\partial_{xx}^2 y + b\partial_x y + cy = f 1_\omega, & \text{in } (0,T)\times\mathcal{E}, \\ y = 0, & \text{on } (0,T)\times\mathcal{V}_\partial, \\ y^{e_i} = y^{e_j}, & \text{on } (0,T)\times\mathcal{V}_0,\ \forall e_i, e_j \in \mathcal{E}(v), \\ \sum\limits_{e\in\mathcal{E}(v)} \mu^e \partial_{n^e} y^e = \gamma y, & \text{on } (0,T)\times\mathcal{V}_0, \\ y(0,\cdot) = y_0, & \text{in } \mathcal{E}. \end{cases} \tag{1}$$

In this model y denotes the flux of the heat on the entire network. Throughout this article, we denote the restriction of a function to an edge e by adding the superscript e. In addition, a and μ are positive coefficients and b and c are coefficients which characterize the properties of the pipes of the network (roughness or properties of the heat flux, for example). Moreover, γ is a real coefficient measuring the flux of the heat on junctions, and $\omega \subset \mathcal{E}$ is the control domain. Here, when writing $\omega \subset \mathcal{E}$ we make a small abuse of notation to mean $\omega \subset \cup_{e\in\mathcal{E}}\, e$.

In addition, by coefficients we mean functions which model the properties of the systems like the heat diffusivity and, unless stated otherwise, depend on the time and spatial variables.

It is trivial to prove in system (1) the usual energy estimations in $L^2(0,T;H^1(\mathcal{E})) \cap C^0([0,T];L^2(\mathcal{E}))$ and regularity result in $L^2(0,T;H^2_{pw}(\mathcal{E})) \cap H^1(0,T;L^2(\mathcal{E}))$ for parabolic equations. This can be done by multiplying the first equation of the system by y and y_t and integrating it in $(0,T)\times\mathcal{E}$ (see [11] for a particular case).

In order to solve the controllability and inverse problems with respect to the parabolic system (1), we studied the observability properties of the adjoint system, which is given by:

$$\begin{cases} -a\partial_t\varphi - \mu\partial_{xx}^2\varphi - \partial_x b\varphi - b\partial_x\varphi + c\varphi = 0, & \text{in } (0,T)\times\mathcal{E}, \\ \varphi = 0, & \text{on } (0,T)\times\mathcal{V}_\partial, \\ \varphi^{e_i} = \varphi^{e_j}, & \text{on } (0,T)\times\mathcal{V}_0,\ \forall e_i, e_j \in \mathcal{E}(v), \\ \sum\limits_{e\in\mathcal{E}(v)} \mu^e\partial_{n^e}\varphi^e = \left(-\sum\limits_{e\in\mathcal{E}(v)} n^e b^e + \gamma\right)\varphi, & \text{on } (0,T)\times\mathcal{V}_0, \\ \varphi(T,\cdot) = \varphi_T, & \text{in } \mathcal{E}. \end{cases} \tag{2}$$

System (2) might not be observable unless the control domain intersects a sufficient number of edges. In particular, in order to avoid some of those non-observable cases, we assumed that the control domain intersects a sufficient number of edges:

Hypothesis 1 (Existence of an Indexing Function) *Let $\mathcal{G} = (\mathcal{V}, \mathcal{E})$ be a network and $\omega \subset \mathcal{E}$ an open subdomain. We suppose that:*

1. *ω intersects all the cycles of $\mathcal{G}$. That is, if $v_1, ..., v_n \in \mathcal{V}$ such that $e_1 := v_1v_2, ..., e_{n-1} := v_{n-1}v_n, e_n := v_nv_1$ satisfy $e_i \in \mathcal{E}$ for all $i = 1, ..., n$, then there is $k \in \{1, ..., n\}$ such that $e_k \cap \omega \neq \emptyset$.*

Moreover, we suppose that there exists a function $u : \{e \in \mathcal{E} : e \cap \omega = \emptyset\} \mapsto \mathcal{V}_0$ such that:

2. *u is injective.*
3. *e is incident to $u(e)$.*

Roughly speaking, the state of the equation in the edge e is controlled by ω if $e \cap \omega \neq \emptyset$, and by $u(e)$ otherwise, which is controlled by the rest of the adjacent edges. Identifying the right hypothesis, in the sense that allows us to prove the results without being too restrictive, is not trivial and is one of the contributions of our work. Indeed, the main breakthrough with respect to the previous work, and notably [15], is to make sure that we do not enter a circular reasoning fallacy. This was done with Hypothesis 1, as the proof of the controllability follows in a fluid way.

Remark 1 (Identification of Edges) Let $\mathcal{G} = (\mathcal{V}, \mathcal{E})$ be a network and let ω be a control domain such that Hypothesis 1 is satisfied with an indexing function u that we fix. In order to identify an edge e such that $\omega \cap e = \emptyset$ with an interval $[0, L^e]$, we established that the end identified with L^e is the vertex $u(e)$. This assignment simplifies some computations in the proof of Proposition 1.

The main observability result in this work is a Carleman inequality (see Proposition 2 in Sect. 2.2). With that inequality, we proved the following null controllability result regarding open-loop control:

Theorem 1 (Null Control of the Heat Equation on Networks with Loops) *Let $\mathcal{G} = (\mathcal{V}, \mathcal{E})$ be a network satisfying Hypothesis 1, $a, \mu \in W^{1,\infty}((0, T); L^\infty(\mathcal{E})) \cap L^\infty((0, T); W^{1,\infty}_{pw}(\mathcal{E}))$ such that $\inf a, \inf \mu > 0$, $b \in L^\infty((0, T); W^{1,\infty}_{pw}(\mathcal{E}))$, $c \in L^\infty((0, T) \times \mathcal{E})$ and $\gamma \in L^\infty((0, T) \times \mathcal{V}_0)$. Then, there exists $C > 0$ such that for all $y_0 \in L^2(\mathcal{E})$ there is $f \in L^2((0, T) \times \omega)$ such that:*

$$\|f\|_{L^2((0,T)\times\omega)} \leq C\|y_0\|_{L^2(\mathcal{E})},$$

and the solution of (1) *satisfies $y(T, \cdot) = 0$.*

Theorem 1 was proved by duality with the results of Sect. 2 (see [1]).

Now, in the next lines, we will focus on feedback control, also known as closed-loop control, in order to show stabilization properties for the simplified system of (1) when $b = c = 0$.

Many applications of feedback type controls can be found in industry and engineering: water level controller, air conditioner, adaptive measurement in quantum systems or servo voltage stabilizer. The study and construction of feedback controls is a well-established research topic (see for instance, [7, 19, 22] and the survey [8]).

In the works [22] and [23] we see the relation between the mild solution of Riccati equation and the null controllability of a system. The purpose of introducing Riccati equations into the study of control theory is to design feedback controls for linear quadratic control problems. We applied Theorem 1 to obtain the existence of feedback controls for a simplified version of system (1), based mainly on the theory of Riccati equations and it is formulated below in Theorem 2.

In order to state our result we need first to define the following concepts:

- A and B are bounded operators defined as:

$$A : H_0^1(\mathcal{E}) \cap H^2(\mathcal{E}) \longrightarrow L^2(\mathcal{E}) \text{ and } B : L^2(\mathcal{E}) \longrightarrow L^2(\mathcal{E}).$$

- $\sum^+(L^2(\mathcal{E}))$ is the Banach space of all symmetric and positive operators acting in $L^2(\mathcal{E})$.
- $C_S\left([0,T); \sum^+(L^2(\mathcal{E}))\right)$ denotes the set of all mappings $S : [0,T) \longrightarrow \sum^+(L^2(\mathcal{E}))$ such that $S(\cdot)z_0$ is continuous in $[0,T)$ for each $z_0 \in L^2(\mathcal{E})$.
- For each $T > 0$ and $g \in L^2((0,T) \times \mathcal{E})$, we denote $(g(t))(x) := g(t,x)$ for $(t,x) \in (0,T) \times \mathcal{E}$.
- A^* and B^* denote the adjoint operators of the above mentioned corresponding operators A and B.
- A function P is called a *mild solution* to the Riccati equation:

$$P'(t) + A^*P(t) + P(t)A - P(t)BB^*P(t) = 0 \text{ on } [0,T), \tag{3}$$

if for each $\delta \in (0,T)$ the function P satisfies:

$$\begin{aligned} P(T-\delta-t)z_0 =& e^{tA^*}P(T-\delta)e^{tA}z_0 \\ &- \int_0^t e^{(t-s)A^*}P(T-\delta-s)BB^*P(T-\delta-s)e^{(t-s)A}z_0 ds, \end{aligned}$$

for each $t \in [0, T-\delta]$ and $z_0 \in L^2(\mathcal{E})$, and the equality $\lim\limits_{(s,z)\to(T,z_0)} \langle P(s)z, z\rangle = +\infty$ holds for each $z_0 \in L^2(\mathcal{E})$ and $z_0 \neq 0$.

Theorem 2 (Existence of Feedback Control) *Let $\mathcal{G} = (\mathcal{V}, \mathcal{E})$ be a network satisfying the hypotheses of Theorem 1. Then, for each $z_0 \in H_0^1(\mathcal{E})$ and $T > 0$, there exists the following feedback control:*

$$f(t,x) := -a^{-1}1_{\omega}P(t)y(t),$$

for the solution y to (1) *with $b = c = 0$ and $a, \mu \in \mathbb{R}$ constant coefficients, where $P \in C_S([0,T]; \sum^+(L^2(\mathcal{E})))$ is the unique mild solution of the Riccati system* (3) *for $A = a^{-1}\mu\partial_{xx}^2$ and $B = a^{-1}1_{\omega}$.*

Theorem 2 was proved by using auxiliary results from [22].

1.3 Application to the Resolution of Inverse Problems

Carleman estimates can also be used to obtain results in the field of inverse problems, which is an additional objective of our work. In fact, the link between Carleman inequalities and their applications is well known. Some important references regarding this topic include [16, 17], and detailed surveys are included in [3, 24].

In this work, we sought to generalize the results of [15] to systems with loops. With that purpose, we considered the system:

$$\begin{cases} \partial_t y - \mu\partial_{xx}^2 y + py = 0, & \text{in } (0,T)\times\mathcal{E}, \\ y = 0, & \text{on } (0,T)\times\mathcal{V}_\partial, \\ y^{e_i} = y^{e_j}, & \text{on } (0,T)\times\mathcal{V}_0,\ \forall e_i, e_j \in \mathcal{E}(v), \\ \sum\limits_{e\in\mathcal{E}(v)} \mu^e\partial_{n^e}y^e = \gamma y, & \text{on } (0,T)\times\mathcal{V}_0, \\ y(0,\cdot) = y_0, & \text{in } \mathcal{E}, \end{cases} \tag{4}$$

for μ a piecewise constant function, γ a real parameter, y_0 the initial state and p the static potential. Moreover, we denoted by $y[p, y_0]$ the solution of (4).

Our objective was to recover the potential p by making measurements on the flux of the heat at a time $t_0 > 0$ and also on the observation domain ω but throughout the whole time interval $(0,T)$. In particular, we proved the following result:

Theorem 3 (Resolution of an Inverse Problem) *Let $\mathcal{G} = (\mathcal{V}, \mathcal{E})$ be a network satisfying Hypothesis 1, $p \in L^\infty(\mathcal{E})$, $r > 0$ and $y_0 \in L^2(\mathcal{E})$ such that $y[p, y_0] \in H^1(0,T; H^2_{pw}(\mathcal{E})) \cap H^2(0,T; L^2_{pw}(\mathcal{E}))$ and such that for some $t_0 \in (0,T)$ the following estimate holds:*

$$|y[p, y_0](t_0,\cdot)| \geq r \quad \text{in } \mathcal{E}. \tag{5}$$

Then, for any $m > 0$, there is a constant $C(m, r, T, \|\partial_t y[p, y_0]\|_{L^\infty((0,T)\times\mathcal{E})})$ such that for any $q \in L^\infty(\mathcal{E})$ satisfying:

$$\|q\|_{L^\infty(\mathcal{E})} \leq m,$$

we have:

$$\|q - p\|_{L^2(\mathcal{E})} \leq C \left(\|y[p, y_0](t_0, \cdot) - y[q, y_0](t_0, \cdot)\|_{H^2(\mathcal{E})} \right.$$
$$\left. + \|y[p, y_0] - y[q, y_0]\|_{H^1(0,T;L^2(\omega))} \right). \tag{6}$$

The proof of Theorem 3 is an easy consequence of the Carleman inequality in Proposition 2 (see [1]).

2 The Observability Problem

In this section we explain the main results and tools we needed in order to prove the observability inequality for system (2). With that purpose, in Sect. 2.1 we show the construction of an auxiliary function of Fursikov-Imanuvilov type, and in Sect. 2.2, using appropriate weights, we show the observability of system (2) with a Carleman inequality.

2.1 Construction of the Auxiliary Function

In this section we construct an auxiliary function that is required to define the Fursikov-Imanuvilov weights in Sect. 2.2. Throughout this section we consider an open subdomain $\tilde{\omega} \subset \omega$ such that $\overline{\tilde{\omega}} \subset \omega$ and such that, for all $e \in \mathcal{E}$, $\tilde{\omega} \cap e \neq \emptyset$ if and only if $\omega \cap e \neq \emptyset$.

We needed to make sure that for all edge e, if $e \cap \tilde{\omega} \neq \emptyset$, the maximum of η^e is achieved in $\tilde{\omega}$ and if $e \cap \tilde{\omega} = \emptyset$, the maximum of η^e is achieved on $u(e)$, being its derivative small near $u(e)$. As the "smallness" depends on the coefficients of the system, we get a family of auxiliary functions whose derivatives near $u(e)$ are as small as needed, and such that they are uniformly bounded in $W^{2,\infty}_{pw}(\mathcal{E})$.

Proposition 1 (Construction of the Auxiliary Function) *Let $\mathcal{G} = (\mathcal{V}, \mathcal{E})$ be a network, and $\tilde{\omega}$ be a domain satisfying Hypothesis 1 with the indexing function u. We identify the edges of $\mathcal{G}$ as in Remark 1. Then, there is $C > 0$ such that for all $\delta \in [0, 1]$ there exists a function η satisfying:*

1. *The function $\eta \in C^0(\overline{\mathcal{E}}) \cap C^2_{pw}(\mathcal{E})$ and $\|\eta\|_{W^{2,\infty}_{pw}} \leq C$.*
2. *For all edges e such that $e \cap \tilde{\omega} = \emptyset$, then:*
   - $\partial_x \eta^e \geq \delta$ *on* e,
   - $\partial_{n^e} \eta^e(0) = -\partial_x \eta^e(0) = -1$,
   - $\partial_{n^e} \eta^e(L^e) = \partial_x \eta^e(L^e) = \delta$.

3. *If an edge e that we identify with $[0, L^e]$ satisfies $e \cap \tilde{\omega} \neq \emptyset$, then:*
   - $|\partial_x \eta| = 1$ *on* $e \setminus \tilde{\omega}$,
   - $\partial_{n^e} \eta^e(0) = -\partial_x \eta^e(0) = -1$,
   - $\partial_{n^e} \eta^e(L^e) = \partial_x \eta^e(L^e) = -1$.

The proof of the existence of such function is based on an induction on the number of edges of $\mathcal{G}$. In order to prove Proposition 1, we needed to study the case of an edge assuming we have some restrictions on the boundary. This was done with Lemmas 1 and 2, whose proofs are standard (see [1, 12]). In Lemma 1 we studied the construction of the auxiliary function for edges that have no intersection with $\tilde{\omega}$.

Lemma 1 (Extension of the Auxiliary Function with One Constraint) *Let $\tilde{\omega} \subset \mathcal{E}$ be an open subdomain, $e \simeq [0, L^e]$ be an edge such that $\tilde{\omega} \cap e = \emptyset$, $R \in \mathbb{R}$, $p \in \{0, L^e\}$ and $\delta \in [0, 1]$. Then, there is a function $\eta^e \in C^2(\overline{e})$ such that:*

- $\partial_x \eta^e \geq \delta$ *on* $[0, L^e]$,
- $\|\eta^e\|_{L^\infty(0,L^e)} \leq |R| + L^e$, $\|\partial_x \eta^e\|_{L^\infty(0,L^e)} \leq 1$, $\|\partial_{xx} \eta^e\|_{L^\infty(0,L^e)} \leq \frac{1}{L^e}$,
- $\partial_{n^e} \eta^e(0) = -\partial_x \eta^e(0) = -1$,
- $\partial_{n^e} \eta^e(L^e) = \partial_x \eta^e(L^e) = \delta$,
- $\eta^e(p) = R$.

Next, in the following Lemma, we studied the construction of the auxiliary function for edges which intersect $\tilde{\omega}$.

Lemma 2 (Extension of the Auxiliary Function with Two Constraints) *Let $\tilde{\omega} \subset \mathcal{E}$ be an open subdomain, $e \simeq [0, L^e]$ be an edge such that $\tilde{\omega} \cap e \neq \emptyset$ and $R_1, R_2 \in \mathbb{R}$. Then, there is a function $\eta^e \in C^2(\overline{e})$ such that:*

- $\|\eta^e\|_{W^{2,\infty}_{pw}(0,L^e)} \leq C(|R_1|, |R_2|, L^e, \tilde{\omega})$, *for C increasing with $|R_1|$ and $|R_2|$ for a fixed L^e and $\tilde{\omega}$,*
- $|\partial_x \eta^e| = 1$ *on* $[0, L^e] \setminus \tilde{\omega}$,
- $\partial_{n^e} \eta^e(0) = -\partial_x \eta^e(0) = -1$,
- $\partial_{n^e} \eta^e(L^e) = \partial_x \eta^e(L^e) = -1$,
- $\eta^e(0) = R_1$, $\eta^e(L^e) = R_2$.

2.2 A New Carleman Inequality

The auxiliary function constructed in the previous section allowed us to define the usual Fursikov-Imanuvilov weights:

$$\alpha(t,x) := \frac{e^{6\lambda\|\eta\|_\infty} - e^{\lambda(4\|\eta\|_\infty + \eta(x))}}{t(T-t)}, \quad \xi(t,x) := \frac{e^{\lambda(4\|\eta\|_\infty + \eta(x))}}{t(T-t)}, \quad \forall (t,x) \in (0,T) \times \mathcal{E}, \tag{7}$$

where η is defined in Proposition 1 for $\tilde{\omega}$ an open domain compactly included in ω such that $e \cap \tilde{\omega} = \emptyset$ if and only if $e \cap \omega = \emptyset$ for all $e \in \mathcal{E}$, and for $\delta > 0$

a sufficiently small parameter that will be defined in the proof of Proposition 2 (see [1]) for absorbing boundary terms. Bearing this in mind, we stated the next Carleman inequality:

Proposition 2 (A New Carleman Inequality) *Let* $a, \mu \in W^{1,\infty}((0,T); L^{\infty}(\mathcal{E})) \cap L^{\infty}((0,T); W^{1,\infty}_{pw}(\mathcal{E}))$ *such that* $\inf a, \inf \mu > 0$, $h \in L^{\infty}((0,T) \times \mathcal{V}_0)$ *and* $g \in L^2(Q)$*. Then, there is* $C > 0$ *depending on* $\mathcal{G}$, ω, a, *and* μ *such that for all* $\varphi_T \in L^2(\mathcal{E})$, $\lambda \geq C$ *and* $s \geq C(T + T^2)$ *the following inequality is satisfied:*

$$s^3\lambda^4 \iint_Q e^{-2s\alpha}\xi^3|\varphi|^2 \mathrm{d}x\mathrm{d}t \leq C\left(s^3\lambda^4 \iint_{Q_\omega} e^{-2s\alpha}\xi^3|\varphi|^2\mathrm{d}x\mathrm{d}t + \iint_Q e^{-2s\alpha}|g|^2\mathrm{d}x\mathrm{d}t\right), \tag{8}$$

for α *and* ξ *the weights defined in* (7), *and* φ *the solution of:*

$$\begin{cases} -a\partial_t\varphi - \mu\partial^2_{xx}\varphi = g, & \text{in } (0,T) \times \mathcal{E}, \\ \varphi = 0, & \text{on } (0,T) \times \mathcal{V}_\partial, \\ \varphi^{e_i} = \varphi^{e_j}, & \text{on } (0,T) \times \mathcal{V}_0,\ \forall e_i, e_j \in \mathcal{E}(v), \\ \sum_{e \in \mathcal{E}(v)} \mu^e \partial_{n^e}\varphi^e = h\varphi, & \text{on } (0,T) \times \mathcal{V}_0, \\ \varphi(T,\cdot) = \varphi_T, & \text{in } \mathcal{E}. \end{cases} \tag{9}$$

Acknowledgments During the time this research work was done, both authors were supported by the grant PID2021-126813NB-I00 funded by MICIU/AEI/10.13039/501100011033 and by "ERDF A way of making Europe" and by the grant IT1615-22 funded by the Basque Government.

References

1. J. Apraiz, J.A. Bárcena-Petisco, Observability and control of parabolic equations on networks with loops. *hal-03501343* (2022)
2. J.A. Bárcena-Petisco, M. Cavalcante, G.M. Coclite, N. de Nitti, E. Zuazua, Control of hyperbolic and parabolic equations on networks and singular limits. *hal-03233211* (2021)
3. M. Bellassoued, M. Yamamoto, *Carleman Estimates and Applications to Inverse Problems for Hyperbolic Systems* (Springer, 2017)
4. V.D. Blondel, E.D. Sontag, M. Vidyasagar, J.C. Willems, *Open Problems in Mathematical Systems and Control Theory*. Communication and Control Engineering Series (Springer, London, 1999)
5. J. Brouwer, I. Gasser, M. Herty, Gas pipeline models revisited: Model hierarchies, non-isothermal models and simulations on networks. Multiscale Model. Simul. **9**, 601–623 (2011)
6. P. Cannarsa, A. Duca, C. Urbani, Exact controllability to eigensolutions of the bilinear heat equation on compact networks. arXiv:2111.02250 (2021)
7. S. Chen, I. Lasiecka, Feedback exact null controllability for unbounded control problems in Hilbert space. J. Optim. Theory App. **74**(2), 191–219 (1992)
8. J.-M. Coron, *Control and Nonlinearity*, vol. 136 (American Mathematical Society, 2007)

9. R. Dager, E. Zuazua, *Wave Propagation, Observation and Control in 1-d Flexible Multistructures*. Volume 50 of Mathematics & Applications (Springer, Berlin, 2006)
10. K. Egger, T. Kugler, Damped wave systems on networks: exponential stability and uniform approximations. Numer. Math. **138**(4), 839–867 (2018)
11. H. Egger, N. Philippi, On the transport limit of singularly perturbed convection-diffusion problems on networks. Math. Methods Appl. Sci. **44**, 5005–5020 (2021)
12. A.V. Fursikov, O.Y. Imanuvilov, *Controllability of Evolution Equations*. Volume 34 of Lecture Notes Series (Seoul National University, Research Institute of Mathematics, Global Analysis Research Center, Seoul, 1996)
13. G. Notarstefano, G. Parlangeli, Controllability and observability of grid graphs via reduction and symmetries. IEEE T. Automat. Contr. **58**(7), 1719–1731 (2013)
14. F.M. Hante, G. Leugering, A. Martin, L. Schewe, M. Schmidt, *Challenges in Optimal Control Problems for Gas and Fluid Flow in Networks of Pipes and Canals: From Modeling to Industrial Applications*. Industrial Mathematics and Complex Systems (Springer, 2017), pp. 77–122
15. L. Ignat, A.F. Pazoto, L. Rosier, Inverse problem for the heat equation and the Schrödinger equation on a tree. Inverse Prob. **28**(1), 015011 (2011)
16. O.Y. Imanuvilov, M. Yamamoto, Lipschitz stability in inverse parabolic problems by the Carleman estimate. Inverse Prob. **14**, 1229–1245 (1998)
17. O.Y. Imanuvilov, M. Yamamoto, Global Lipschitz stability in an inverse hyperbolic problem by interior observations. Inverse Prob. **17**(4), 717 (2001)
18. L.E. Lagnese, G. Leugering, E.J.P.G. Schmidt, *Modeling, Analysis and Control of Dynamic Elastic Multi-Link Structures*. Volume 19 of Systems Control: Foundations Applications (Springer Science+Business Media, New York, 1994)
19. P. Lin, Global blowup controllability of heat equation with feedback control. Commun. Contemp. Math. **20**(5), 1750062–1–11 (2018)
20. V. Mehandiratta, M. Mehra, G. Leugering, Optimal control problems driven by time-fractional diffusion equations on metric graphs: optimality system and finite difference approximation. SIAM J. Control and Optim. **59**(6), 4216–4242 (2021)
21. M. Newman, A.L. Barabási, D.J. Watts, *The Structure and Dynamics of Networks*. Volume 19 of Princeton Studies in Complexity (Princeton University Press, Princeton, 2011)
22. M. Sîrbu, A Riccati equation approach to the null controllability of linear systems. Comm. Appl. Anal., 164–177(2) (2002)
23. M. Sîrbu, G. Tessitore, Null controllability of an infinite dimensional sde with state and control-dependent noise. Syst. Control Lett. **44**(5), 385–394(44) (2001)
24. M. Yamamoto, Carleman estimates for parabolic equations and applications. Inverse Prob. **25**(12), 123013 (2009)

Some Results on the Energy Decay of Solutions for a Wave Equation with a General Internal Feedback of Diffusive Type

Abbes Benaissa and Abdelkader Boudaouad

2010 Mathematics Subject Classification 93D15, 35B40, 47D03, 74D05

1 Introduction

In this paper, we investigate the existence and stabilization of wave equation with general internal feedback of diffusive type. The system is given by:

$$(P)\begin{cases} u_{tt}(x,t) - \Delta u(x,t) + \zeta \displaystyle\int_{-\infty}^{+\infty} \mu(\xi)\phi(x,\xi,t)d\xi = 0 & \text{in } \Omega \times (0,+\infty), \\ u(x,t) = 0 & \text{on } \partial\Omega \times (0,+\infty), \\ \phi_t(x,\xi,t) + (\xi^2+\eta)\phi(x,\xi,t) - u_t(x,t)\mu(\xi) = 0 & \text{in } \Omega \times (-\infty,+\infty) \\ & \times(0,+\infty), \\ u(x,0) = u_0(x), u_t(x,0) = u_1(x) & \text{in } \Omega, \\ \phi(x,\xi,0) = \phi_0(x,\xi) & \text{on } \Omega \times (-\infty,+\infty), \end{cases}$$

where Ω is a bounded open domain in $\mathbb{R}^n$ $(n \geq 1)$, with a smooth boundary $\partial\Omega$, $\zeta > 0$, $\eta \geq 0$ and μ is a general measure density and the initial data are taken in suitable spaces. When $\mu(\xi) = |\xi|^{\frac{2\alpha-1}{2}}$, $\zeta = \gamma\pi^{-1}\sin(\alpha\pi)$ and $\phi_0 \equiv 0$ $(P)_1$ becomes $u_{tt}(x,t) - \Delta u(x,t) + \gamma\partial_t^{\alpha,\eta}u(x,t) = 0$ in $\Omega \times (0,+\infty)$, where $\partial_t^{\alpha,\eta}$ denoted the generalized Caputo's fractional derivative of order α, $0 < \alpha < 1$ with respect to the time variable. It is defined by

$$\partial_t^{\alpha,\eta} w(t) = \frac{1}{\Gamma(1-\alpha)}\int_0^t (t-s)^{-\alpha}e^{-\eta(t-s)}\frac{dw}{ds}(s)\,ds, \ \eta \geq 0.$$

In [7] Mbodje studied the energy decay of the wave equation with a boundary control of fractional derivative type, that is

A. Benaissa (✉) · A. Boudaouad
Laboratory of Analysis and Control of PDEs, Djillali Liabes University, Sidi Bel Abbes, Algeria

K. Ammari et al. (eds.), *Advances in Partial Differential Equations and Control*, Trends in Mathematics, https://doi.org/10.1007/978-3-031-62265-6_4

$$\begin{cases} u_{tt}(x,t) - u_{xx}(x,t) = 0 & \text{in } (0,L) \times (0,+\infty), \\ u(0,t) = 0 & \text{on } (0,+\infty), \\ u_x(L,t) + \rho\partial_t^{\alpha,\eta} u(L,t) = 0 & \text{on } (0,+\infty), \\ u(x,0) = u_0(x), \quad u_t(x,0) = u_1(x) & \text{on } (0,L). \end{cases}$$

In order to reduce the difficulties, he used a new approach called "diffusive representation." The original model is transformed into an augmented system which can be more easily tackled by the energy method. The author showed strong asymptotic stability of solutions when $\eta = 0$ and polynomial type decay rate $\mathcal{E}(t) \leq C/t$ for $t > 0$ when $\eta \neq 0$.

Recently in [5], Benaissa and Rafa extended the result of Mbodje to higher-space dimension and boundary control of diffusive type. The authors established a less decay estimate by adopting the multiplier method. Very recently, in [1], Ammari et al., studied the wave equation with internal fractional damping. The system considered is as follows:

$$\begin{cases} u_{tt}(x,t) - \Delta u(x,t) + \gamma\partial_t^{\alpha,\eta} u(x,t) = 0 & \text{in } \Omega \times (0,+\infty), \\ u(x,t) = 0 & \text{on } \partial\Omega \times (0,+\infty), \\ u(x,0) = u_0(x), \quad u_t(x,0) = u_1(x) & \text{on } \Omega. \end{cases}$$

The authors proved that the energy decays polynomially as $t^{-2/(1-\alpha)}$.

The main result of this paper is to obtain a precise and optimal energy decay estimate for a general internal control of diffusive type, from which the usual control of fractional derivative type is a special case.

This paper is organized as follows. In Sect. 3, existence and uniqueness of strong and weak solutions of the system are proved, using the Hille–Yosida Theorem. In Sect. 4, we show the lack of exponential stability by spectral analysis. In Sect. 5, we study asymptotic stability of the above model and the main results are Theorems 4, 7, 8 and 11. In Theorem 7, we show a general decay rate depending on the form of the measure density μ. The proof heavily relies on the multiplier method.

2 Preliminaries Results

Let us state the precise hypotheses on function μ. μ is an even nonnegative measurable function such that

$$\int_{-\infty}^{\infty} \frac{\mu(\xi)^2}{1+\xi^2}\, d\xi < \infty. \tag{1}$$

Now, we recall a definition which will be needed in Sect. 5 for application.

Definition 1 Given $a \geq 0$ and a measurable function $M : [a,+\infty) \to (0,+\infty)$ we say that M has positive increase if there exist strictly positive constants $\alpha > 0$, $c \in$

(0, 1] and $s_0 \geq a$ such that

$$\frac{M(\lambda s)}{M(s)} \geq c\lambda^{\alpha}, \quad \lambda \geq 1, \quad s \geq s_0. \tag{2}$$

3 Global Existence

In this section, we give an existence and uniqueness result for the problem (P) using semigroup theory. We define the following space

$$H_0^1(\Omega) = \{u \in H^1(\Omega) : u|_{\partial\Omega} = 0\}$$

and the energy space

$$\mathcal{H} = H_0^1(\Omega) \times L^2(\Omega) \times L^2(\Omega \times (-\infty, +\infty))$$

equipped with the following inner product

$$\langle U, \tilde{U}\rangle_{\mathcal{H}} = \int_{\Omega} \left(v\bar{\tilde{v}} + \nabla u \nabla \bar{\tilde{u}}\right) dx + \zeta \int_{\Omega} \int_{-\infty}^{+\infty} \phi\bar{\tilde{\phi}} d\xi dx, \tag{3}$$

where $U = (u, v, \phi)^T, \tilde{U} = (\tilde{u}, \tilde{v}, \tilde{\phi})^T$, In order to transform the problem (P) to an abstract problem on the Hilbert space $\mathcal{H}$. We introduce the vector function $U = (u, v, \phi)^T$, where $v = u_t$. Then problem (P) can be rewritten as

$$U_t = \mathcal{A}U, \quad U(0) = U_0, \tag{4}$$

where $U_0 = (u_0, u_1, \phi_0)^T$ and $\mathcal{A} : D(\mathcal{A}) \subset \mathcal{H} \to \mathcal{H}$ is defined as follows

$$\mathcal{A}(u, v, \phi) = \left(v, \Delta u - \zeta \int_{-\infty}^{+\infty} \mu(\xi)\phi(\xi) d\xi, -(\xi^2 + \eta)\phi + v(x)\mu(\xi)\right) \tag{5}$$

and its domain is given by

$$D(\mathcal{A}) = \left\{ \begin{array}{l} (u, v, \phi)^T \text{ in } \mathcal{H} : u \in H^2(\Omega) \cap H_0^1(\Omega), v \in H_0^1(\Omega), \\ \Delta u(x) - \zeta \int_{-\infty}^{+\infty} \mu(\xi)\phi(\xi) d\xi \in L^2(\Omega), \\ -(\xi^2 + \eta)\phi + v(x)\mu(\xi) \in L^2(\Omega \times (-\infty, +\infty)), \\ |\xi|\phi \in L^2(\Omega \times (-\infty, +\infty)) \end{array} \right\}. \tag{6}$$

The energy associated with the solution of the problem (P) is given by

$$\mathcal{E}(t) = \frac{1}{2}\|u_t\|_2^2 + \frac{1}{2}\|\nabla u\|_2^2 + \frac{\zeta}{2}\int_\Omega \int_{-\infty}^{+\infty} |\phi(x,\xi,t)|^2 \, d\xi dx. \tag{7}$$

Differentiating $\mathcal{E}$ in a formal way, using (P) and integrating by parts, we obtain after a straightforward computation the following Lemma.

Lemma 1 *Let(u, ϕ) be a regular solution of problem (P). Then, the energy functional defined by Eq. (7) satisfies*

$$\frac{d}{dt}\mathcal{E}(t) = -\zeta \int_\Omega \int_{-\infty}^{+\infty} (\xi^2 + \eta)|\phi(x,\xi,t)|^2 \, d\xi dx \leq 0. \tag{8}$$

We have the following result (see [6]).

Theorem 1 (Existence and Uniqueness) *If $U_0 \in \mathcal{H}$, then system Eq. (4) has a unique weak solution $U \in C^0(\mathbb{R}_+, \mathcal{H})$. Moreover if $U_0 \in D(\mathcal{A})$, then system Eq. (4) has a unique strong solution $U \in C^0(\mathbb{R}_+, D(\mathcal{A})) \cap C^1(\mathbb{R}_+, \mathcal{H})$.*

4 Lack of Exponential Stability

Theorem 2 ([8]) *Let $S(t) = e^{\mathcal{A}t}$ be a C_0-semigroup of contractions on Hilbert space X. Then $S(t)$ is exponentially stable if and only if*

$$\rho(\mathcal{A}) \supseteq \{i\beta : \beta \in \mathbb{R}\} \equiv i\mathbb{R} \tag{9}$$

and

$$\overline{\lim_{|\beta|\to\infty}} \|(i\beta I - \mathcal{A})^{-1}\|_{\mathcal{L}(X)} < \infty. \tag{10}$$

Our main result in this part is the following theorem.

Theorem 3 *The semigroup generated by the operator $\mathcal{A}$ is not exponentially stable.*

Proof of Theorem 3 Let $-\beta_n^2 = (i\beta_n)^2$ be a sequence of eigenvalues corresponding to the sequence of normalized eigenfunctions u_n of the operator Δ such that $|\beta_n| \longrightarrow$ as $n \longrightarrow \infty$ and

$$\begin{cases} \Delta u_n = -\beta_n^2 u_n & \text{in } \Omega, \\ u_n = 0 & \text{on } \partial\Omega. \end{cases} \tag{11}$$

Our aim is to prove, under some conditions, that if $i\beta_n$ satisfies Eq. (9) then Eq. (10) does not hold. In other words we want to show that an infinite number of eigenvalues of $\mathcal{A}$ approach the imaginary axis which prevents the wave system (P) from being exponentially stable. Indeed we first compute the characteristic equation that gives the eigenvalues of $\mathcal{A}$. Let λ be an eigenvalue of $\mathcal{A}$ with associated eigenvector $U = (u, v, \phi)^T$. Then $\mathcal{A}U = \lambda U$ is equivalent to

$$\begin{cases} \lambda u - v = 0, \\ \lambda v - \Delta u + \zeta \int_{-\infty}^{+\infty} \mu(\xi)\phi(\xi)d\xi = 0, \\ \lambda\phi + (\xi^2 + \eta)\phi - v(x)\mu(\xi) = 0. \end{cases} \tag{12}$$

From Eq. $(12)_1$, we have

$$v = \lambda u. \tag{13}$$

Inserting Eq. (13) in Eq. $(12)_2$, we get

$$\lambda^2 u - \Delta u + \zeta \int_{-\infty}^{+\infty} \mu(\xi)\phi(\xi)d\xi = 0. \tag{14}$$

Then from Eq. (13), Eq. $(12)_3$ and Eq. (14) we get

$$\lambda^2 u - \Delta u + \zeta \int_{-\infty}^{+\infty} \lambda u(x) \tfrac{\mu^2(\xi)}{\xi^2+\eta+\lambda} d\xi = 0, \tag{15}$$

it follows that

$$\Delta u = \left(\lambda^2 + \zeta \int_{-\infty}^{+\infty} \lambda \tfrac{\mu^2(\xi)}{\xi^2+\eta+\lambda} d\xi\right) u. \tag{16}$$

From Eq. (11) and Eq. (16) we get the existence of a sequence of eigenvalues λ_n of $\mathcal{A}$ corresponding to the sequence β_n, such that

$$-\beta_n^2 u_n = \Delta u_n = \left(\lambda_n^2 + \zeta \int_{-\infty}^{+\infty} \lambda_n \frac{\mu^2(\xi)}{\xi^2+\eta+\lambda_n} d\xi\right) u_n, \tag{17}$$

then we get

$$\beta_n^2 = -\lambda_n^2 - \zeta \int_{-\infty}^{+\infty} \lambda_n \frac{\mu^2(\xi)}{\xi^2+\eta+\lambda_n} d\xi. \tag{18}$$

By taking $\phi_n = \frac{\mu(\xi)}{\xi^2+\eta+i\beta_n} u_n$ and the vector $U_n = (\frac{u_n}{i\beta_n}, u_n, \phi_n)^T$. We have $U_n \in D(\mathcal{A})$. Then a direct computation gives

$$\mathcal{A}\begin{pmatrix} \frac{u_n}{i\beta_n} \\ u_n \\ \phi_n \end{pmatrix} = \begin{pmatrix} u_n \\ i\beta_n u_n - \zeta \int_{-\infty}^{+\infty} \mu(\xi)\phi_n(\xi)d\xi \\ i\beta_n\phi_n \end{pmatrix}. \tag{19}$$

It follows that

$$(i\beta_n I - \mathcal{A})U_n = \begin{pmatrix} 0 \\ \zeta u_n \int_{-\infty}^{+\infty} \frac{\mu^2(\xi)}{\xi^2+\eta+i\beta_n} d\xi \\ 0 \end{pmatrix}. \tag{20}$$

Proving

$$\left\|(i\beta_n I - \mathcal{A})^{-1}\right\|_{\mathcal{L}(\mathcal{H})} \longrightarrow \infty \quad \text{as} \quad |\beta_n| \longrightarrow \infty \quad \left(\text{i.e. as} \quad n \longrightarrow \infty\right) \tag{21}$$

reduces to show that

$$\left\|\zeta u_n \int_{-\infty}^{+\infty} \frac{\mu^2(\xi)}{\xi^2+\eta+i\beta_n} d\xi\right\|_{L^2(\Omega)} \longrightarrow 0 \quad \text{as} \quad n \longrightarrow \infty. \tag{22}$$

Indeed, using the fact that

$$\left|\int_{-\infty}^{+\infty} \frac{\mu^2(\xi)}{\xi^2+\eta+i\beta_n} d\xi\right| \longrightarrow 0 \quad \text{as} \quad n \longrightarrow \infty \tag{23}$$

(see Lemma 4.3 in [5]) and the fact that u_n is a normalized eigenfunction of the operator Δ, for each $n \in \mathbb{N}$, we get the desired limit.

5 Asymptotic Stability

5.1 Strong Stability of the System

In this part, we use a general Theorem of Arendt-Batty in [2] to show the strong stability of the C_0-semigroup $e^{t\mathcal{A}}$ associated with the wave system (P). Our main result is the following (see [6]).

Theorem 4 *The C_0-semigroup $e^{t\mathcal{A}}$ is strongly stable in $\mathcal{H}$; i.e, for all $U_0 \in \mathcal{H}$, the solution of Eq. (4) satisfies*

$$\lim_{t\to\infty} \|e^{t\mathcal{A}} U_0\|_{\mathcal{H}} = 0.$$

For the proof of Theorem 4, we need the following two lemmas.

Lemma 2 *$\mathcal{A}$ does not have eigenvalues in $i\mathbb{R}$.*

Lemma 3 *We have*

$$i\mathbb{R} \subset \rho(\mathcal{A}) \text{ if } \eta \neq 0, \quad i\mathbb{R}^* \subset \rho(\mathcal{A}) \text{ if } \eta = 0.$$

5.2 General Decay Estimate

Since the energy of system (P) has no uniform decay rate, we will look for non-uniform decay rate for smooth data. To do this, we shall use the following two important abstract results connecting resolvent estimates of generators to the decay rate of their corresponding semigroups.

Theorem 5 ([3]) *Let X be a Banach space and let $\mathcal{A}$ be the generator of a bounded C_0-semigroup $(S(t))_{t\geq 0}$ on X. If*

$$i\mathbb{R} \subset \rho(\mathcal{A}) \text{ and } \|(i\beta I - \mathcal{A})^{-1}\|_{\mathcal{L}(X)} \leq M(|\beta|),$$

where $M : \mathbb{R}_+ \to (0, \infty)$ is a continuous non-decreasing function, then there exists constants $C, c > 0$ such that

$$\|e^{\mathcal{A}t} U_0\| \leq \frac{C}{M_{\log}^{-1}(ct)} \|U_0\|_{D(\mathcal{A})},$$

where $M_{\log} : \mathbb{R}_+ \to (0, \infty)$ is defined by $M_{\log}(s) = M(s)(\log(1+M(s))+\log(1+s)), \quad s \geq 0$.

In particular, we have the following important theorem when the resolvent estimate is a function of positive increase.

Theorem 6 ([9]) *Let X be a Hilbert space and let $\mathcal{A}$ be the generator of a bounded C_0-semigroup $(S(t))_{t\geq 0}$ on X. If*

$$i\mathbb{R} \subset \rho(\mathcal{A}) \text{ and } \|(i\beta I - \mathcal{A})^{-1}\|_{\mathcal{L}(X)} \leq M(|\beta|),$$

where $M : \mathbb{R}_+ \to (0, \infty)$ is a continuous non-decreasing function of positive increase, then there exists a positive constant $C > 0$ such that

$$\|e^{\mathcal{A}t} U_0\| \leq C \frac{1}{M^{-1}(t)} \|U_0\|_{D(\mathcal{A})}, \quad t \to \infty.$$

Our result reads as follows

Theorem 7 *Let*

$$\mathcal{M}(\lambda) = c|\lambda|^{-3/2} \mathcal{S}^{-2}$$

for a suitable positive constant c and where $\mathcal{S} = \int_{-\infty}^{+\infty} \frac{|\xi||\mu(\xi)}{(|\lambda|+\xi^2+\eta)^2} d\xi$. Then the semigroup $S_{\mathcal{A}}(t)_{t\geq 0}$ satisfies the following decay estimate

(1) If $\mathcal{M}$ be a non-decreasing function of positive increase, then

$$\|e^{\mathcal{A}t}U_0\| \le C\frac{1}{\mathcal{M}^{-1}(t)}\|U_0\|_{D(\mathcal{A})}, \quad t \to \infty,$$

where $\mathcal{M}^{-1}$ is any asymptotic inverse of $\mathcal{M}$.

(2) *If*

$$\mathcal{M}(\lambda) \sim cl(|\lambda|), |\lambda| \to \infty,$$

where l is a non-decreasing slowly varying function, then

$$\|e^{\mathcal{A}t}U_0\| \le \frac{1}{l_{\log}^{-1}(ct)}\|U_0\|_{D(\mathcal{A})},$$

where $l_{\log}(s) = l(s)(\log(1 + l(s)) + \log(1 + s)), \quad s \ge 0.$

Proof We will need to study the resolvent equation $(i\lambda I - \mathcal{A})U = F$, for $\lambda \in \mathbb{R}$, namely

$$\begin{cases} i\lambda u - v = f_1, \\ i\lambda v - \Delta u + \zeta \int_{-\infty}^{+\infty} \mu(\xi)\phi(x,\xi)d\xi = f_2, \\ i\lambda\phi + (\xi^2 + \eta)\phi - v(x)\mu(\xi) = f_3, \end{cases} \tag{24}$$

where $F = (f_1, f_2, f_3)^T$. Taking inner product in $\mathcal{H}$ with $U = (u, v, \phi)^T$, using the fact that $\mathcal{E}(t) = \frac{1}{2}\|U\|_{\mathcal{H}}^2$ and from (7), we get

$$|Re\langle \mathcal{A}U, U\rangle| \le \|U\|_{\mathcal{H}}\|F\|_{\mathcal{H}}. \tag{25}$$

This implies that

$$\zeta \int_{\Omega}\int_{-\infty}^{+\infty} (\xi^2 + \eta)|\phi(x,\xi)|^2\, d\xi\, dx \le \|U\|_{\mathcal{H}}\|F\|_{\mathcal{H}}. \tag{26}$$

From Eq. $(24)_3$, we obtain

$$v(x)\mu(\xi) = (i\lambda + \xi^2 + \eta)\phi(x,\xi) - f_3(x,\xi), \quad \forall (x,\xi) \in \Omega \times (-\infty, +\infty). \tag{27}$$

By multiplying Eq. (27) by $(i\lambda + \xi^2 + \eta)^{-2}|\xi|$, we get

$$(i\lambda+\xi^2+\eta)^{-2}v(x)|\xi|\mu(\xi) = (i\lambda+\xi^2+\eta)^{-1}|\xi|\phi - (i\lambda+\xi^2+\eta)^{-2}|\xi| f_3(x,\xi), x \in \Omega. \tag{28}$$

Integrating over the interval $]-\infty, +\infty[$ with respect to the variable ξ, then taking absolute values of both sides of Eq. (28) and applying triangle inequality we obtain

$$\left|\int_{-\infty}^{+\infty} \frac{|\xi|\,\mu(\xi)}{(i\lambda+\xi^2+\eta)^2}\,d\xi\right|\,|v(x)| \le \int_{-\infty}^{+\infty} \frac{|\xi|\,|\phi|}{|i\lambda+\xi^2+\eta|}\,d\xi + \int_{-\infty}^{+\infty} \frac{|\xi|\,|f_3(x,\xi)|}{|i\lambda+\xi^2+\eta|^2}\,d\xi. \tag{29}$$

Applying Cauchy–Schwartz inequality to the Eq. (29) and integrating over Ω, we get

$$\begin{aligned} &\left(\int_\Omega |v(x)|^2\,dx\right)\left|\int_{-\infty}^{+\infty} \frac{|\xi|\mu(\xi)}{(i\lambda+\xi^2+\eta)^2}\,d\xi\right|^2 \\ &\le 2\int_\Omega \int_{-\infty}^{+\infty} (|\xi|^2+\eta)|\phi|^2\,d\xi\,dx\left(\int_{-\infty}^{+\infty} \frac{d\xi}{|i\lambda+\xi^2+\eta|^2}\right) \\ &+2\left(\int_\Omega \int_{-\infty}^{+\infty} |f_3(x,\xi)|^2\,d\xi\,dx\right)\left(\int_{-\infty}^{+\infty} \frac{|\xi|^2}{|i\lambda+\xi^2+\eta|^4}\,d\xi\right). \end{aligned} \tag{30}$$

Now from the following inequality (see proposition 2.4 in [4])

$$\begin{aligned} \left|\int_{-\infty}^{+\infty} (i\lambda+\eta+\xi^2)^{-2}|\xi|\mu(\xi)\,d\xi\right| &\ge \frac{\sqrt{1+\cos\theta}}{\sqrt{2}}\int_{-\infty}^{+\infty} (|i\lambda+\eta|+\xi^2)^{-2}|\xi|\mu(\xi)\,d\xi \\ &\ge \frac{\sqrt{1+\cos\theta}}{\sqrt{2}}\int_{-\infty}^{+\infty} (|\lambda|+\xi^2+\eta)^{-2}|\xi|\mu(\xi)\,d\xi, \end{aligned} \tag{31}$$

where $\cos\theta = \eta/\sqrt{\lambda^2+\eta^2}$. We obtain

$$\left|\int_{-\infty}^{+\infty} \frac{|\xi|\mu(\xi)}{(i\lambda+\xi^2+\eta)^2}\,d\xi\right| \ge \frac{1}{\sqrt{2}}\int_{-\infty}^{+\infty} \frac{|\xi|\mu(\xi)}{(|\lambda|+\xi^2+\eta)^2}\,d\xi\,. \tag{32}$$

Denoting $\mathcal{S} = \int_{-\infty}^{+\infty} \frac{|\xi|\mu(\xi)}{(|\lambda|+\xi^2+\eta)^2}\,d\xi$, and by using Eq. (26), Eq. (30) and Eq. (32) we obtain

$$\begin{aligned} &\mathcal{S}^2\|v\|^2_{L^2(\Omega)} \\ &\le 4\left(\int_{-\infty}^{+\infty} \frac{d\xi}{|i\lambda+\xi^2+\eta|^2}\right)\|U\|_{\mathcal{H}}\,\|F\|_{\mathcal{H}}+4\|f_3\|^2_{L^2(\Omega\times(-\infty,+\infty))}\left(\int_{-\infty}^{+\infty} \frac{|\xi|^2}{|i\lambda+\xi^2+\eta|^4}\,d\xi\right) \\ &\le 8\left(\int_{-\infty}^{+\infty} \frac{d\xi}{(|\lambda|+\xi^2+\eta)^2}\right)\|U\|_{\mathcal{H}}\,\|F\|_{\mathcal{H}}+16\|f_3\|^2_{L^2(\Omega\times(-\infty,+\infty))}\left(\int_{-\infty}^{+\infty} \frac{|\xi|^2}{(|\lambda|+\xi^2+\eta)^4}\,d\xi\right). \end{aligned} \tag{33}$$

Now we use the classical multiplier method. Let us now introduce the following notations

$$\mathcal{I}_u(x) = |v(x)|^2 + |\nabla u(x)|^2 \text{ and } \mathcal{E}_u = \int_\Omega \left(|v(x)|^2 + |\nabla u(x)|^2\right)dx.$$

Lemma 4 *We have that*

$$\mathcal{E}_u \le C\int_\Omega \left|\int_{-\infty}^{+\infty} \mu(\xi)\phi(x,\xi)d\xi\right|^2 dx + 2\|v\|_2^2 + C'\|F\|^2_{\mathcal{H}}.$$

Proof Multiplying Eq. $(24)_2$ by $\bar{u}$, using Eq. $(24)_1$, integrating over Ω, and applying Cauchy–Schwartz inequality, we get the result.

Now, using

$$\left|\int_{-\infty}^{+\infty}\mu(\xi)\phi(x,\xi)d\xi\right|^2 \le \Big(\int_{-\infty}^{+\infty}\frac{\mu^2(\xi)}{\xi^2+\eta}d\xi\Big)\Big(\int_{-\infty}^{+\infty}(\xi^2+\eta)|\phi|^2d\xi\Big),$$

then we get

$$\mathcal{E}_u \le c\|v\|_2^2 + c'\|F\|_{\mathcal{H}}^2 + c''\|U\|_{\mathcal{H}}\|F\|_{\mathcal{H}}$$

and from Eq. (26) it follows that

$$\begin{aligned}\|\phi\|_{L^2(\Omega\times(-\infty,+\infty))} &= \int_\Omega\int_{-\infty}^{+\infty}|\phi|^2d\xi\,dx\\ &\le C\int_\Omega\int_{-\infty}^{+\infty}(\xi^2+\eta)|\phi|^2d\xi\,dx \le c\|U\|_{\mathcal{H}}\|F\|_{\mathcal{H}}.\end{aligned}$$

We conclude that

$$\|U\|_{\mathcal{H}}^2 \le c\|v\|_{L^2(0,1)}^2 + c'\|F\|_{\mathcal{H}}^2 + c''\|U\|_{\mathcal{H}}\|F\|_{\mathcal{H}}. \tag{34}$$

Inserting Eq. (33) into Eq. (34), we obtain

$$\begin{aligned}\|U\|_{\mathcal{H}}^2 \le c\mathcal{S}^{-2}\Big(\int_{-\infty}^{+\infty}\frac{d\xi}{(|\lambda|+\xi^2+\eta)^2}\Big)\|U\|_{\mathcal{H}}\|F\|_{\mathcal{H}}+c'\mathcal{S}^{-2}\|F\|_{\mathcal{H}}^2\Big(\int_{-\infty}^{+\infty}\frac{|\xi|^2}{(|\lambda|+\xi^2+\eta)^4}d\xi\Big)\\ +c''\|F\|_{\mathcal{H}}^2 + c'''\|U\|_{\mathcal{H}}\|F\|_{\mathcal{H}}.\end{aligned}$$

Then we get

$$\|U\|_{\mathcal{H}}^2 \le \mathcal{M}^2(\lambda)\|F\|_{\mathcal{H}}^2, \tag{35}$$

where

$$\mathcal{M}(\lambda) = c\mathcal{S}^{-2}\Big(\int_{-\infty}^{+\infty}\frac{d\xi}{(|\lambda|+\xi^2+\eta|)^2}\Big) \equiv c|\lambda|^{-3/2}\mathcal{S}^{-2}.$$

It follows that

$$\frac{1}{\mathcal{M}(\lambda)}\|(i\lambda I-\mathcal{A})^{-1}\|_{\mathcal{L}(\mathcal{H})} \le C, \quad \forall\lambda\in\mathbb{R}$$

for a positive constant C. The conclusion then follows by applying Theorems 5 and 6 following the form of $\mathcal{M}$.

Example

$$\mu(\xi) = |\xi|^\alpha(\ln|\xi|)^\beta,\ \alpha,\beta\in\mathbb{R}.$$

We obtain the following optimal energy decay

- If $|\alpha| < \frac{1}{2}$, then

$$\mathcal{E}(t) \leq C(\mathcal{E}(0))t^{-\frac{4}{1-2\alpha}}(\ln t)^{-\frac{8\beta}{1-2\alpha}}.$$

- If $\alpha = -\frac{1}{2}$, then

$$\mathcal{E}(t) \leq C(\mathcal{E}(0))t^{-2}(\ln t)^{-4(2\beta+1)} \text{ if } 2\beta < -1.$$

- If $\alpha = \frac{1}{2}$ and $2\beta < -1$, then

$$\|(i\lambda I - \mathcal{A})^{-1}\|_{\mathcal{H}} \leq C(\ln|\lambda|)^{-(2\beta+1)}.$$

Then $l_{log}(s) \sim C(\ln|\lambda|)^{-2\beta}$. Hence

$$\mathcal{E}(t) \leq C(\mathcal{E}(0))e^{-\omega t^{-\frac{1}{2\beta}}}.$$

Now, using a variant of Watson Lemma, our system take a form

$$u_{tt} - \Delta u + \int_0^t h(t-s)u_t(s)\,ds = 0,$$

where

$$h(t) \sim c\frac{1}{t^{\alpha+1/2}}\left(\ln\frac{1}{t}\right)^{2\beta}e^{-\eta t} \text{ as } t \to \infty, \quad \text{if } \alpha > -\frac{1}{2}.$$

$$h(t) \sim c\left(\ln\frac{1}{t}\right)^{2\beta+1}e^{-\eta t} \qquad \text{as } t \to \infty, \quad \text{if } \alpha = -\frac{1}{2},\ 2\beta < -1.$$

5.3 Spectral Analysis

Let λ be an eigenvalue of $\mathcal{A}$ associated with eigenvector $U = (u, v, \phi)^T$. Then $\mathcal{A}U = \lambda U$. then we get

$$\beta_k^2 = -\lambda^2 - \zeta\int_{-\infty}^{+\infty}\lambda\frac{\mu^2(\xi)}{\xi^2+\eta+\lambda}d\xi. \tag{36}$$

Our purpose in the sequel is to prove, thanks to Rouché's Theorem, that there is a subsequence of eigenvalues for which their real part tends to 0.

In the sequel, since $\mathcal{A}$ is dissipative, we study the asymptotic behavior of the large eigenvalues λ of $\mathcal{A}$ in the strip $-\alpha_0 \leq \Re(\lambda) \leq 0$, for some $\alpha_0 > 0$ large enough and for such λ, we remark that e^{t_i}, $i = 1, 2$ remains bounded.

Let us set

$$S(\lambda) = \int_{-\infty}^{+\infty} \frac{\mu^2(\xi)}{\xi^2 + \eta + \lambda} d\xi.$$

We have the following Lemma.

Lemma 5 *There exists $N \in \mathbb{N}$ such that*

$$\{\lambda_k\}_{k \in \mathbb{Z}^*, |k| \geq N} \subset \sigma(\mathcal{A}), \tag{37}$$

where

$$\begin{cases} \lambda_k = i\beta_k - \zeta \Re S(\lambda_k^0) - i\zeta \Im S(\lambda_k^0) + o(\Re S(\lambda_k^0)) + i\, o(\Im S(\lambda_k^0)) & k \geq N, \\ \lambda_k = \overline{\lambda_{-k}} \text{ if } k \leq -N. \end{cases} \tag{38}$$

Moreover for all $|k| \geq N$, the eigenvalues λ_k are simple.

In particular, when μ is chosen such that

$$\int_0^s \frac{\mu^2(\sqrt{\xi})}{\sqrt{\xi}} d\xi \sim s^{1-\sigma} l(s) \text{ as } s \to \infty,$$

where l is a slowly varying function on $\mathbb{R}_+$, and either $0 < \sigma \leq 1$, or $\sigma = 0$ and $\int_0^\infty l(s)/(s+1)\, ds$ is finite, we get the following expansion of eigenvalues of $\mathcal{A}$

$$\lambda_k = i\beta_k + \frac{\tilde{\alpha}}{l(\beta_k)^{-1}\beta_k^\sigma} + \frac{\beta}{l(k)^{-1}k^\sigma} + o\left(\frac{1}{l(\beta_k)^{-1}\beta_k^\sigma}\right), k \geq N, \tilde{\alpha} \in i\mathbb{R}, \beta \in \mathbb{R}, \beta < 0 \text{ if } 0 < \sigma \leq 1,$$

$$\lambda_k = \overline{\lambda_{-k}} \text{ if } k \leq -N,$$

and for $\sigma = 0$,

$$\lambda_k = i\beta_k + \frac{\beta}{\tilde{l}(|\beta_k|)^{-1}} + o\left(\frac{1}{\tilde{l}(|\beta_k|)^{-1}}\right), k \geq N, \beta \in \mathbb{R}, \beta < 0 \text{ if } \sigma = 0,$$

$$\lambda_k = \overline{\lambda_{-k}} \text{ if } k \leq -N.$$

Proof We can write

$$\lambda_k = i\beta_k + \varepsilon_k. \tag{39}$$

Using Roché Theorem and substituting (39) into (36), we get

$$\lambda_k = i\beta_k - \zeta \int_{-\infty}^{+\infty} \frac{\mu^2(\xi)}{\xi^2 + \eta + i\beta_k} d\xi + o(\Re S(i\beta_k)) + o(\Im S(i\beta_k)).$$

Now, as an application, if μ is chosen such that

$$g(s) \sim s^{1-\sigma} l(s) \text{ as } s \to \infty,$$

where l is a slowly varying function on $\mathbb{R}_+$, and either $0 < \sigma \leq 1$, or $\sigma = 0$ and $\int_0^\infty l(s)/(s+1)\, ds$ is finite and

$$g(s) = \int_0^s \frac{\mu^2(\sqrt{\xi})}{\sqrt{\xi}} d\xi. \tag{40}$$

We have

$$\lambda_k = i\beta_k - \frac{\zeta \Gamma(\sigma)\Gamma(2-\sigma)}{l\,(\beta_k)^{-1}\,(\beta_k)^{\sigma}} \left(\cos\sigma\frac{\pi}{2} - i\sin\sigma\frac{\pi}{2}\right) + o\left(\frac{1}{l(\beta_k)^{-1}\beta_k^{\sigma}}\right) \text{ for } k \succeq 0. \tag{41}$$

From (41) we have in that case $l(|\beta_k|)^{-1}|\beta_k|^{\sigma}\Re\lambda_k \sim \beta$, with

$$\beta = -\zeta\Gamma(\sigma)\Gamma(2-\sigma)\cos\sigma\frac{\pi}{2}.$$

If $\sigma = 0$, there exists a slowly varying function $\tilde{l}$ such that $\tilde{l}(|\beta_k|)^{-1}\Re\lambda_k \sim \beta$, with $\beta < 0$. The operator $\mathcal{A}$ has a non exponential decaying branch of eigenvalues.

Now, setting $\tilde{U}_k = (\lambda_k^0 - \mathcal{A})U_k$, where U_k is a normalized eigenfunction associated with λ_k. We then have

$$\begin{aligned}
\|(\lambda_k^0 - \mathcal{A})^{-1}\|_{\mathcal{L}(\mathcal{H})} &= \sup_{U\in\mathcal{H}, U\neq 0} \frac{\|(\lambda_k^0 - \mathcal{A})^{-1}U\|_{\mathcal{H}}}{\|U\|_{\mathcal{H}}} \\
&\geq \frac{\|(\lambda_k^0 - \mathcal{A})^{-1}\tilde{U}_k\|_{\mathcal{H}}}{\|\tilde{U}_k\|_{\mathcal{H}}} \\
&\geq \frac{\|U_k\|_{\mathcal{H}}}{\|(\lambda_k^0 - \mathcal{A})U_k\|_{\mathcal{H}}}.
\end{aligned}$$

Hence, by (38), we deduce that

$$\|(\lambda_k^0 - \mathcal{A})^{-1}\|_{\mathcal{L}\mathcal{H}} \geq \frac{c}{|S(\lambda_k^0)|},$$

which implies that (10) does not hold and the proof is thus complete.

This shows an expected optimal behavior of the resolvent like

$$\|(i\lambda I - \mathcal{A})^{-1}\|_{\mathcal{H}} \equiv (\Re S(i\lambda))^{-1} \text{ as } |\lambda| \to \infty.$$

□

5.4 Optimality of Energy Decay in One-Dimensional Case

But, it is interesting to remark that, in the one-dimensional case, it is possible to use an explicit representation of the resolvent to obtain the optimal estimate. Indeed, we take $\Omega = (0, 1)$.

Theorem 8 *Let*

$$\Lambda(\lambda) = \frac{1}{(\Re S(i\lambda))}.$$

Assume that Λ has a positive increase. Then the semigroup $S_{\mathcal{A}}(t)_{t\geq 0}$ satisfies the following decay estimate

$$\|e^{\mathcal{A}t}U_0\| \leq C\frac{1}{\Lambda^{-1}(t)}\|U_0\|_{D(\mathcal{A})}, \quad t \to \infty,$$

where Λ^{-1} is any asymptotic inverse of Λ. Moreover, the rate of energy decay is optimal for any initial data in $D(\mathcal{A})$.

Proof Let $\lambda \in \mathbb{R}$. Let $F = (f_1, f_2, f_3)^T \in \mathcal{H}$ be given, and let $X = (u, v, \phi)^T \in D(\mathcal{A})$ be such that

$$(i\lambda I - \mathcal{A})X = F. \tag{42}$$

Equivalently, we have

$$\begin{cases} i\lambda u - v = f_1, \\ i\lambda v - u_{xx} + \zeta \displaystyle\int_{-\infty}^{+\infty} \mu(\xi)\phi(x,\xi)\,d\xi = f_2, \\ i\lambda\phi + (\xi^2 + \eta)\phi - v(x)\mu(\xi) = f_3. \end{cases} \tag{43}$$

From $(43)_3$,we have

$$\phi(x,\xi) = \frac{v(x)\mu(\xi) + f_3(x,\xi)}{i\lambda + \xi^2 + \eta}. \tag{44}$$

From $(43)_1$, $(43)_2$ and (44), we have

$$\lambda^2 u + u_{xx} - i\zeta\lambda \int_{-\infty}^{+\infty} \frac{\mu^2(\xi)}{i\lambda+\xi^2+\eta}\, d\xi = -(f_2+i\lambda f_1) + \zeta\lambda \int_{-\infty}^{+\infty} \frac{\mu(\xi) f_3(x,\xi)}{i\lambda+\xi^2+\eta}\, d\xi \\ -\zeta f_1(x) \int_{-\infty}^{+\infty} \frac{\mu^2(\xi)}{i\lambda+\xi^2+\eta}\, d\xi .$$

Then

$$u_{xx} + (\lambda^2 - i\zeta\lambda S(i\lambda))u = F(x),$$

where

$$S(i\lambda) = \int_{-\infty}^{+\infty} \frac{\mu^2(\xi)}{i\lambda+\xi^2+\eta}\, d\xi,$$

$$F(x) = -(f_2+i\lambda f_1) + \zeta\lambda \int_{-\infty}^{+\infty} \frac{\mu(\xi) f_3(x,\xi)}{i\lambda+\xi^2+\eta}\, d\xi - \zeta f_1(x) \int_{-\infty}^{+\infty} \frac{\mu^2(\xi)}{i\lambda+\xi^2+\eta}\, d\xi,$$

with $u(0) = 0$. Suppose that $\lambda \neq 0$. Then

$$u(x) = c_1 \sin\tilde{\lambda}x + \frac{1}{\tilde{\lambda}} \int_0^x F(\sigma) \sin\tilde{\lambda}(x-\sigma)\, d\sigma, \tag{45}$$

$$u_x(x) = c_1\tilde{\lambda} \cos\tilde{\lambda}x + \int_0^x F(\sigma) \cos\tilde{\lambda}(x-\sigma)\, d\sigma, \tag{46}$$

where

$$\tilde{\lambda} = \sqrt{\lambda^2 - i\zeta\lambda S(i\lambda)}.$$

As $u(1) = 0$, then we have

$$c_1 \sin\tilde{\lambda} + \frac{1}{\tilde{\lambda}} \int_0^1 F(\sigma) \sin\tilde{\lambda}(1-\sigma)\, d\sigma = 0. \tag{47}$$

For λ large enough, we have

$$\tilde{\lambda} = \lambda - \frac{i\zeta}{2} S(i\lambda) + o(S(i\lambda)).$$

Then

$$\sin\tilde{\lambda} = \sin\lambda - \frac{i\zeta}{2} S(i\lambda)\cos\lambda + O(S(i\lambda)^2).$$

We deduce that

$$|\sin\tilde{\lambda}| \geq c(\Re S(i\lambda)) \text{ for large } \lambda. \tag{48}$$

The constant c_1 in (47) satisfies

$$|c_1||\sin\tilde{\lambda}| \leq \frac{1}{|\tilde{\lambda}|}\left|\int_0^1 F(\sigma)\sin\tilde{\lambda}(1-\sigma)\,d\sigma\right|. \tag{49}$$

Using integration by parts and the fact that $f_1 \in H_0^1(0,1)$. Then, from (49) and (48) we conclude that

$$|c_1| \leq c\frac{1}{|\lambda|(\Re S(i\lambda))}.$$

From (46), we deduce that

$$\|u_x\|_{L^2(0,1)} \leq c\frac{1}{\Re S(i\lambda)}\left(\|f_1\|_{H^1(0,1)} + \|f_2\|_{L^2(0,1)}\right).$$

From $(43)_1$ and (45), we get

$$\|v\|_{L^2(0,1)} \leq c\frac{1}{\Re S(i\lambda)}\left(\|f_1\|_{H^1(0,1)} + \|f_2\|_{L^2(0,1)}\right).$$

Moreover from (26), we get

$$\begin{aligned}\int_{-\infty}^{+\infty}|\phi(\xi)|^2\,d\xi &\leq C\int_{-\infty}^{+\infty}(\xi^2+\eta)|\phi(\xi)|^2\,d\xi \\ &\leq C\|U\|_{\mathcal{H}}\|F\|_{\mathcal{H}}.\end{aligned}$$

Thus, we conclude that

$$\|(i\lambda I - \mathcal{A})^{-1}\|_{\mathcal{H}} \leq c\frac{1}{\Re S(i\lambda)} \quad \text{as } |\lambda| \to \infty.$$

Now, in order to prove the inverse inequality we argue by contradiction. Suppose that $O(\Lambda(\lambda)^{1/2})$ is not optimal. This means that there exists a function $\tilde{\Lambda}$ such that

$$\|(i\lambda I - \mathcal{A})^{-1}\|_{\mathcal{H}} \leq \tilde{\Lambda}(\lambda) \text{ and } \lim_{\lambda\to\infty}\frac{\tilde{\Lambda}(\lambda)}{(\Re S(i\lambda))^{-1}} = 0, \tag{50}$$

which implies that, for all $F \in \mathcal{H}$, there exists $c > 0$ such that

$$\frac{1}{\tilde{\Lambda}(\lambda)}\|U\|_{\mathcal{H}} \leq c\|F\|_{\mathcal{H}} \quad \forall \lambda \in \mathbb{R}, \lambda \neq 0,$$

where $U \in \mathcal{H}$ is the solution of the resolvent equation $i\lambda U - \mathcal{A}U = F$. This is a contradiction because, taking advantage of Lemma 5, let λ_k be an eigenvalue of the operator $\mathcal{A}$ and $U_k \in D(\mathcal{A})$ the associated normalized eigenfunction. Moreover, we introduce the following sequence

$$\beta_k = Im\lambda_k, \quad |k| \geq k_1.$$

Next, using Lemma 5, we have

$$\|(i\beta_k - \mathcal{A})U_k\|_{\mathcal{H}} = \|(i\beta_k - \lambda_k)U_k\|_{\mathcal{H}} \leq \frac{1}{(\Re S(i\beta_k))^{-1}}\|U_k\|_{\mathcal{H}}.$$

This contradicts the asymptotic equation (50) and the proof is thus complete.

Remark 1 From the Theorem 7, we have obtained that:

$$\|(i\lambda I - \mathcal{A})^{-1}\|_{\mathcal{H}} \leq c|\lambda|^{-3/2}\left(\int_{-\infty}^{+\infty}\frac{|\xi|\mu(\xi)}{(\xi^2+\eta+|\lambda|)^2}d\xi\right)^{-2} \quad \text{as } |\lambda| \to \infty$$

and from theorem 8, we have obtained that:

$$\|(i\lambda I - \mathcal{A})^{-1}\|_{\mathcal{H}} \leq c\frac{1}{\Re S(i\lambda)} \quad \text{as } |\lambda| \to \infty.$$

But, it is interesting to see that we have same behavior of the resolvent in the case $\mu(\xi) = |\xi|^{\frac{2\alpha-1}{2}}$ (i.e fractional feedback: $\partial_t^{\alpha,\eta}u$).

5.5 Polynomial Stability (for $\eta = 0$)

When $\eta = 0$, 0 may be a spectral point. Therefore it is convenient to have the following generalization (of Borichev–Tomilov Theorem) at hand:

Theorem 9 ([9]) *Let X be a Hilbert space and let $\mathcal{A}$ be the generator of a bounded C_0-semigroup $(S(t))_{t\geq 0}$ on X. Assume that $\sigma(\mathcal{A}) \cap i\mathbb{R} = \{0\}$ and that there exist $M_0, M_\infty : [1, \infty) \to (0, \infty)$ are continuous non-decreasing functions such that*

$$\|(isI - \mathcal{A})^{-1}\|_{\mathcal{L}(\mathcal{H})} = \begin{cases} O(M_0(s)), & s \to 0, \\ O(M_\infty(s)), & |s| \to \infty. \end{cases}$$

Let $M : [1, \infty) \to (0, \infty)$ be defined by $M(s) = \max\{M_0(s); M_\infty(s)\}$, $s \geq 1$ and suppose that M has positive increase, then there exists a positive constant $C > 0$

such that

$$\|e^{\mathcal{A}t}U_0\| \leq C\frac{1}{M^{-1}(t)}\|U_0\|_{D(\mathcal{A})\cap R(\mathcal{A})}, \quad t \to \infty.$$

In particular, we have the following theorem.

Theorem 10 ([4]) *Let $S(t)$ be a bounded C_0-semigroup on a Hilbert space $\mathcal{H}$ with generator $\mathcal{A}$. Assume that $\sigma(\mathcal{A}) \cap i\mathbb{R} = \{0\}$ and that there exist $\vartheta > 1$ and $\upsilon > 0$ such that*

$$\|(isI - \mathcal{A})^{-1}\|_{\mathcal{L}(\mathcal{H})} = \begin{cases} O(|s|^{-\vartheta}), & s \to 0, \\ O(|s|^{\upsilon}), & |s| \to \infty. \end{cases}$$

Then there exist constants $C, t_0 > 0$ such that for all $t \geq t_0$ and $U_0 \in D(\mathcal{A}) \cap R(\mathcal{A})$ we have

$$\|e^{\mathcal{A}t}U_0\|^2 \leq C\frac{1}{t^{\frac{2}{\varsigma}}}\|U_0\|^2_{D(\mathcal{A})\cap R(\mathcal{A})},$$

where $\varsigma = \max\{\vartheta, \upsilon\}$.

Our main result is the following.

Theorem 11 *The semigroup $S_{\mathcal{A}}(t)_{t\geq 0}$ satisfies the following decay estimate*

$$\mathcal{E}(t) = \|S_{\mathcal{A}}(t)U_0\|^2_{\mathcal{H}} \leq C\frac{1}{M^{-1}(t)}\|U_0\|^2_{D(\mathcal{A})\cap R(\mathcal{A})},$$

where

$$M(s) = \max\{M_0(s), M_\infty(s)\},$$

$$M_\infty(s) = cS^{-2}\Big(\int_{-\infty}^{+\infty}\frac{d\xi}{(|\lambda| + \xi^2 + \eta)^2}\Big),$$

$$M_0(s) = C\Bigg[|\lambda|^{-3}S^{-4} + |\lambda|^{-5/2}S^{-2} + |\lambda|^{-3}S^{-4}\left(\int_{-\infty}^{+\infty}\frac{\mu^2(\xi)}{(|\lambda| + \xi^2)^2}\,d\xi\right)^2$$
$$+|\lambda|^{-5/2}S^{-2}\left(\int_{-\infty}^{+\infty}\frac{\mu^2(\xi)}{(|\lambda| + \xi^2)^2}\,d\xi\right) + |\lambda|^{-2} + \left(\int_{-\infty}^{+\infty}\frac{\mu^2(\xi)}{(|\lambda| + \xi^2)^2}\,d\xi\right)$$
$$+|\lambda|^{-5/2}S^{-2}\left(\int_{-\infty}^{+\infty}\frac{\mu^2(\xi)}{(|\lambda| + \xi^2)}\,d\xi\right)^2 |\lambda|^{-3}S^{-4}\left(\int_{-\infty}^{+\infty}\frac{\mu^2(\xi)}{(|\lambda| + \xi^2)}\,d\xi\right)^4\Bigg].$$

In particulat when $\mu(\xi) = |\xi|^{\frac{2\alpha-1}{2}}$, we have

Theorem 12 *The semigroup $S_{\mathcal{A}}(t)_{t\geq 0}$ is polynomially stable and*

$$\mathcal{E}(t) = \|S_{\mathcal{A}}(t)U_0\|^2_{\mathcal{H}} \leq C\frac{1}{t}\|U_0\|^2_{D(\mathcal{A})\cap R(\mathcal{A})}.$$

Proof of Theorem 11 First for λ large enough, from the estimation in the proof of Theorem 7, we have

$$\|(i\lambda I - \mathcal{A})^{-1}\|_{\mathcal{H}} \leq M_\infty(|\lambda|).$$

For λ near 0, we have from $(24)_3$

$$v(x)\mu(\xi) = (i\lambda + \xi^2)\phi - f_3(x,\xi). \tag{51}$$

By multiplying (51) by $(i\gamma + \xi^2)^{-2}|\xi|$, we get

$$(i\lambda + \xi^2)^{-2}v(x)\mu(\xi)|\xi| = (i\lambda + \xi^2)^{-1}|\xi|\phi - (i\lambda + \xi^2)^{-2}|\xi|f_3(x,\xi). \tag{52}$$

Hence, by taking absolute values of both sides of (52), integrating over the interval $]-\infty, +\infty[$ with respect to the variable ξ and applying Cauchy–Schwartz inequality, we obtain

$$\mathcal{S}|v(x)| \leq \sqrt{2}\mathcal{U}\left(\int_{-\infty}^{+\infty} \xi^2|\phi|^2\,d\xi\right)^{\frac{1}{2}} + 2\mathcal{V}\left(\int_{-\infty}^{+\infty} |f_3(x,\xi)|^2\,d\xi\right)^{\frac{1}{2}}, \tag{53}$$

where

$$\mathcal{S} = \left|\int_{-\infty}^{+\infty} (i\lambda + \xi^2)^{-2}|\xi|\mu(\xi)\,d\xi\right|,$$

$$\mathcal{U} = \left(\int_{-\infty}^{+\infty} (|\lambda| + \xi^2)^{-2}\,d\xi\right)^{\frac{1}{2}} = (\frac{\pi}{2})^{1/2}|\lambda|^{-\frac{3}{4}},$$

$$\mathcal{V} = \left(\int_{-\infty}^{+\infty} (|\lambda| + \xi^2)^{-4}|\xi|^2\,d\xi\right)^{\frac{1}{2}} = \frac{\pi}{16}|\lambda|^{-\frac{5}{4}}.$$

Thus, by using the inequality $2PQ \leq P^2 + Q^2$, $P \geq 0$, $Q \geq 0$, again, we get

$$\begin{aligned}\mathcal{S}^2\int_\Omega |v(x)|^2\,dx &\leq 2\mathcal{U}^2\left(\int_\Omega\int_{-\infty}^{+\infty} \xi^2|\phi|^2\,d\xi\,dx\right)\\ &+4\mathcal{V}^2\left(\int_\Omega\int_{-\infty}^{+\infty} |f_3(x,\xi)|^2\,d\xi\,dx\right).\end{aligned} \tag{54}$$

We deduce that for λ near 0, we have from (54)

$$\int_\Omega |v(x)|^2\,dx \le c|\lambda|^{-3/2}S^{-2}\|U\|_{\mathcal{H}}\|F\|_{\mathcal{H}} + c|\lambda|^{-5/2}S^{-2}\|F\|^2_{\mathcal{H}}. \tag{55}$$

$$\begin{aligned}\|\phi\|^2 &\le 2\int_\Omega |v(x)|^2\,dx \int_{-\infty}^{+\infty}\frac{\mu^2(\xi)}{|i\lambda+\xi^2|^2}\,d\xi + 2\int_\Omega\int_{-\infty}^{+\infty}\frac{|f_3(x,\xi)|^2}{|i\lambda+\xi^2|^2}\,d\xi\,dx\\ &\le 4\int_\Omega |v(x)|^2\,dx \int_{-\infty}^{+\infty}\frac{\mu^2(\xi)}{(|\lambda|+\xi^2)^2}\,d\xi + 4\int_\Omega\int_{-\infty}^{+\infty}\frac{|f_3(x,\xi)|^2}{(|\lambda|+\xi^2)^2}\,d\xi\,dx\\ &\le 4\int_{-\infty}^{+\infty}\frac{\mu^2(\xi)}{(|\lambda|+\xi^2)^2}\,d\xi\,\|v\|^2_{L^2(\Omega)} + 4|\lambda|^{-2}\|f_3\|^2_{L^2(\Omega\times(-\infty,+\infty))}.\end{aligned} \tag{56}$$

$$\begin{aligned}&\left|\int_{-\infty}^{+\infty}\mu(\xi)\mu(\xi)\,d\xi\right|^2\\ &= \left|v(x)\int_{-\infty}^{+\infty}\frac{\kappa^2(\xi)}{i\lambda+\xi^2}\,d\xi + \int_{-\infty}^{+\infty}\frac{\mu(\xi)f_3(\xi)}{i\lambda+\xi^2}\,d\xi\right|^2\\ &\le 2|v(x)|^2\left|\int_{-\infty}^{+\infty}\frac{\mu^2(\xi)}{i\lambda+\xi^2}\,d\xi\right|^2 + 2\int_{-\infty}^{+\infty}\frac{\mu^2(\xi)}{|i\lambda+\xi^2|^2}\,d\xi\,\|f_3\|^2_{L^2(-\infty,+\infty)}\\ &\le 4|v(x)|^2\left|\int_{-\infty}^{+\infty}\frac{\mu^2(\xi)}{|\lambda|+\xi^2}\,d\xi\right|^2 + 4\int_{-\infty}^{+\infty}\frac{\mu^2(\xi)}{(|\lambda|+\xi^2)^2}\,d\xi\,\|f_3\|^2_{L^2(-\infty,+\infty)}.\end{aligned} \tag{57}$$

Then

$$\begin{aligned}\int_\Omega\left|\int_{-\infty}^{+\infty}\mu(\xi)\mu(\xi)\,d\xi\right|^2 dx &\le 4\left(\int_{-\infty}^{+\infty}\frac{\mu^2(\xi)}{|\lambda|+\xi^2}\,d\xi\right)^2\|v\|^2_{L^2(\Omega)}\\ &\quad +4\left(\int_{-\infty}^{+\infty}\frac{\mu^2(\xi)}{(|\lambda|+\xi^2)^2}\,d\xi\right)\|F\|^2_{\mathcal{H}}.\end{aligned}$$

Finally, we deduce that

$$\begin{aligned}\|U\|^2_{\mathcal{H}} &\le C\Bigg[|\lambda|^{-3}S^{-4} + |\lambda|^{-5/2}S^{-2} + |\lambda|^{-3}S^{-4}\left(\int_{-\infty}^{+\infty}\frac{\mu^2(\xi)}{(|\lambda|+\xi^2)^2}\,d\xi\right)^2\\ &+|\lambda|^{-5/2}S^{-2}\left(\int_{-\infty}^{+\infty}\frac{\mu^2(\xi)}{(|\lambda|+\xi^2)^2}\,d\xi\right) + |\lambda|^{-2} + \left(\int_{-\infty}^{+\infty}\frac{\mu^2(\xi)}{(|\lambda|+\xi^2)^2}\,d\xi\right)\\ &+|\lambda|^{-5/2}S^{-2}\left(\int_{-\infty}^{+\infty}\frac{\mu^2(\xi)}{(|\lambda|+\xi^2)}\,d\xi\right)^2|\lambda|^{-3}S^{-4}\left(\int_{-\infty}^{+\infty}\frac{\mu^2(\xi)}{(|\lambda|+\xi^2)}\,d\xi\right)^4\Bigg]\|F\|^2_{\mathcal{H}}\\ &\equiv M_0^2(|\lambda|)\|F\|^2_{\mathcal{H}}\end{aligned}.$$

Hence

$$\|(i\lambda I - \mathcal{A})^{-1}\|_{\mathcal{H}} \le CM_0(|\lambda|) \text{ as } \lambda \to 0.$$

In particular for $\mu(\xi) = |\xi|^{\frac{2\alpha-1}{2}}$, we have

$$\|(i\lambda I - \mathcal{A})^{-1}\|_{\mathcal{H}} \leq \frac{C}{|\lambda|^2} \text{ as } \lambda \to 0.$$

Applying Theorem 10, we obtain that

$$\mathcal{E}(t) \leq C\frac{1}{t}\|U_0\|^2_{D(\mathcal{A})\cap R(\mathcal{A})}.$$

References

1. K. Ammari, H. Fathi, L. Robbiano, Fractional-feedback stabilization for a class of evolution systems. J. Differ. Equ. **268**, 5751–5791 (2020)
2. W. Arendt, C.J.K. Batty, Tauberian theorems and stability of one-parameter semigroups. Trans. Am. Math. Soc., **306**, 837–852 (1988)
3. C.J.K. Batty, T. Duyckaerts, Non-uniform stability for bounded semi-groups on Banach spaces. J. Evol. Equ. **8**, 765–780 (2008)
4. C.J.K. Batty, R. Chill, Y. Tomilov, Fine scales of decay of operator semigroups. J. Eur. Math. Soc. (JEMS) **18**, 853–929 (2016)
5. A. Benaissa, S. Rafa, Well-posedness and energy decay of solutions to a wave equation with a general boundary control of diffusive type. Mathematische Nachrichten **292**, 1644–1673 (2019)
6. A. Boudaouad, A. Benaissa, Stabilization of a wave equation with a general internal control of diffusive type. Discontin. Nonlinear. Complex. **12**, 879–891 (2023)
7. B. Mbodje, Wave energy decay under fractional derivative controls. IMA J. Math. Control Inf. **23**, 237–257 (2006)
8. J. Pruss, On the spectrum of C_0-semigroups. Trans. Am. Math. Soc. **284**, 847–857 (1984)
9. J. Rozendaal, D. Seifert, R. Stahn, Optimal rates of decay for operator semigroups on Hilbert spaces. Adv. Math. **346**, 359–388 (2019)

Numerical Approximation of the Boundary Control for the Wave Equation with Periodic Oscillating Coefficients

Somia Boumimez and Carlos Castro

2010 Mathematics Subject Classification 93B05, 35B30, 93C20

1 Introduction

Let $\rho(x)$ be a periodic function with $0 < \rho_m \leq \rho(x) \leq \rho_M < \infty$ and $\varepsilon > 0$ is a small parameter. Consider the one-dimensional wave equation in the interval $x \in (0,1)$ with a periodic oscillating density $\rho^\varepsilon(x) = \rho(x/\varepsilon) \in L^\infty(0,1)$ and a boundary control at the extreme $x = 1$,

$$\begin{cases} \rho^\varepsilon(x)u''(t,x) - u_{xx}(t,x) = 0 & t \in (0,T),\ x \in (0,1) \\ u(t,0) = 0, \qquad u(t,1) = f(t), & t \in (0,T) \\ u(0,x) = u^0(x),\ u'(0,x) = u^1(x) & x \in (0,1), \end{cases} \tag{1}$$

where $(u^0, u^1) \in L^2(0,1) \times H^{-1}(0,1)$ are the initial data. Here $H^{-1}(0,1)$ represents the dual space of $H_0^1(0,1)$. To fix ideas and without loss of generality we take ρ of period 1 though our analysis is independent of the period. Note that the function ρ^ε oscillates more and more as $\varepsilon \to 0$ and then (1) constitutes a simplified model for the wave propagation in a medium with highly heterogeneous density.

Assuming some extra smoothness, $\rho^\varepsilon \in W^{1,\infty}(0,1)$, the following controllability result is known to hold (see [16] or [5]): Given $T > 2\sqrt{\overline{\rho}} = 2\sqrt{\int_0^1 \rho(s)\, ds}$, for any initial data $(u^0, u^1) \in L^2 \times H^{-1}(0,1)$ there exists a control $f^\varepsilon \in L^2(0,T)$ such that the solution u^ε of system (1) satisfies

$$u^\varepsilon(x,T) = u_t^\varepsilon(x,T) = 0, \quad x \in (0,1).$$

S. Boumimez · C. Castro (✉)
M^{2}ASAI. Universidad Politécnica de Madrid, ETSI Caminos, Calle Profesor Aranguren, Madrid, Spain
e-mail: carlos.castro@upm.es

K. Ammari et al. (eds.), *Advances in Partial Differential Equations and Control*, Trends in Mathematics, https://doi.org/10.1007/978-3-031-62265-6_5

This is usually known as the null controllability problem where the boundary control f^ε is chosen to drive the system to the equilibrium. Note that, due to the time reversibility of the wave equation, this is equivalent to control any initial data to any other final one, not necessarily zero (see for example [14] or [18]).

Another important remark is that the controls, when they exist, are not unique in general. To fix ideas we always refer as the control with minimal L^2-norm, which is unique and can be easily characterized with the Hilbert Uniqueness Method (HUM) introduced by J.-L. Lions [14].

The following estimate is known to hold ([16]) for the control of (1),

$$\|f^\varepsilon\|_{L^2(0,T)} \le C_2 \exp\left(C_3 T V(\rho^\varepsilon)\right) \|(u_0, u_1)\|_{L^2(0,1)\times H^{-1}(0,1)}, \tag{2}$$

where $TV(\rho^\varepsilon)$ denotes the total variation norm, i.e., $TV(\rho^\varepsilon) = \|(\rho^\varepsilon)_x\|_{L^1(0,1)}$, and C_2, C_3 are positive constants which only depend on ρ_m, ρ_M and T. Note that, in our particular case, even if ρ^ε is smooth this estimate does not guarantee a uniform bound of the controls when $\varepsilon \to 0$. In fact, we have

$$\|f^\varepsilon\|_{L^2(0,T)} \le C_2 e^{C_3/\varepsilon} \|(u_0, u_1)\|_{L^2(0,1)\times H^{-1}(0,1)}. \tag{3}$$

It turns out that this estimate is sharp and the control f^ε cannot be chosen uniformly bounded in the L^2-norm, as $\varepsilon \to 0$, in general [17]. This is due to the presence of a class of solutions exponentially locallized on one of the extremes that interact with the microstructure in a resonant way. These solutions are also the reason why the TV-norm in (2) is optimal, as proved in [6].

On the other hand, when $\varepsilon \to 0$, the coefficient $\rho^\varepsilon(x) = \rho(x/\varepsilon)$ converges weakly to its average $\overline{\rho} = \int_0^1 \rho(s)ds$ and system (1) converges in some sense to the so-called homogenized limit (see [2, 3]). More precisely, if $(u^0, u^1) \in H_0^1 \times L^2$ and there exists a subsequence, still denoted by $\varepsilon \to 0$, such that

$$\rho^\varepsilon u_1 \rightharpoonup u_1^*, \text{ weakly in } L^2(0,1), \tag{4}$$

then the solution of (1) converges to the solution of the following system:

$$\begin{cases} \overline{\rho} u_{tt} - u_{xx} = 0, & 0 < x < 1,\ 0 < t < T, \\ u(0,t) = 0,\ u(1,t) = f(t), & 0 < t < T, \\ u(x,0) = u_0(x),\ u_t(x,0) = u_1^*(x)/\overline{\rho},\ 0 < x < 1, & \end{cases} \tag{5}$$

when there is no control, i.e., $f = 0$. For this constant coefficients limit system we can apply the classical theory in [14] and the null controllability holds.

The relation between f^ε, the minimal L^2-norm control of system (1) and f, the one associated with (5) for the same initial data can be established through the projection of the solutions in the finite dimensional space generated by the first eigenfunctions of the associated eigenvalue problem, as described in [5] and [9]. To be more precise we introduce the eigenvalue problem,

$$\begin{cases} w'' + \lambda\rho^{\varepsilon}(x)w = 0, & x \in (0,1), \\ w(0) = w(1) = 0. \end{cases} \tag{6}$$

For each $\varepsilon > 0$, there exists a sequence of eigenvalues

$$0 < \lambda_1^{\varepsilon} < \lambda_2^{\varepsilon} < \ldots < \lambda_n^{\varepsilon} < \ldots \to \infty \tag{7}$$

and a sequence of associated eigenfunctions $(w_n^{\varepsilon})_{n\in N}$ which can be chosen to constitute an orthonormal basis in $L^2(0,1)$ with the norm

$$||w||_{\varepsilon}^2 = \int_0^1 \rho^{\varepsilon}(x)|w(x)|^2 dx,$$

which is equivalent to the usual $L^2(0,1)$-norm. Observe that

$$L^2(0,1) = span\{(w_n^{\varepsilon})_{n\in N}\}. \tag{8}$$

Multiplying the equation in (6) by any eigenfunction and integrating by parts we easily deduce that the eigenfunctions are also orthogonal in $H_0^1(0,1)$ with the norm

$$||w||_{H_0^1(0,1)}^2 = \int_0^1 |w'(x)|^2 dx.$$

In fact, $(w_n^{\varepsilon}/\sqrt{\lambda_n^{\varepsilon}})_{n\in N}$ constitutes a orthonormal basis in $H_0^1(0,1)$.

Given a continuous increasing functions $K(\varepsilon)$ such that $K(\varepsilon) \to \infty$ as $\varepsilon \to 0$, we introduce the space

$$L_{K(\varepsilon)} = span\{(w_n^{\varepsilon})_{n\le K(\varepsilon)}\}, \tag{9}$$

and its associated projection $\Pi_{K(\varepsilon)} : X \to L_{K(\varepsilon)}$ defined on any space X for which linear combinations of eigenfunctions are dense as follows

$$\Pi_{K(\varepsilon)}\left(\sum_{n=1}^{\infty} c_n^{\varepsilon} w_n^{\varepsilon}\right) = \sum_{n=1}^{K(\varepsilon)} c_n^{\varepsilon} w_n^{\varepsilon}, \qquad c_n \in \mathbb{R}.$$

When $X = H_0^1(0,1)$ or $L^2(0,1)$ this corresponds to the orthogonal projection with respect to the scalar products associated with the norms $\|\cdot\|_{H_0^1}$ and $\|\cdot\|_{\varepsilon}$ respectively.

Note that $L_{K(\varepsilon)}$ is a finite dimensional space generated by the low frequencies. The number of frequencies $K(\varepsilon)$ is made explicit below and depends on ε in such a way that $K(\varepsilon) \to \infty$ as $\varepsilon \to 0$. Thus, we consider more and more frequencies for smaller ε.

The following result is proved in [5]:

Theorem 1 *Let $\rho \in L^\infty(\mathbb{R})$ be a periodic function such that $0 < \rho_m \leq \rho(x) \leq \rho_M < \infty$ a.e. $x \in R$, and consider $T > 2\sqrt{\overline{\rho}} = 2\sqrt{\int_0^1 \rho(s)ds}$. Then, if $K(\varepsilon) \leq D\varepsilon^{-1}$ with D small enough, for any (u_0, u_1) with $(u_0, \rho^\varepsilon u_1) \in L^2(0,1)\times H^{-1}(0,1)$ and $0 < \varepsilon < 1$ there exists $f_L^\varepsilon \in L^2(0,T)$ such that the solution u^ε of (1) verifies*

$$(\Pi_{K(\varepsilon)} u^\varepsilon(T), \Pi_{K(\varepsilon)} u_t^\varepsilon(T)) = 0. \tag{10}$$

Moreover, there exist constants $C_1, C_2 > 0$ independent of ε such that

$$\|f_L^\varepsilon\|_{L^2(0,T)} \leq C_1 \|(\Pi_{K(\varepsilon)} u_0, \rho^\varepsilon \Pi_{K(\varepsilon)} u_1)\|_{L^2(0,1)\times H^{-1}(0,1)}, \tag{11}$$

$$\|(\Pi_{K(\varepsilon)} u^\varepsilon, \rho^\varepsilon \Pi_{K(\varepsilon)} u_t^\varepsilon)\|_{L^\infty(0,T;L^2(0,1)\times H^{-1}(0,1))} \leq C_2 \|f_L^\varepsilon\|_{L^2(0,T)}, \tag{12}$$

for any (u_0, u_1) and $0 < \varepsilon < 1$.

Assume that there exists a subsequence, still denoted $\varepsilon \to 0$, such that

$$\rho^\varepsilon u_1 \rightharpoonup u_1^* \text{ weakly in } H^{-1}(0,1). \tag{13}$$

Then, the sequence of solutions $(u^\varepsilon, u_t^\varepsilon)$ and the sequence of controls $f_L^\varepsilon(t)$ verify

$$(\Pi_{K(\varepsilon)} u^\varepsilon, \rho^\varepsilon \Pi_{K(\varepsilon)} u_t^\varepsilon) \rightharpoonup (u, \overline{\rho} u_t) \text{ weakly in } L^\infty(0,T; L^2(0,1) \times H^{-1}(0,1)) \tag{14}$$

$$f_L^\varepsilon \rightharpoonup f \text{ weakly in } L^2(0,T), \tag{15}$$

where (u, u_t) and $f(t)$ are the solution and the control respectively of the limit system (5). Moreover, if $\rho^\varepsilon u_1 \to u_1^*$ strongly in $H^{-1}(0,1)$, then

$$(\Pi_{K(\varepsilon)} u^\varepsilon, \rho^\varepsilon \Pi_{K(\varepsilon)} u_t^\varepsilon) \to (u, \overline{\rho} u_t) \text{ strongly in } L^\infty(0,T; L^2(0,1) \times H^{-1}(0,1)) \tag{16}$$

$$f_L^\varepsilon \to f \text{ strongly in } L^2(0,T). \tag{17}$$

Remark 1 Theorem 1 establishes a convergence result for a particular class of controls f_L^ε of (1) that affects only the projection of the solution on the low frequencies. The subindex L in the control is included in the notation to make this explicit.

The optimality of the limit value $K(\varepsilon)$ in Theorem 1 is due to the singular behavior of the eigenfunctions w_n^ε with wavelength of the order of ε, i.e., $n \sim \varepsilon^{-1}$. More precisely, the eigenfunctions which can exhibit a singular behavior are those for which its associated eigenvalues λ_n^ε verify

$$\varepsilon\sqrt{\lambda_n^\varepsilon} \in I, \tag{18}$$

where I is the set of all so-called instability intervals of the associated Hill equation

$$w''(y) + \lambda\rho(y)w(y) = 0, \quad y \in \mathbb{R}. \tag{19}$$

Using the Rayleigh formula (see for example [15]) it is easy to check that

$$\frac{n\pi}{\sqrt{\rho_M}} \leq \sqrt{\lambda_n^\varepsilon} \leq \frac{n\pi}{\sqrt{\rho_m}}$$

for any $0 < \varepsilon < 1$ and $n \in \mathbb{N}$. On the other hand, as I is constituted by non-empty closed intervals we deduce that (18) only happens when $\sqrt{\lambda_n^\varepsilon}\varepsilon \geq \min_{x\in I} x > 0$, i.e.,

$$n \geq (\min_{x\in I} x)\sqrt{\rho_m}\pi^{-1}\varepsilon^{-1}. \tag{20}$$

We refer to [10] for a detailed description of the instability intervals of the Eq. (19).

The singular eigenfunctions w_n^ε when $n \sim \varepsilon^{-1}$ are used also to construct solutions of the wave equation for which estimate (3) is sharp with respect to $\varepsilon \to 0$.

A possible application of Theorem 1 is to approximate the control f_L^ε by the one associated with the limit system f. However, it is not clear how good is this approximation. For example, it does not say anything about the rate of convergence or how we can include correctors to f if we want to improve the approximation.

In this work we propose a numerical method to give accurate approximations of f_L^ε. There are several difficulties associated with this approximation. The natural approach based on substituting the continuous control problem by a consistent discrete approximation, depending on a new discretization parameter $h \to 0$, solve the discrete control problem and use the discrete solution as an approximation of the continuous control is delicate. Note that the discrete system will depend on two parameters ε and h that could interact to give a wrong approximation.

For example, the presence of the oscillating coefficient makes any numerical method based on finite differences or finite elements difficult to implement since it usually requires a discretization parameter much smaller than the period of the coefficient, even for direct simulation of the uncontrolled wave equation.

On the other hand, even if we can use a sufficiently small discretization parameter, the discrete system is not uniformly controllable with respect to h when considering usual discretization of the wave equation based on finite differences or classical finite elements. This has been observed by a number of authors in different situations (see [18] or [11] and the references therein for a complete description and number of examples).

This has a dramatic consequence in the applications, even for a fixed ε not two small. For example, in the context of the boundary controllability, if we compute the boundary controls of the discrete approximation, their norms grow exponentially as the discretization parameter $h \to 0$, for some initial data, and do not produce any approximation of the control for the continuous system.

These phenomena were first described in a series of papers by R. Glowinski and J.-L Lions (see the review paper [13] and the references therein). They also proposed a number of cures to deal with this lack of uniformity, as bigrid approximations or Tychonoff regularization methods. Later on, other methods have been proposed: filtering, projection of the controlled solution, etc. (see the review [18]).

To overcome all the difficulties associated with the natural approach described above we adopt a different idea. First we characterize f_L^ε as the control of a finite dimensional control problem obtained by a projection method on the eigenfunctions of the associated eigenvalue problem, in the spirit of the result in Theorem 1 (see also [1, 4]). This gives a finite dimensional control problem for the Fourier coefficients of the eigenfunctions. Following the idea in HUM the control with minimal L^2-norm can be characterized as the solution of a suitable linear system that involves the solution of an adjoint problem, which is a wave equation with the same oscillating coefficient. As we are projecting in the set generated by a finite number of eigenfunctions, the solution of the adjoint wave equation can be easily computed and the method provides an efficient way to compute the control. The only obstacle is that we do not know the eigenfunctions and eigenvalues of the associated eigenvalue problem. To obtain an approximate problem we use the asymptotic formulas for the spectra in [7]. We focus on the numerical implementation of the method and give some experiments that show its efficiency. This provides numerical evidences of the convergence. The analytical details of this convergence will be analyzed in a future publication.

Note that the projection method described in this paper is not restricted to the one-dimensional case and it can be extended to several dimensions, as long as we can compute efficiently the eigenfunctions of the underlying eigenvalue problem. In some special domains as balls or intervals the method could be adapted. It can also be applied to other equations as the heat model with oscillating coefficients (see [8]).

The rest of the paper is divided as follows. In Sect. 2 we characterize the control associated with the low frequencies f_L^ε in terms of the spectra. In Sect. 3 we use the known results on the asymptotic of the spectra to obtain numerical approximations of the control. In Sect. 4 we show some numerical experiments that illustrate the efficiency of the method.

2 Characterization of the Control

In this section we introduce a numerical approximation of the control problem (1). In fact, we propose an approximate finite dimensional discrete control problem that we can solve using the finite dimensional control theory and provides an approximation of f_L^ε.

For numerical reasons we focus on controls f_L^ε that are compactly supported in $t \in (0, T)$. To guarantee this, we introduce a weighted L^2-norm for the controls,

$$\|f\|^2_{L^2_w(0,T)} = \int_0^T \eta(t)|f(t)|^2\, dt, \tag{21}$$

where η is a $C^\infty(0, T)$ compactly supported bump function which satisfies $0 \le \eta \le 1$ and $\eta(t) = 1$ for $t \in (\delta, T - \delta)$, with sufficiently small $\delta > 0$.

Minimal L^2-weighted controls were introduced in [12] to recover smooth controls from smooth data. Here we adopt the same idea but in order to have compactly supported controls.

According to [5] we have the following variational characterization of the control f_L^ε,

Proposition 1 *Any control f_L^ε for which the solution of system* (1) *vanishes at time $t = T$, i.e.,* (10)*, verifies the variational identity,*

$$\begin{aligned} &- < \rho^\varepsilon u_1(x), v^\varepsilon(x,0) >_{H^{-1},H_0^1} + \int_0^1 \rho^\varepsilon(x) u_0(x) v_t^\varepsilon(x,0)dx \\ &+ \int_0^T f_L^\varepsilon(t) v_x^\varepsilon(1,t)dt = 0 \end{aligned} \tag{22}$$

for all $(v_0, v_1) \in L_{K(\varepsilon)}$, where v^ε is the solution of the adjoint system

$$\begin{cases} \rho^\varepsilon(x) v_{tt} - v_{xx} = 0, & 0 < x < 1,\ 0 < t < T, \\ v(t,0) = v(t,1) = 0, & 0 < t < T, \\ v(0,x) = v_0(x),\ \ v_t(0,x) = v_1(x),\ 0 < x < 1. \end{cases} \tag{23}$$

We include the proof of this result for completeness.

Proof Multiplying the Eqs. (1) by v^ε, the solution of (23), and integrating by parts we easily obtain the following identity:

$$\begin{aligned} &< \rho^\varepsilon(x) u_t^\varepsilon(x,T), v_0(x) >_{H^{-1},H_0^1} - \int_0^1 \rho^\varepsilon(x) u^\varepsilon(x,T) v_1(x)dx \\ &\quad - < \rho^\varepsilon(x) u_0(x), v^\varepsilon(x,0) >_{H^{-1},H_0^1} + \int_0^1 \rho^\varepsilon(x) u_0(x) v_t^\varepsilon(x,0)dx \\ &\quad + \int_0^T f_L^\varepsilon(t) v_x^\varepsilon(1,t) = 0. \end{aligned} \tag{24}$$

Here $< \cdot, \cdot >_{H^{-1},H_0^1}$ represents the duality product between H_0^1 and its dual H^{-1}.

As the eigenfunctions are orthogonal in $L^2(0,1)$ and $v_1 \in L_{K(\varepsilon)}$ which is a subspace generated by the first $K(\varepsilon)$ eigenfunctions we can simplify the second term in (24) as follows

$$\int_0^1 \rho^\varepsilon(x)u^\varepsilon(x,T)v_1(x)dx = \int_0^1 \rho^\varepsilon(x)\Pi_{K(\varepsilon)}(u^\varepsilon(x,T))v_1(x)dx. \tag{25}$$

In the same way, we can simplify the first term in (24). To do this we observe that both $u_t^\varepsilon(x,T)$ and v_0 can be spanned in terms of the eigenfunctions w_n^ε and that

$$< \rho^\varepsilon w_n^\varepsilon, w_m^\varepsilon >_{H^{-1},H_0^1} = \int_0^1 \rho^\varepsilon(x)w_n^\varepsilon(x)w_m^\varepsilon(x)dx \tag{26}$$

which is nonzero if and only if $m = n$. So, as $v_0 \in L_{K(\varepsilon)}$ we easily obtain

$$< \rho^\varepsilon(x)u_t^\varepsilon(x,T), v_0(x) >_{H^{-1},H_0^1} = < \rho^\varepsilon(x)\Pi_{K(\varepsilon)}(u_t^\varepsilon(x,T)), v_0(x) >_{H^{-1},H_0^1} \tag{27}$$

Using (25) and (27) we can simplify (24) into

$$< \rho^\varepsilon(x)\Pi_{K(\varepsilon)}(u_t^\varepsilon(x,T)), v_0(x) >_{H^{-1},H_0^1} - \int_0^1 \rho^\varepsilon(x)\Pi_{K(\varepsilon)}(u^\varepsilon(x,T))v_1(x)dx$$
$$- < \rho^\varepsilon(x)u_0(x), v^\varepsilon(x,0) >_{H^{-1},H_0^1} + \int_0^1 \rho^\varepsilon(x)u_0(x)v_t^\varepsilon(x,0)dx$$
$$+ \int_0^T f_L^\varepsilon(t)v_x^\varepsilon(1,t) = 0. \tag{28}$$

Now, observe that the condition (10) is equivalent to the following one:

$$< \rho^\varepsilon(x)\Pi_{K(\varepsilon)}(u_t^\varepsilon(x,T)), v_0(x) >_{H^{-1},H_0^1} - \int_0^1 \rho^\varepsilon(x)\Pi_{K(\varepsilon)}(u^\varepsilon(x,T))v_0(x)dx = 0,$$
$$\forall (v_0, v_1) \in L_{K(\varepsilon)}. \tag{29}$$

From (28) and (29) we deduce that the statement in Proposition 1 is equivalent to the existence of $f_L^\varepsilon \in L^2(0,T)$ such that (22) holds. This concludes the proof.

Now, following HUM, we consider the Hilbert space $V = H_0^1 \times L^2$ and the following bilinear form on V,

$$A^\varepsilon((v^0, v^1), (\psi^0, \psi^1)) = \int_0^T \eta(t)v_x^\varepsilon(t,1)\psi_x^\varepsilon(t,1)dt,$$

where v, ψ are the solutions of the adjoint system (23) with final data (v^0, v^1), (ψ^0, ψ^1) respectively. We also consider the linear form on V,

$$L^\varepsilon((\psi^0,\psi^1)) = -\left\langle \psi(0,\cdot), \rho^\varepsilon(\cdot)u^1(\cdot)\right\rangle_{H^{-1},H_0^1} + \int_0^1 \rho^\varepsilon(x)\psi_t(0,x)u^0(x)\,dx.$$

Then, any solution of the variational equation

$$A^\varepsilon((\hat{v}^{0,\varepsilon},\hat{v}^{1,\varepsilon}),(\psi^0,\psi^1)) = L^\varepsilon(\psi^0,\psi^1), \quad \text{for all } (\psi^0,\psi^1) \in L_{K(\varepsilon)} \times L_{K(\varepsilon)} \subset V, \tag{30}$$

gives a control

$$f_L^\varepsilon(t) = \eta(t)\hat{v}_x^\varepsilon(t,1)$$

that satisfies (22). Moreover, this control is the one with the minimal L^2-weighted norm, defined in (21).

3 Matrix Formulation of the Control Problem

We assume that the eigenfunctions $\{w_k^\varepsilon\}_{n=1}^\infty$ are normalized with the H_0^1 norm. Let us define

$$W_k^\varepsilon = \frac{1}{\sqrt{2}}(w_{|k|}^\varepsilon, i\mu_k^\varepsilon w_{|k|}^\varepsilon), \text{ for } k \in \mathbb{Z}^*, \text{ and } \mu_k = sign(k)\sqrt{\lambda_{|k|}}.$$

Note that W_k^ε is normalized in $V = H_0^1 \times L^2$, when considering the L^2-norm $\|\cdot\|_\varepsilon$ (which is equivalent to the usual L^2-norm).

Any $(v^0, v^1) \in L_{K(\varepsilon)} \times L_{K(\varepsilon)}$ can be written as follows

$$(v^0,v^1) = \sum_{|k|\le K(\varepsilon)} c_k^\varepsilon W_k^\varepsilon, \quad \|(v^0,v^1)\|_{H_0^1\times L^2}^2 = \sum_{|k|\le K(\varepsilon)} |c_k^\varepsilon|^2,$$

where $c_k^\varepsilon \in \mathbb{C}$ can be computed from (v^0, v^1) as follows,

$$c_n = \frac{a_n + ib_n}{2}, \quad c_{-n} = \overline{c_n}, \quad n \in \mathbb{N} \tag{31}$$

$$a_n = \int_0^1 (u^0)'(x)(w_n^\varepsilon)'(x)\,dx, \quad b_n = \lambda_n^\varepsilon \int_0^1 \rho^\varepsilon u^1(x) w_n^\varepsilon(x)\,dx. \tag{32}$$

Note also that the solution $(v^\varepsilon, v_t^\varepsilon)$ of the adjoint system (23) with initial data (v^0, v^1) is given by

$$(v^\varepsilon, v_t^\varepsilon) = \sum_{|k|\le K(\varepsilon)} c_k^\varepsilon e^{i\mu_k^\varepsilon(T-t)} W_k^\varepsilon.$$

In particular

$$v_x^\varepsilon(t,1)=\sum_{|k|\le K(\varepsilon)} c_k^\varepsilon e^{i\mu_k^\varepsilon(T-t)}(w_k^\varepsilon)'(1). \tag{33}$$

Therefore, we can write (30) as a linear system for the Fourier coefficients $c^\varepsilon = \{c_k^\varepsilon\}_{|k|\le K(\varepsilon)}$, i.e.,

$$M^\varepsilon c^\varepsilon = d^\varepsilon, \quad M^\varepsilon \in \mathcal{M}_{2K(\varepsilon)\times 2K(\varepsilon)}, \quad c^\varepsilon,\ d^\varepsilon \in \mathbb{C}^{2K(\varepsilon)}. \tag{34}$$

Here, the matrix components are

$$\begin{aligned}
M_{jk}^\varepsilon &= A^\varepsilon(W_j^\varepsilon, W_k^\varepsilon) = \int_0^T \eta(t) e^{i(\mu_j^\varepsilon-\mu_k^\varepsilon)(T-t)}dt\ (w_j^\varepsilon)'(1)(w_k^\varepsilon)'(1),\\
d_j^\varepsilon &= L^\varepsilon(W_j^\varepsilon)\\
&= -e^{i\mu_j^\varepsilon T}\left\langle w_j^\varepsilon(\cdot), \rho^\varepsilon(\cdot)u^1(\cdot)\right\rangle_{H^{-1},H_0^1} - i\mu_j^\varepsilon e^{i\mu_j^\varepsilon T}\int_0^1 \rho^\varepsilon(x) w_j^\varepsilon(x) u^0(x)\,dx.
\end{aligned}$$

The control is given by $\eta(t)\hat{v}_x^\varepsilon(t,1)$ where $\hat{v}_x(t,1)$ is given by (33) with the Fourier coefficients $c^\varepsilon = \{c_k^\varepsilon\}_{|k|\le K(\varepsilon)}$ obtained by solving the linear system (34).

4 Numerical Approximation of the Control Problem

To obtain numerical approximations of the control we consider an approximation of the linear system (34) obtained by approximating their coefficients with the asymptotic expansions of the eigenvalues and eigenfunctions of (6). The full expansion below the critical size $k \sim \varepsilon^{-1}$ is given in [7]. This expansion requires an infinite number of correctors that are computed by a recurrence formula. To illustrate the process we focus on the first terms in the approximation.

The following result is proved in [7].

Theorem 2 *Consider $\rho \in L^\infty(\mathbb{R})$ a 1-periodic function such that $0 < \rho_m \le \rho(x) \le \rho_M < \infty$ a. e. in $\mathbb{R}$. Then, given $N \in \mathbb{N}$ the eigenvalues λ_k^ε of (6) verify the following,*

$$\begin{aligned}
\sqrt{\lambda_k^\varepsilon} &= \frac{k\pi}{\sqrt{\bar{\rho}}} - \sum_{n=1}^N \varepsilon^{2n}(k\pi)^{2n+1}\frac{d_{2n-1}}{\bar{\rho}^{n+1}}\\
&- \sum_{n=1}^N \varepsilon^{2n+1}(k\pi)^{2n+1}\frac{c_{2n}(\varepsilon^{-1})}{\bar{\rho}^{n+1}} + \mathcal{O}(\varepsilon^{2N+2}k^{2N+3}),
\end{aligned} \tag{35}$$

where the constants d_{2n-1} are given by

$$d_{2n-1} = \frac{1}{2\sqrt{\bar{\rho}}} \int_0^1 \sum_{i+j=2n,\ i,j\geq 1} S_t^i(t) S_t^j(t) dt, \qquad n \geq 1, \tag{36}$$

and $c_{2n}(s)$ are 1-periodic functions given by

$$c_{2n}(s) = -i \int_0^s \left[\int_0^{t_1} S_{tt}^{2n}(t_2) dt_2 - \int_0^1 \int_0^t S_{tt}^{2n}(t_2) dt_2 dt \right] dt_1, \quad n \geq 1. \tag{37}$$

The functions $S_t^n(t)$ are defined for $t \in [0, \varepsilon^{-1}]$. They depend only on ρ and can be computed explicitly from the following recursion formula:

$$S_t^0(t) = i\sqrt{\bar{\rho}}, \quad S_t^1(t) = -\int_0^t (\rho(t_2) - \bar{\rho}) dt_2 + \gamma_1, \tag{38}$$

$$S_t^n(t) = -\int_0^t \sum_{i+j=n-1} S_t^i(t_2) S_t^j(t_2) dt_2 + \gamma_n, \quad n \geq 2, \tag{39}$$

where

$$\gamma_1 = \int_0^1 \int_0^{t_1} (\rho(t_2) - \bar{\rho}) dt_2 dt_1,$$

$$\gamma_n = \frac{1}{2i\sqrt{\bar{\rho}}} \int_0^1 \sum_{i+j=n,\ i,j\geq 1} S_t^i(t_1) S_t^j(t_1) dt_1 + \int_0^1 \int_0^{t_1} \sum_{i+j=n-1} S_t^i(t_2) S_t^j(t_2) dt_2 dt_1. \tag{40}$$

The coefficients S_t^n verify the following properties:

(a) *$S_t^n(t)$ are 1-periodic functions.*
(b) *$S_t^{2n-1}(t)$ is a real function while $S_t^{2n}(t)$ is purely imaginary.*

For the eigenfunctions we have

$$w_k^\varepsilon(x) = \sqrt{\frac{2}{\lambda_k^\varepsilon \bar{\rho}}} \sin\left(\sqrt{\lambda_k^\varepsilon} \sqrt{\bar{\rho}} x\right) + \mathcal{O}(\varepsilon k). \tag{41}$$

According to the number of terms that we consider in the expansion for the eigenvalues above we have different order approximations.

Zero Order Approximation This corresponds to consider $N = 0$ in Theorem 2. In this case,

$$\sqrt{\lambda_k^\varepsilon} = \frac{k\pi}{\sqrt{\bar\rho}} + \mathcal{O}(\varepsilon^2 k^3) = \mu_k^0 + \mathcal{O}(\varepsilon^2 k^3)$$

$$w_k^\varepsilon(x) = \frac{\sqrt{2}}{k\pi}\sin(k\pi x) + \mathcal{O}(\varepsilon^2 k^3) = w_k^0(x) + \mathcal{O}(\varepsilon^2 k^3).$$

Therefore, if we assume that $K(\varepsilon) \le C\varepsilon^{-3/2}$ with a sufficiently small constant $C > 0$, the spectra can be approximated uniformly by the zero order approximation and we can, at least formally, approximate the linear system (34) by the following approximation

$$M^0 c^{0,\varepsilon} = d^{0,\varepsilon},$$

$$M_{jk}^0 = \int_0^T \eta(t) e^{i(\mu_j^0 - \mu_k^0)(T-t)} dt \; (w_j^0)'(1)(w_k^0)'(1),$$

$$d_j^\varepsilon = -e^{i\mu_j^0 T}\left\langle w_j^0(\cdot), \rho^\varepsilon(\cdot) u^1(\cdot)\right\rangle_{H^{-1},H_0^1} - i\mu_k e^{i\mu_j^0 T}\int_0^1 \rho^\varepsilon(x) w_j^0(x) u^0(x)\, dx.$$

First Order Approximation This corresponds to consider $N = 1$ in Theorem 2. In this case,

$$\sqrt{\lambda_k^\varepsilon} = \frac{k\pi}{\sqrt{\bar\rho}} - \varepsilon^2 (k\pi)^3 \frac{d_1}{\bar\rho^2} - \varepsilon^3 (k\pi)^3 \frac{c_2(\varepsilon^{-1})}{\bar\rho^2} + \mathcal{O}(\varepsilon^4 k^5) = \mu_k^{1,\varepsilon} + \mathcal{O}(\varepsilon^4 k^5), \quad (42)$$

$$w_k^\varepsilon(x) = \frac{\sqrt{2}}{\mu_k^{1,\varepsilon}\sqrt{\bar\rho}}\sin\left(\mu_k^{1,\varepsilon}\sqrt{\bar\rho} x\right) + \mathcal{O}(\varepsilon^2 k^3) = w_k^{1,\varepsilon}(x) + \mathcal{O}(\varepsilon^2 k^3). \quad (43)$$

Therefore, if we assume that $K(\varepsilon) \le C\varepsilon^{-5/4}$ with a sufficiently small constant $C > 0$, the spectra can be approximated uniformly by the first order approximation and we can, at least formally, approximate the linear system (34) by the following approximation

$$M^{1,\varepsilon} c^{1,\varepsilon} = d^{1,\varepsilon},$$

$$M_{jk}^{1,\varepsilon} = \int_0^T \eta(t) e^{i(\mu_j^{1,\varepsilon} - \mu_k^{1,\varepsilon})(T-t)} dt \; (w_j^{1,\varepsilon})'(1)(w_k^{1,\varepsilon})'(1),$$

$$d_j^\varepsilon = -e^{i\mu_j^{1,\varepsilon} T}\left\langle w_j^{1,\varepsilon}(\cdot), \rho^\varepsilon(\cdot) u^1(\cdot)\right\rangle_{H^{-1},H_0^1} - i\mu_k e^{i\mu_j^{1,\varepsilon} T}\int_0^1 \rho^\varepsilon(x) w_j^{1,\varepsilon}(x) u^0(x)\, dx.$$

Higher Order Approximations Considering more terms in the uniform approximation of the spectrum provides an approximation of the linear system (34) valid for more frequencies. In particular this will allow us to control a larger projection of the solution up to the order $\varepsilon^{1-1/(2N+1)}$ for all $N \ge 1$.

On the other hand, if we are interested in the projection of a finite number of frequencies, including more terms in the asymptotic expansion give a better order of approximation in ε.

5 Numerical Experiments

In this section we show some experiments that illustrate the method. We consider the following periodic function:

$$\rho(s) = 2 + \sin(2\pi s).$$

In this case, we can easily compute the first functions $S^n(t)$ in (38)–(38),

$$S_t^0(t) = -i\sqrt{2}, \qquad S_t^1(t) = \frac{\cos 2\pi t}{2\pi}, \quad S_t^2(t) = i\frac{8\sqrt{2}\sin 2\pi t}{16\pi^2}.$$

This allows us to compute d_1 and $c_2(s)$:

$$d_1 = \frac{1}{16\pi^2}, \qquad c_2(s) = \sqrt{2}\frac{\cos 2s\pi\ - 1}{4\pi^3}.$$

Therefore, substituting in formula (42) we have

$$\sqrt{\lambda_k^\varepsilon} = \frac{k\pi}{\sqrt{2}} - \varepsilon^2 k^2 \frac{\pi}{64} - \varepsilon^3 k^3 \frac{\sqrt{2}}{16}(\cos(2\pi/\varepsilon) - 1) + \mathcal{O}(\varepsilon^4 k^5). \tag{44}$$

The zero order approximation corresponds to consider

$$\mu_k^0 = \frac{k\pi}{\sqrt{2}}, \qquad w_k^0 = \frac{\sqrt{2}}{k\pi}\sin(k\pi x).$$

The first order approximation is,

$$\mu_k^{1,\varepsilon} = \frac{k\pi}{\sqrt{2}} - \varepsilon^2 k^2 \frac{\pi}{64} - \varepsilon^3 k^3 \frac{\sqrt{2}}{16}(\cos(2\pi/\varepsilon) - 1),$$

$$w_k^{1,\varepsilon} = \frac{\sqrt{2}}{\mu_k^{1,\varepsilon}\sqrt{\bar{\rho}}}\sin\left(\mu_k^{1,\varepsilon}\sqrt{\bar{\rho}}x\right).$$

Note that the oscillating coefficient produce a dispersion in the homogenized limit. In fact the eigenvalues and eigenfunctions of the oscillating coefficient must be corrected if we want to consider either higher values of k or larger values of ε.

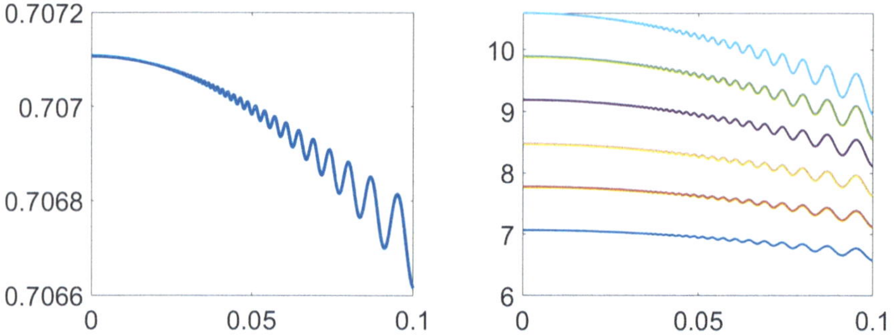

Fig. 1 Asymptotic behavior of the first eigenvalue $\sqrt{\lambda_1^\varepsilon}$ with respect to ε (left), and the same for the eigenvalues $\sqrt{\lambda_{10}^\varepsilon}, \ldots, \sqrt{\lambda_{15}^\varepsilon}$

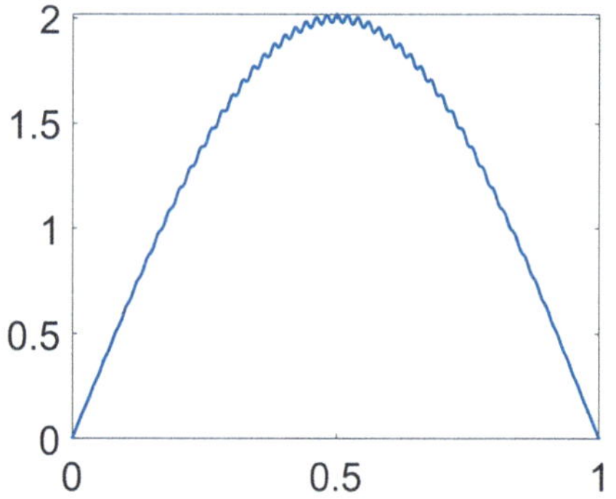

Fig. 2 Asymptotic behavior of the first eigenfunction w_1^ε with respect to ε

Figure 1 shows the behavior of the first eigenvalue with respect to ε (left) and how the correctors are more important for larger eigenvalues (right).

The eigenfunctions exhibit also an oscillatory behavior but this corrector is small below the critical size $k \sim \varepsilon^{-1}$ and therefore it does not appear in the approximation formula (41) (see Fig. 2).

We now show two examples with different initial data where we illustrate the behavior of the control.

Experiment 1 Here we consider as initial data the bump function

$$u^0(x) = \max\{0, e^{-100(x-1/2)^2}\}, \qquad u^1(x) = 0,$$

which is smooth. The final time is chosen to be $T = 3$, the time-step is set to $dt = 10^{-2}$ and the space discretization to compute the Fourier coefficients $dx = 10^{-4}$. We also consider 100 Fourier coefficients and $\varepsilon = 0.01$.

In Fig. 3 we compare the second order control $f^{2,\varepsilon}$ with the correctors corresponding to the zero order approximation f^0 and the first order approximation $f^{1,\varepsilon}$. We appreciate how the dispersion phenomena in the spectrum affects to the

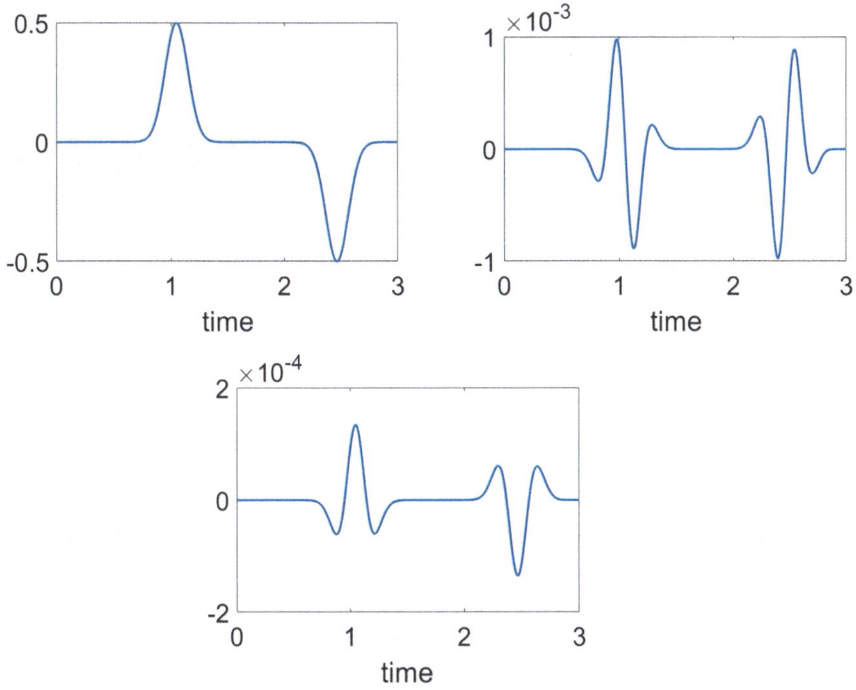

Fig. 3 Experiment 1. Second order control $f^{2,\varepsilon}(t)$ (upper left), correction to the zero order control $f^{2,\varepsilon}(t) - f^0(t)$ (upper right) and correction to the first order control $f^{2,\varepsilon}(t) - f^{1,\varepsilon}(t)$ (lower graph)

corrections of the control. We also observe that each correction is smaller than the previous one.

Experiment 2 Now we consider as initial data the function

$$u^0(x) = \max\{0, 1 - 4|x - 1/2|\}, \qquad u^1(x) = 0,$$

which is Lipschitz continuous but not in $C^1[0, 1]$. In this case we set $\varepsilon = 10^{-3}$ and the rest of the parameters are chosen as in the previous experiment.

As in the previous experiment, we show in Fig. 3 the comparison between the second order control $f^{2,\varepsilon}$ and the correctors corresponding to the zero order approximation f^0 and the first order approximation $f^{1,\varepsilon}$. As the solution now is less regular, Fourier coefficients associated with larger eigenvalues are more important and the dispersion effect is larger at these frequencies. This produces very oscillating corrections as shown in Fig. 4.

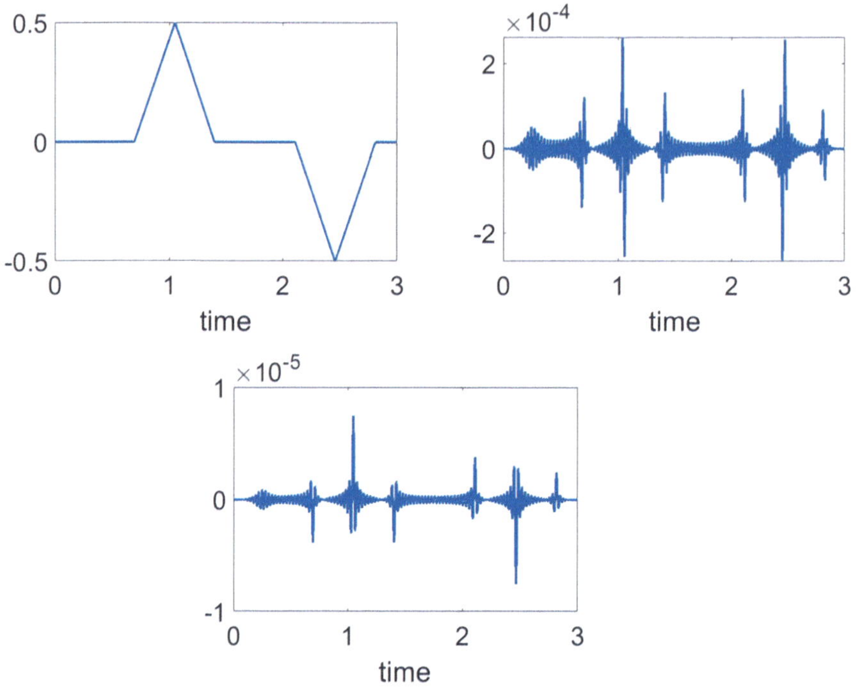

Fig. 4 Experiment 2. Second order control $f^{2,\varepsilon}(t)$ (upper left), correction to the zero order control $f^{2,\varepsilon}(t) - f^{0}(t)$ (upper right) and correction to the first order control $f^{2,\varepsilon}(t) - f^{1,\varepsilon}(t)$ (lower graph)

Acknowledgments The authors acknowledges the support of Ministerio de Ciencia, Innovación y Universidades of the Spanish government through the grant PID2021-124195NB-C31.

References

1. M. Abdelli, C. Castro, Numerical approximation of the averaged controllability for the wave equation with unknown velocity of propagation. ESAIM COCV **27**(61), 64 (2021)
2. A. Bensoussan, J.-L. Lions, G. Papanicolau, *Asymptotic Analysis for Periodic Structures* (North Holland, 1978)
3. S. Brahim-Otsmane, G. Francfort, F. Murat, Correctors for the homogenization of the wave and heat equations. J. de Mathématiques Pures et Appliquées **71**, 197–231 (1992)
4. C. Castro, *A Spectral Numerical Method to Approximate the Boundary Controllability of the Wave Equation with Variable Coefficients*, in Research in PDEs and Related Fields, The 2019 Spring School, Sidi Bel Abbès, Algeria (Tutorials, Schools, and Workshops in the Mathematical Sciences, Birkhäuser, 2022)
5. C. Castro, Boundary controllability of the one-dimensional wave equation with rapidly oscillating density. Asymptot. Anal. **20**(3–4), 317–350 (1999)
6. C. Castro, E. Zuazua, Concentration and lack of observability of waves in highly heterogeneous media. Arch. Rat. Mech. Anal. **164**(1), 39–72 (2002)

7. C. Castro, E. Zuazua, Low frequency asymptotic analysis of a string with rapidly oscillating density. SIAM J. Appl. Math. **60**, 1205–1233 (2000)
8. C. Castro, E. Zuazua, High frequency asymptotic analysis of a string with rapidly oscillating density. Eur. J. Appl. Math. **11**, 595–622, (2000)
9. C. Castro, E. Zuazua, *Control and homogenization of wave equations*, Homogenization 2001, in *Proceedings of the First HMS2000 International School and Conference on Homogenization*, ed. by L. Carbone, R. De Arcangelis. GAKUTO Internat. Ser. Math. Sci. Appl., vol. 18. (Gakkotosho, Tokyo, Naples, 2002), pp. 45–94
10. M.S.P. Eastham, *The Spectral Theory of Periodic Differential Equations* (Scottish Academic Press, Edinburgh, 1973)
11. S. Ervedoza, E. Zuazua, *Numerical Approximation of Exact Controls for Waves* (Springer, 2013)
12. S. Ervedoza, E. Zuazua, A systematic method for building smooth controls for smooth data. Discrete Contin. Dyn. Syst. Ser. B **14**, 1375–1401 (2010)
13. R. Glowinski, J.-L. Lions, Exact and approximate controllability for distributed parameter systems. Acta Numerica **3**, 269–378 (1994)
14. J.-L. Lions, *Contrôlabilité exacte, perturbations et stabilisation de systèmes distribués. Tome 1*, vol. 8 of Recherches en Mathématiques Appliquées [Research in Applied Mathematics], Masson, Paris, 1988. Contrôlabilité exacte. [Exact controllability], With appendices by E. Zuazua, C. Bardos, G. Lebeau and J. Rauch
15. P.-A. Raviart, J.-M. Thomas, *Introduction à l'analyse numérique des équations aux dérivées partielles* (Dunod, Paris, 1998)
16. E. Zuazua, Exact controllability for semilinear wave equations in one space dimension. Ann. de 'IHP, sction C **10**(1), 109–129 (1993)
17. M. Avellaneda, C. Bardos, and J. Rauch, *Contrôlabilité exacte, homogénéisation et localisation d'ondes dans un milieu non-homogène*, Asymptotic Anal., **5**, 481–494 (1992)
18. E. Zuazua, Propagation, observation, and control of waves approximated by finite difference methods. SIAM Rev. **47**(2), 197–243 (2005)

Numerical Impulse Controllability for Parabolic Equations by a Penalized HUM Approach

Salah-Eddine Chorfi, Ghita El Guermai, Lahcen Maniar, and Walid Zouhair

2022 Mathematics Subject Classification 93C27; 49N25; 35R12

1 Introduction and Main Results

Parabolic equations, where the heat equation is the prototype, constitute a class of Partial Differential Equations (PDEs) that describe the evolution of physical quantities over time and space. The heat equation is particularly important in the study of heat transfer and diffusion processes.

Impulsive systems in the context of PDEs refer to systems whose behavior changes abruptly or impulsively at certain points in space or time. These impulsive changes can be modeled mathematically using PDEs with discontinuities or Dirac delta functions, which are used to present concentrated impulses at specific spatial or temporal locations. Impulsive systems are encountered in various fields, from biological models to fluid dynamics and economics, among others. They manifest as sudden boundary conditions in switched control inputs or shock waves in compressible flows. Controlling impulsive systems via impulse controls can be challenging (due to the presence of delta functions) requiring specific techniques, such as logarithmic convexity, Carleman commutator approach, and optimal impulse control theory, see, e.g., [1, 2, 11].

Impulsive controllability is a concept in control theory that deals with the ability to control a dynamical system by applying control inputs at specific discrete instants or time intervals, often referred to as "impulse times" (or intervals). In impulsive controllability, the control actions are not continuously applied but occur at distinct

S.-E. Chorfi · G. El Guermai · L. Maniar
LMDP, UMMISCO (IRD-UPMC), Cadi Ayyad University, Faculty of Sciences Semlalia, Marrakesh, Morocco
e-mail: s.chorfi@uca.ac.ma; maniar@uca.ac.ma

W. Zouhair (✉)
Department of mathematics, Ibn Zohr University, Faculty of Applied Sciences Ait Melloul, Azrou, Morocco

K. Ammari et al. (eds.), *Advances in Partial Differential Equations and Control*, Trends in Mathematics, https://doi.org/10.1007/978-3-031-62265-6_6

time points. Impulsive systems can model situations where control inputs change abruptly. They make it possible to handle systems in situations where continuous control may not be feasible or practical. For instance, in switched control systems, the control input can change instantaneously at a specific instant. In this context, the impulse approximate controllability was studied for a linear heat equation with homogeneous Dirichlet and Neumann boundary conditions in [5, 13], using a new strategy combining the logarithmic convexity method and the Carleman commutator approach. In [4], the authors have established a Lebeau–Robbiano-type spectral inequality for a degenerate one-dimensional elliptic operator with application to impulse control and finite-time stabilization. It should be pointed out that this method is a new approach to steer the solution to zero using impulse control as a stabilizer in finite time. Recently, in [6–8], the authors have established new results of impulse controllability for a general type of dynamic boundary conditions which introduce mathematical issues that require sophisticated estimates due to the boundary terms. We refer to the seminal paper [12] for more details on the nonimpulsive control case.

Boundary conditions play a crucial role in solving PDEs as they have a significant effect on the behavior of the solutions. They describe the interaction of the system with its surroundings. For example, in heat transfer problems, the temperature on the boundary may represent an insulated or constant temperature boundary, reflecting the physical properties of the system. The choice of appropriate boundary conditions is a fundamental step in the analysis of PDEs in various fields of science and engineering. From a numerical perspective, boundary conditions are fundamental for ensuring the accuracy, stability, and convergence of numerical solutions to PDEs. They guide how the spatial domain should be discretized. In this chapter, we propose an algorithm designed for the numerical computation of impulse controls with minimal energy. This approach involves an adaptation of the penalized HUM and a CG method to the impulsive case. For further information, we recommend the book [10] and the paper [3]. Our investigation encompasses both static boundary conditions (Dirichlet and Neumann) and dynamic boundary conditions. To validate and compare theoretical findings regarding impulse controllability, we conduct several numerical experiments using the algorithm we have developed. Finally, it should be emphasized that the numerical computation of impulse controls has not been considered before for static boundary conditions. We refer to the "Future works" section of the thesis [15].

The remainder of this chapter is structured as follows: In Sect. 2, we provide a review of various results related to impulse controllability for the heat equation with different boundary conditions. Section 3 is devoted to the algorithm for computing impulse optimal controls, accompanied by numerical simulations for illustration. Finally, we conclude with a comparative analysis of numerical outcomes across various boundary conditions.

2 Preliminary Results

Let $\Omega \subset \mathbb{R}^n$ be a bounded domain with smooth boundary $\Gamma := \partial\Omega$. Let $T > 0$ be an arbitrary control time and $\tau \in (0, T)$ be an arbitrary fixed impulsion time. We consider the following impulse-controlled system:

$$\begin{cases} \partial_t \Psi - \mathbf{A}\Psi = 0, & \text{in } (0,T)\backslash\{\tau\}, \\ \psi(\cdot,\tau) = \psi\left(\cdot,\tau^-\right) + \mathbb{1}_\omega h(\cdot,\tau), & \text{in } \Omega, \\ \Psi(0) = \Psi^0, \end{cases} \tag{1}$$

where $\psi(\cdot,\tau^-)$ denotes the left limit of the function ψ at time τ, the control region $\omega \Subset \Omega$ is a nonempty open subset, $\mathbb{1}_\omega$ stands for the characteristic function of ω, and $h(\cdot,\tau)$ is an impulsive control acting at the impulse instant τ. The notation $\mathbf{A}$ designates a linear operator on an L^2-space with norm $\|\cdot\|$, and the state Ψ might be a couple (ψ, ψ_Γ), depending on the type of boundary conditions (see the next subsections).

If the operator $\mathbf{A}$ generates a C_0-semigroup. Then, for every initial datum Ψ_0, the system (1) has a unique mild solution given by

$$\Psi(t) = \mathrm{e}^{t\mathbf{A}}\Psi_0 + \mathbb{1}_{\{t\geq\tau\}}(t)\,\mathrm{e}^{(t-\tau)\mathbf{A}}(\mathbb{1}_\omega h(\tau), 0), \qquad t \in (0,T).$$

Definition 1 System (1) is null approximate impulse controllable at time T if for any $\varepsilon > 0$ and any Ψ^0, there exists a control function $h \in L^2(\omega)$, such that the associated state at final time satisfies

$$\|\Psi(\cdot,T)\| \leq \varepsilon \left\|\Psi^0\right\|.$$

This means that for every $\varepsilon > 0$ and every initial datum Ψ^0, the set

$$\mathcal{R}_{T,\Psi^0,\varepsilon} := \left\{h \in L^2(\omega) : \text{ the solution of (1) satisfies } \|\Psi(\cdot,T)\| \leq \varepsilon \left\|\Psi^0\right\|\right\},$$

is nonempty, which leads to the definition of the cost of null approximate impulse controllability.

Definition 2 The quantity

$$K(T,\varepsilon) := \sup_{\|\Psi^0\|=1} \inf_{h\in\mathcal{R}_{T,\Psi^0,\varepsilon}} \|h\|_{L^2(\omega)}$$

is called the cost of null approximate impulse controllability at time T.

2.1 Dirichlet Case

In this subsection, we recall the impulsive controllability result for the heat equation with the Dirichlet boundary condition:

$$\begin{cases} \partial_t \psi - \Delta \psi = 0, & \text{in } \Omega \times (0,T)\backslash\{\tau\}, \\ \psi(\cdot,\tau) = \psi\left(\cdot,\tau^-\right) + \mathbb{1}_\omega h(\cdot,\tau), & \text{in } \Omega, \\ \psi = 0, & \text{on } \Gamma \times (0,T), \\ \psi(\cdot,0) = \psi^0, & \text{on } \Omega. \end{cases} \tag{2}$$

To obtain the null approximate impulse controllability of the above equation, the key ingredient is the following logarithmic convexity estimate:

Lemma 1 ([13]) *For any $T > 0$ and any ω nonempty open subset of Ω,*

$$\|u(\cdot,T)\|_{L^2(\Omega)} \le e^{C\frac{K}{T}} \|u(\cdot,T)\|^{\beta}_{L^2(\omega)} \|u(\cdot,0)\|^{1-\beta}_{L^2(\Omega)}, \tag{3}$$

where $\beta \in (0,1)$, $C, K > 0$ are constants, and u is the solution of the following homogeneous system:

$$\begin{cases} \partial_t u - \Delta u = 0, & \text{in } \Omega \times (0,T), \\ u = 0, & \text{on } \Gamma \times (0,T), \\ u(\cdot,0) = u^0, & \text{in } \Omega. \end{cases} \tag{4}$$

The previous lemma reflects an observability estimate at a single instant of time. This estimate has been proven using the weight function

$$\Phi(x,t) = \frac{-|x-x_0|^2}{4(T-t+\rho)}, \qquad (x,t) \in \overline{\Omega} \times (0,T),$$

where $x_0 \in \omega$ and $\rho > 0$ is suitably chosen.

Consequently, the following result on impulse controllability of the Eq. (2) was established:

Theorem 1 ([14, 16]) *The heat equation (2) is null approximate impulse controllable at time T. Moreover, we have the following upper bound for the control cost:*

$$K_1(T,\varepsilon) \le \frac{M_1 e^{\frac{M_2}{T-\tau}}}{\varepsilon^\delta},$$

where M_1, M_2, and δ are positive constants depending on Ω and ω.

2.2 Neumann Case

Here, we recall the impulsive controllability result for the heat equation with the Neumann boundary condition:

$$\begin{cases} \partial_t \psi - \Delta \psi = 0, & \text{in } \Omega \times (0,T)\backslash\{\tau\}, \\ \psi(\cdot,\tau) = \psi\left(\cdot,\tau^-\right) + \mathbb{1}_\omega h(\cdot,\tau), & \text{in } \Omega, \\ \partial_\nu \psi = 0, & \text{on } \Gamma \times (0,T), \\ \psi(\cdot,0) = \psi^0, & \text{in } \Omega, \end{cases} \tag{5}$$

where ν is the unit outward normal vector to Γ, and $\partial_\nu \psi$ denotes the normal derivative. As before, the key ingredient is the following logarithmic convexity estimate:

Lemma 2 ([5]) *For any $T > 0$ and any ω nonempty open subset of Ω,*

$$\|u(\cdot,T)\|_{L^2(\Omega)} \leq \left(e^{C\left(1+\frac{1}{T}\right)} \|u(\cdot,T)\|_{L^2(\omega)}\right)^{\beta} \|u(\cdot,0)\|^{1-\beta}_{L^2(\Omega)},$$

where $\beta \in (0,1)$, $C > 0$ are constants only depending on Ω and ω, and u is the solution of the homogeneous system

$$\begin{cases} \partial_t u - \Delta u = 0, & \textit{in } \Omega \times (0,T), \\ \partial_\nu u = 0, & \textit{on } \Gamma \times (0,T), \\ u(\cdot,0) = u^0, & \textit{in } \Omega. \end{cases} \tag{6}$$

This result has been recently extended to a general parabolic equation with variable diffusion and drift coefficients in [9].

Being different from the Dirichlet case, the above lemma has been established by introducing a small parameter $s \in (0,1)$ in the weight function

$$\Phi_s(x,t) = \frac{-s\,|x-x_0|^2}{4(T-t+\rho)}, \qquad (x,t) \in \overline{\Omega} \times (0,T),$$

where $x_0 \in \omega$ and $\rho > 0$ is suitably chosen.

Then, one can prove the following impulse controllability for the Eq. (5):

Theorem 2 ([5]) *The heat equation* (5) *is null approximate impulse controllable at time T. Moreover, we have the following upper bound for the control cost:*

$$K_2(T,\varepsilon) \leq \frac{N_1 e^{\frac{N_2}{T-\tau}}}{\varepsilon^\sigma},$$

where N_1, N_2, and σ are positive constants depending on Ω and ω.

2.3 Dynamic Case

Now, we consider the following heat equation with dynamic boundary conditions:

$$\begin{cases} \partial_t \psi - \Delta \psi = 0, & \text{in } \Omega \times (0,T)\backslash\{\tau\}, \\ \psi(\cdot,\tau) = \psi\left(\cdot,\tau^-\right) + \mathbb{1}_\omega h(\cdot,\tau), & \text{in } \Omega, \\ \partial_t \psi_\Gamma - \Delta_\Gamma \psi_\Gamma + \partial_\nu \psi = 0, & \text{on } \Gamma \times (0,T)\backslash\{\tau\}, \\ \psi_\Gamma(\cdot,\tau) = \psi_\Gamma\left(\cdot,\tau^-\right), & \text{on } \Gamma, \\ \psi_\Gamma(x,t) = \psi_{|\Gamma}(x,t), & \text{on } \Gamma \times (0,T), \\ \left(\psi(\cdot,0), \psi_\Gamma(\cdot,0)\right) = \left(\psi^0, \psi_\Gamma^0\right), & \text{on } \Omega \times \Gamma, \end{cases} \tag{7}$$

where $\left(\psi^0, \psi_\Gamma^0\right) \in L^2(\Omega) \times L^2(\Gamma)$ denotes the initial condition. Again, the key result is the logarithmic convexity estimate.

Lemma 3 ([6, 7]) *For any $T > 0$ and any ω nonempty open subset of Ω, the following estimate holds:*

$$\|U(\cdot,T)\|_{L^2(\Omega)\times L^2(\Gamma)} \le \left(\mu e^{\frac{K}{T}} \|u(\cdot,T)\|_{L^2(\omega)}\right)^\beta \|U(\cdot,0)\|^{1-\beta}_{L^2(\Omega)\times L^2(\Gamma)}, \tag{8}$$

where $\mu, K > 0$, $\beta \in (0,1)$ are constants, and $U = (u, u_\Gamma)$ is the solution of the following homogeneous system:

$$\begin{cases} \partial_t u - \Delta u = 0, & \text{in } \Omega \times (0,T), \\ \partial_t u_\Gamma - \Delta_\Gamma u_\Gamma + \partial_\nu u = 0, & \text{on } \Gamma \times (0,T), \\ u_\Gamma(x,t) = u_{|\Gamma}(x,t), & \text{on } \Gamma \times (0,T), \\ \left(u(\cdot,0), u_\Gamma(\cdot,0)\right) = \left(u^0, u_\Gamma^0\right), & \text{on } \Omega \times \Gamma. \end{cases} \tag{9}$$

In this dynamic case, several new boundary terms occur and should be absorbed. This has been done thanks to the small parameter s introduced in the weight function Φ_s inspired by the Neumann case.

Consequently, we obtained the following impulse controllability result:

Theorem 3 ([7]) *The system (7) is null approximate impulse controllable at any time $T > 0$. Moreover, we have the following upper bound for the control cost:*

$$K_3(T,\varepsilon) \le \frac{L_1 e^{\frac{L_2}{T-\tau}}}{\varepsilon^\kappa},$$

where L_1, L_2, and κ are positive constants depending on Ω and ω.

3 Algorithm for Calculating HUM Impulse Controls

In this section, we propose a numerical algorithm designed for calculating the HUM impulse controls. This method employs a penalized HUM approach along with a CG algorithm. We refer to [10] and [3] for more details on such a method.

Notations
We introduce the following notations to encapsulate various boundary conditions and give a general algorithm:

$$\mathbb{L}^2 := \begin{cases} L^2(\Omega), & \text{(Dirichlet and Neumann cases)},\\ L^2(\Omega)\times L^2(\Gamma), & \text{(Dynamic case)}, \end{cases}$$

with the inner product

$$\langle\cdot,\cdot\rangle := \begin{cases} \langle\cdot,\cdot\rangle_{L^2(\Omega)}, & \text{(Dirichlet and Neumann cases)},\\ \langle\cdot,\cdot\rangle_{L^2(\Omega)} + \langle\cdot,\cdot\rangle_{L^2(\Gamma)}, & \text{(Dynamic case)} \end{cases}$$

and the norm

$$\|\cdot\| := \begin{cases} \|\cdot\|_{L^2(\Omega)}, & \text{(Dirichlet and Neumann cases)},\\ \|\cdot\|_{L^2(\Omega)\times L^2(\Gamma)}, & \text{(Dynamic case)}. \end{cases}$$

Any capital letter as ϑ will stand for the couple $(\upsilon, \upsilon_\Gamma) \in L^2(\Omega)\times L^2(\Gamma)$. In particular, we will identify $(\upsilon, \upsilon_\Gamma) \in L^2(\Omega)\times L^2(\Gamma)$ with $\upsilon \in L^2(\Omega)$ in Dirichlet and Neumann cases. We denote by **BC** one of the boundary conditions: Dirichlet condition, Neumann condition, or dynamic condition. In each case, the operator $\mathbf{A}$ stands for the governing linear operator, and $e^{t\mathbf{A}}$ designates its associated C_0-semigroup on $\mathbb{L}^2$.

3.1 The HUM Impulse Controls

Let $\varepsilon > 0$ be fixed, and let Ψ^0 be an initial datum to be controlled. Without loss of generality, we may assume that $\|\Psi^0\| = 1$. We define the cost functional $J_\varepsilon : \mathbb{L}^2 \to \mathbb{R}$ by

$$J_{\varepsilon}\left(\vartheta^{0}\right)=\frac{1}{2}\|\upsilon(\cdot,T-\tau)\|_{L^{2}(\omega)}^{2}+\frac{\varepsilon}{2}\left\|\vartheta^{0}\right\|^{2}+\left\langle\Psi^{0},\vartheta(\cdot,T)\right\rangle,$$

where ϑ is the solution of the homogeneous heat equation with **BC** corresponding to ϑ^0. Note that the functional J_ε is strictly convex, of class C^1, and coercive, i.e., $J_\varepsilon\left(\vartheta^0\right) \to \infty$ as $\|\vartheta^0\| \to \infty$. Then the unique minimizer $\tilde{\vartheta}_\varepsilon^0 \in \mathbb{L}^2$ of J_ε is characterized by the Euler–Lagrange equation

$$\int_{\omega}\tilde{\upsilon}_{\varepsilon}(x,T-\tau)z(x,T-\tau)\mathrm{d}x+\varepsilon\left\langle\tilde{\vartheta}_{\varepsilon}^{0},Z^{0}\right\rangle+\left\langle\Psi^{0},Z(\cdot,T)\right\rangle=0 \tag{10}$$

for all $Z^0 \in \mathbb{L}^2$, where $\tilde{\vartheta}_\varepsilon$ and Z are, respectively, the solutions of the homogeneous heat equation with **BC** corresponding to $\tilde{\vartheta}_\varepsilon^0$ and Z^0. We introduce the control operator $\mathcal{B}\colon \mathbb{L}^2 \to \mathbb{L}^2$ defined by

$$\mathcal{B}\vartheta=(\mathbb{1}_{\omega}\upsilon,0),$$

and we consider the nonnegative symmetric operator (the Gramian operator)

$$\Lambda_{\tau}:\mathbb{L}^{2}\to\mathbb{L}^{2},$$

given by

$$\Lambda_{\tau}\varrho=\mathrm{e}^{(T-\tau)\mathbf{A}}\mathcal{B}\,\mathrm{e}^{(T-\tau)\mathbf{A}}\varrho.$$

Thus, the HUM impulse control is given by

$$\widehat{h}=\mathcal{B}\,\mathrm{e}^{(T-\tau)\mathbf{A}}\tilde{\vartheta}_{\varepsilon}^{0},$$

and the identity (10) can be rewritten as

$$\left(\Lambda_{\tau}+\varepsilon\mathbf{I}_{\mathbb{L}^{2}}\right)\tilde{\vartheta}_{\varepsilon}^{0}=-\mathrm{e}^{T\mathbf{A}}\Psi^{0}, \tag{11}$$

where $\mathbf{I}_{\mathbb{L}^2}$ denotes the identity operator. To resolve the above operator equation, we propose the following CG algorithm:

Algorithm 1 HUM with CG Algorithm

Set $k = 0$ and choose an initial guess $\mathbf{f}_0 = \left(f_0, f_{0,\Gamma}\right) \in \mathbb{L}^2$. Solve the problem

$$\begin{cases} \partial_t p_0(x,t) - \Delta p_0(x,t) = 0, & (x,t) \in \Omega_T, \\ \mathbf{BC}, & t \in (0,T), \\ \left(p_0(x,0), p_{0,\Gamma}(x,0)\right) = \mathbf{f}_0(x), & x \in \Omega, \end{cases}$$

and set $\mathbf{u}_0(x) = \mathcal{B}\,\mathbf{p}_0(T-\tau, x)$. Solve the problem

$$\begin{cases} \partial_t y_0(x,t) - \Delta y_0(x,t) = 0, & (x,t) \in \Omega_T, \\ \mathbf{BC}, & t \in (0,T), \\ \left(y_0(x,0), y_{0,\Gamma}(x,0)\right) = \mathbf{u}_0(x), & x \in \Omega, \end{cases}$$

compute $\mathbf{g}_0 = \varepsilon \mathbf{f}_0 + Y_0(T-\tau) + \mathrm{e}^{T\mathbf{A}}\Psi^0$, and set $\mathbf{w}_0 = \mathbf{g}_0$. For $k = 1, 2, \ldots,$ until convergence, solve the problem

$$\begin{cases} \partial_t p_k(x,t) - \Delta p_k(x,t) = 0, & (x,t) \in \Omega_T, \\ \mathbf{BC}, & t \in (0,T), \\ \left(p_k(x,0), p_{k,\Gamma}(x,0)\right) = \mathbf{w}_{k-1}(x), & x \in \Omega, \end{cases}$$

and set $\mathbf{u}_k(x) = \mathcal{B}\,\mathbf{p}_k(T-\tau, x)$. Solve the problem

$$\begin{cases} \partial_t y_k(x,t) - \Delta y_k(x,t) = 0, & (x,t) \in \Omega_T, \\ \mathbf{BC}, & t \in (0,T), \\ \left(y_k(x,0), y_{k,\Gamma}(x,0)\right) = \mathbf{u}_k(x), & x \in \Omega, \end{cases}$$

and compute

$$\bar{\mathbf{g}}_k = \varepsilon \mathbf{w}_{k-1} + Y_k(T-\tau) \qquad \text{and} \qquad \rho_k = \frac{\|\mathbf{g}_{k-1}\|^2}{\langle \bar{\mathbf{g}}_k, \mathbf{w}_{k-1}\rangle},$$

then

$$\mathbf{f}_k = \mathbf{f}_{k-1} - \rho_k \mathbf{w}_{k-1} \qquad \text{and} \qquad \mathbf{g}_k = \mathbf{g}_{k-1} - \rho_k \bar{\mathbf{g}}_k.$$

If $\dfrac{\|\mathbf{g}_k\|}{\|\mathbf{g}_0\|} \leq tol$, stop the algorithm, set $\mathbf{g} = \mathbf{f}_k$, and solve the problem

$$\begin{cases} \partial_t p_k(x,t) - \Delta p_k(x,t) = 0, & (x,t) \in \Omega_T, \\ \mathbf{BC}, & t \in (0,T), \\ \left(p_k(x,0), p_{k,\Gamma}(x,0)\right) = \mathbf{g}(x), & x \in \Omega, \end{cases}$$

and set $\mathbf{u}_k(x) = \mathcal{B}\,\mathbf{p}_k(T-\tau, x)$.
Else compute

$$\gamma_k = \frac{\|\mathbf{g}_k\|^2}{\|\mathbf{g}_{k-1}\|^2} \qquad \text{and then} \qquad \mathbf{w}_k = \mathbf{g}_k + \gamma_k \mathbf{w}_{k-1}.$$

3.2 Numerical Experiments

Now, we conduct several numerical tests to demonstrate the theoretical findings and to highlight the effectiveness of the above CG algorithm.

In all main numerical experiments, we will take the following values:

$$T = 0.02, \quad \tau = 0.01, \quad \Omega = (0,1), \quad \omega = (0.3, 0.7) \Subset (0,1),$$

and the initial datum to be controlled is given by

$$\psi_0(x) = \sqrt{2}\sin(\pi x), \qquad x \in [0,1].$$

We employ the method of lines to numerically solve diverse parabolic equations subject to different boundary conditions in Algorithm 1. In this approach, we use the uniform spatial grid given by $x_j = j\Delta x$ for $j = \overline{0, N_x}$, with $\Delta x = \frac{1}{N_x}$. Next, we denote by $u_j(t) := u(t, x_j)$. The second-order derivative of u is approximated by

$$u_{xx}(t, x_j) \approx \frac{u_{j-1}(t) - 2u_j(t) + u_{j+1}(t)}{(\Delta x)^2}, \qquad j = \overline{1, N_x - 1}.$$

The first-order derivatives on the boundary are approximated by

$$u_x(t,0) \approx \frac{u_1(t) - u_0(t)}{\Delta x}$$
$$u_x(t,1) \approx \frac{u_{N_x}(t) - u_{N_x-1}(t)}{\Delta x}.$$

Thus, it suffices to resolve the resulting system of ordinary differential equations.

For our computations, we take $N_x = 25$ for the spatial mesh parameter. The initial guess in the algorithm is taken as $\mathbf{f}_0 = 0$. We also choose $\varepsilon = 10^{-2}$ and the stopping parameter $tol = 10^{-3}$ for the plots.

3.3 Dirichlet Case

We plot the uncontrolled and the controlled solutions (Figs. 1 and 2).

The algorithm stops at the iteration number $k_* = 10$ (Figs. 3 and 4).

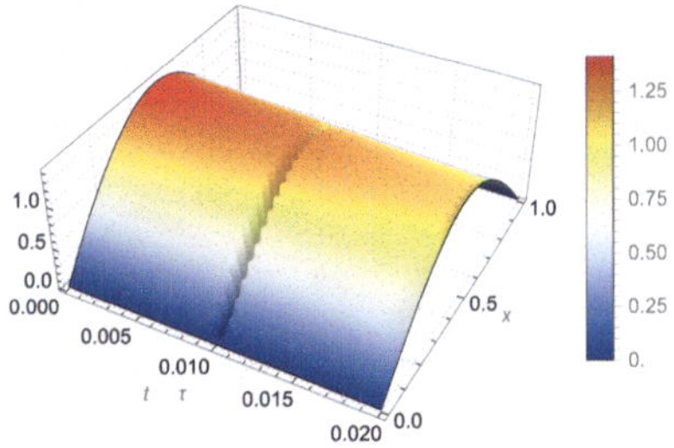

Fig. 1 The uncontrolled solution in Dirichlet case

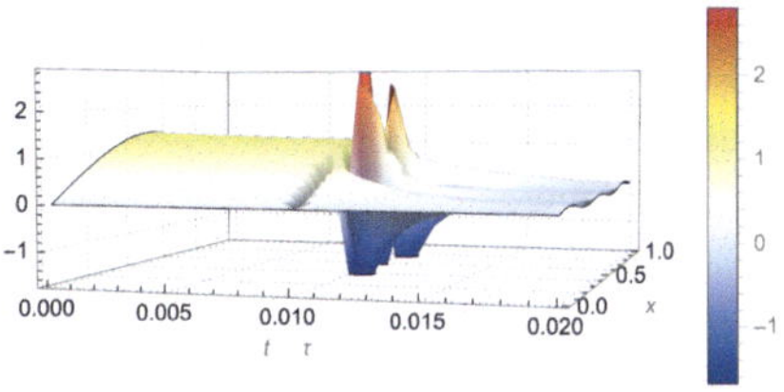

Fig. 2 The controlled solution in Dirichlet case

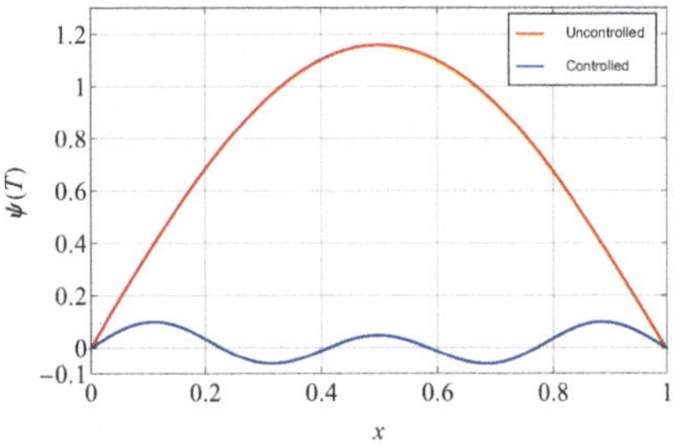

Fig. 3 The final state for uncontrolled and controlled solutions in Dirichlet case

3.4 Neumann Case

Next, we plot the uncontrolled and the controlled solutions (Figs. 5 and 6).

The algorithm stops at the iteration number $k_* = 29$ (Figs. 7 and 8).

3.5 Dynamic Case

Next, we plot the uncontrolled and the controlled solutions (Figs. 9 and 10).

The algorithm stops at the iteration number $k_* = 11$ (Figs. 11 and 12).

By analyzing the previous experiments, some comments and remarks are in order:

- From Figs. 2, 6, and 10, we notice the impact of the impulse controls at time $\tau = 0.01$ on the state.
- Tables 1, 2, and 3 show that when we fix the value of the penalization parameter ε, the Dirichlet case requires fewer iterations. The Dynamic case comes afterward

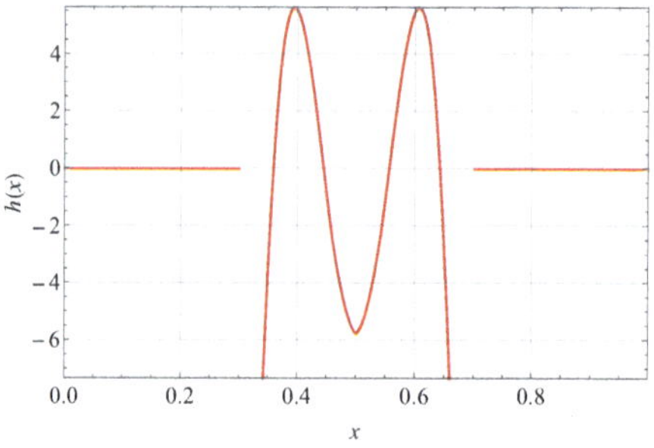

Fig. 4 The computed control h in Dirichlet case

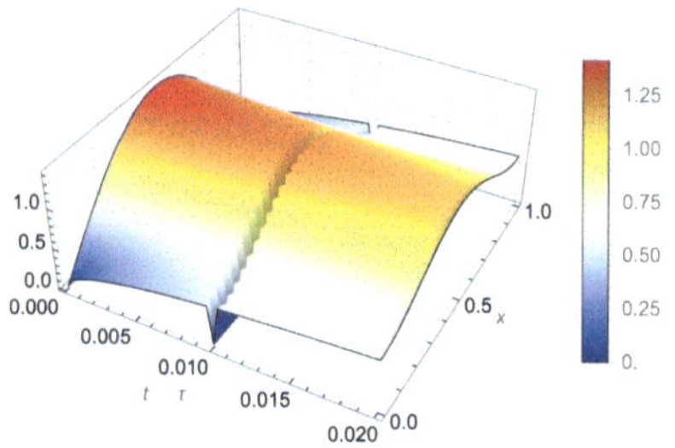

Fig. 5 The uncontrolled solution in Neumann case

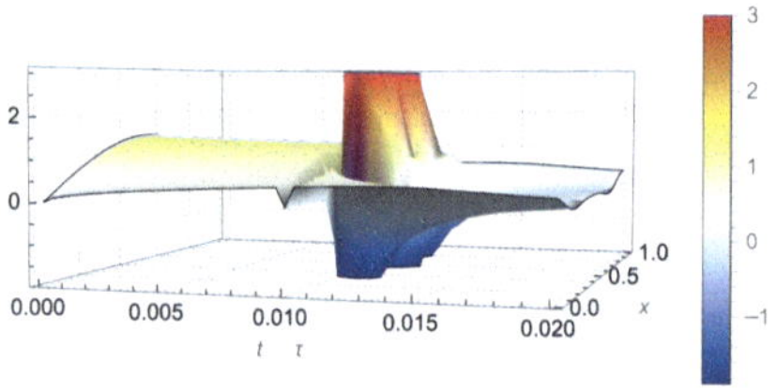

Fig. 6 The controlled solution in Neumann case

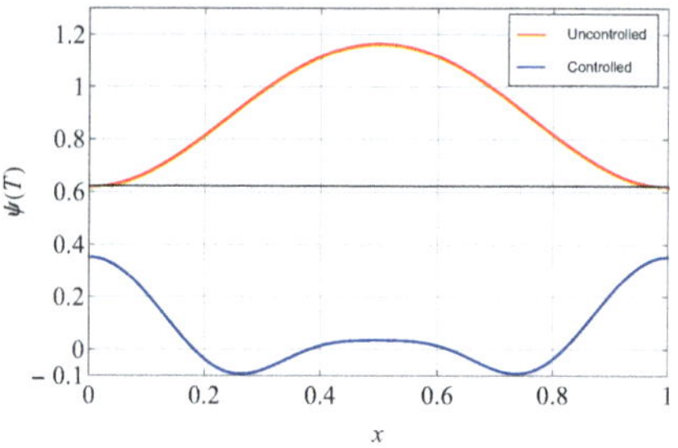

Fig. 7 The final state for uncontrolled and controlled solutions in Neumann case

with more needed iterations than the Dirichlet case. In contrast, the Neumann case requires more iterations than both previous cases.

- The tables also show that the norms $\|\Psi(T)\|$ decrease and the norms of the impulse controls $\|h\|_{L^2(\omega)}$ increase as ε tends to zero. Moreover, for a fixed ε, we have

$$\|\Psi_{\mathrm{D}}(T)\| < \|\Psi_{\mathrm{Dyn}}(T)\| < \|\Psi_{\mathrm{N}}(T)\|,$$

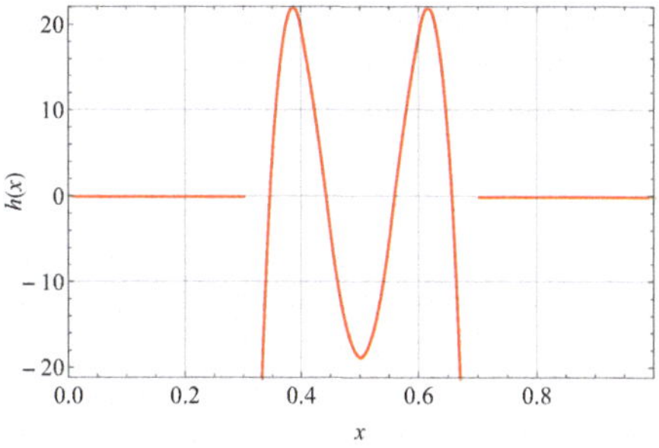

Fig. 8 The computed control h in Neumann case

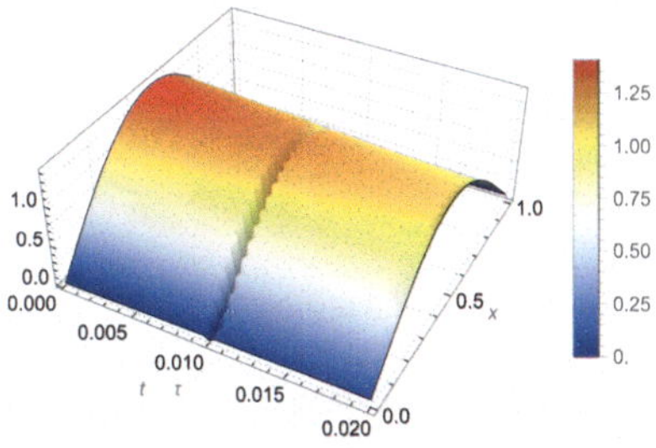

Fig. 9 The uncontrolled solution in dynamic case

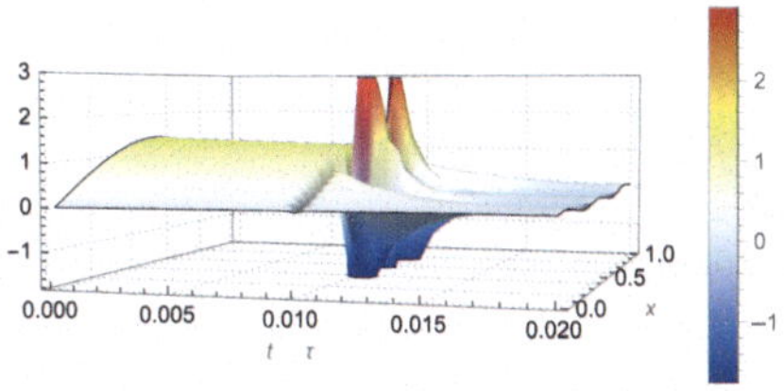

Fig. 10 The controlled solution in dynamic case

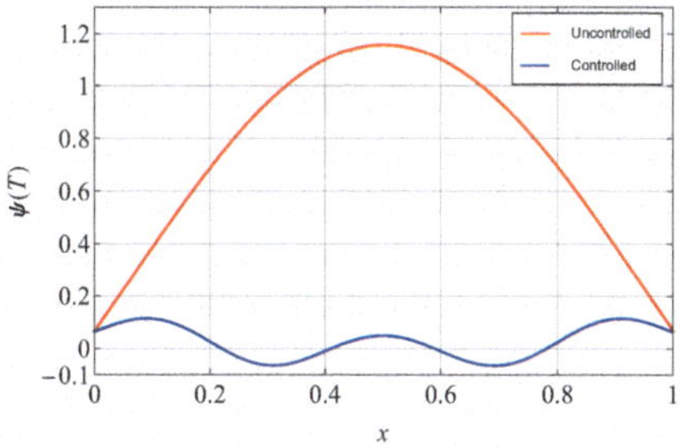

Fig. 11 The final state for uncontrolled and controlled solutions in dynamic case

and

$$\|h_{\mathrm{D}}\|_{L^2(\omega)} < \|h_{\mathrm{Dyn}}\|_{L^2(\omega)} < \|h_{\mathrm{N}}\|_{L^2(\omega)}.$$

These are relevant numerical observations that deserve further theoretical investigation to better understand why the above comparison holds.

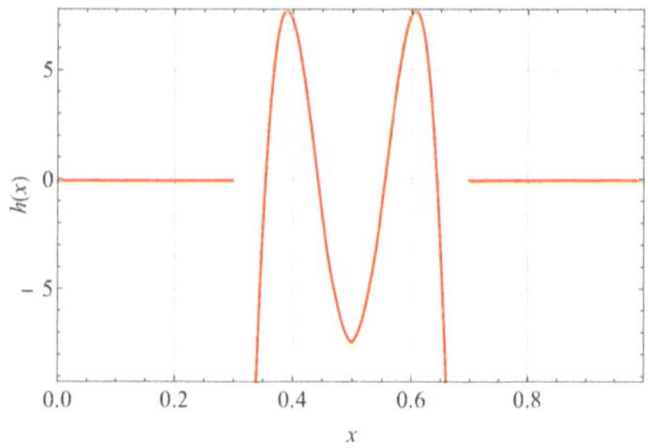

Fig. 12 The computed control h in dynamic case

Table 1 Numerical results for $T = 0.02$, $\tau = 0.01$, and $tol = 10^{-3}$ for Dirichlet condition

ε	10^{-1}	10^{-2}	10^{-3}
N_{iter}	4	10	20
$\\|\psi_{\text{D}}(T)\\|$	1.148×10^{-1}	5.63×10^{-2}	1.73×10^{-2}
$\\|h_{\text{D}}\\|_{L^2(\omega)}$	1.684	7.3014	28.6994

Table 2 Numerical results for $T = 0.02$, $\tau = 0.01$, and $tol = 10^{-3}$ for Neumann condition

ε	10^{-1}	10^{-2}	10^{-3}
N_{iter}	4	29	100
$\\|\psi_{\text{N}}(T)\\|$	2.651×10^{-1}	15.28×10^{-2}	11.16×10^{-2}
$\\|h_{\text{N}}\\|_{L^2(\omega)}$	2.128	17.2777	96.8253

Table 3 Numerical results for $T = 0.02$, $\tau = 0.01$, and $tol = 10^{-3}$ for dynamic condition

ε	10^{-1}	10^{-2}	10^{-3}
N_{iter}	4	11	65
$\\|\psi_{\text{Dyn}}(T)\\|$	1.598×10^{-1}	1.135×10^{-1}	9.02×10^{-2}
$\\|h_{\text{Dyn}}\\|_{L^2(\omega)}$	1.7491	9.234	63.6215

The numerical simulations show that the HUM algorithm yields accurate results for the numerical approximation of impulse controls at one single instant τ for the heat equation with static boundary conditions (Dirichlet and Neumann) and also with dynamic boundary conditions. The developed algorithm deserves more investigation in the context of discrete systems and their convergence analysis in terms of discrete impulse controls. This will be investigated in future research.

References

1. C. Bardos, K.D. Phung, Observation estimate for kinetic transport equations by diffusion approximation. Comptes Rendus Mathematique **355**, 640–664 (2017)
2. A. Ben Aissa, W. Zouhair, Qualitative properties for the $1-D$ impulsive wave equation: controllability and observability. Quaestiones Mathematicae (2021)
3. F. Boyer, On the penalised HUM approach and its applications to the numerical approximation of null-controls for parabolic problems. ESAIM: Proc. **41**, 15–58 (2013)
4. R. Buffe, K.D. Phung, A spectral inequality for degenerate operators and applications. C. R. Math. Acad. Sci. **356**, 1131–1155 (2018)

5. R. Buffe, K.D. Phung, Observation estimate for the heat equations with Neumann boundary condition via logarithmic convexity. J. Evol. Equ. **22**, 86 (2022)
6. S.E. Chorfi, G.El. Guermai, L. Maniar, W. Zouhair, Finite-time stabilization and impulse control of heat equation with dynamic boundary conditions. Dyn. Control Syst., 1–31 (2023)
7. S.E. Chorfi, G. El. Guermai, L. Maniar, W. Zouhair, Impulsive null approximate controllability for heat equation with dynamic boundary conditions. Math. Control Relat. Fields **13**, 1023–1046 (2023)
8. S.E. Chorfi, G. El. Guermai, L. Maniar, W. Zouhair, Logarithmic convexity and impulsive controllability for the one-dimensional heat equation with dynamic boundary conditions, IMA J. Math. Control. Inf. **39**, 861–891 (2022)
9. Y. Duan, L. Wang, C. Zhang, Quantitative unique continuation for parabolic equations with Neumann boundary conditions. Math. Control Relat. Fields **14**, 108–132 (2024)
10. R. Glowinski, J.-L. Lions, J. He, *Exact and Approximate Controllability for Distributed Parameter Systems: a Numerical Approach*, vol. 117. Encyclopedia of Mathematics and Its Applications (Cambridge University Press, Cambridge, UK; New York, 2008)
11. X. Li, S. Song, *Impulsive Systems with Delays* (Springer, Singapore, 2022)
12. L. Maniar, M. Meyries, R. Schnaubelt, Null controllability for parabolic equations with dynamic boundary conditions. Evol. Equ. Cont. Theo. **6**, 381–407 (2017)
13. K.D. Phung, Carleman commutator approach in logarithmic convexity for parabolic equations. Math. Control Rel. Fields **8**, 899–933 (2018)
14. K.D. Phung, G. Wang, Y. Xu, Impulse output rapid stabilization for heat equations, J. Differ. Equ. **263**, 5012–5041 (2017)
15. T.M.N. Vo, Construction of a control and reconstruction of a source for linear and nonlinear heat equations. PhD thesis, Orléans University, 2018
16. T.M.N. Vo, The local backward heat problem (2017). arXiv:1704.05314

Part II
Related Fields

Decoding the Kramers–Fokker–Planck Operator: An Overview

Mona Ben Said

2010 Mathematics Subject Classification 35Q84, 35H20, 35P05, 47A10, 14P10

1 Introduction

The Fokker–Planck equation was first derived in the 1910s by Andriaan Fokker [13] and subsequently by Max Planck [36] to describe the time evolution of the probability density function of the velocity of a particle undergoing Brownian motion [38].

The history of Brownian motion can be traced back to the observations made by Robert Brown, a Scottish botanist, in 1827. While looking through a microscope at pollen grains immersed in water, Brown noticed that the grains exhibited erratic and random movements. This phenomenon came to be known as Brownian motion. At the time, the cause of Brownian motion was not fully understood, and it was not until many years later that the underlying explanation was provided by Albert Einstein in 1905 in his paper [12]. Einstein's explanation, based on the kinetic theory of gases and statistical mechanics, established that the random motion of particles in a fluid is due to the thermal motion of molecules colliding with the suspended particles. In addition to Albert Einstein's work, the mathematical description and explanation of Brownian motion can also be attributed to Marian Smoluchowski [44] and Paul Langevin [28]. In contrast to Einstein–Smoluchowski's diffusion models, Paul Langevin introduced a stochastic differential equation known as the Langevin equation. The Langevin equation incorporates the concept of particle inertia, which was not considered in the earlier models. Later, the problem was generalized to the Brownian motion in an external force field by Hendrik Kramer in 1940 [27]. The so-called Kramers–Fokker–Planck equation emerged then as a variant of the Fokker–Planck equation that incorporates the effects of both random

M. Ben Said (✉)
LR Analysis and Control of PDEs, UR 22ES03, Department of Mathematics, Faculty of Sciences of Monastir, University of Monastir, Monastir, Tunisia

K. Ammari et al. (eds.), *Advances in Partial Differential Equations and Control*, Trends in Mathematics, https://doi.org/10.1007/978-3-031-62265-6_7

forces (such as Brownian motion) and external force fields. This equation has the form

$$\partial_t u(t;q,p) + K_V u(t;q,p) = 0, \qquad (q,p) \in \mathbb{R}^{2d},\ t > 0, \tag{1}$$

with initial data

$$u(0;q,p) = u_0(q,p). \tag{2}$$

Here, u corresponds to the probability density function, q and p represent, respectively, the position and the velocity of the particle, $x.y = \sum_{j=1}^{d} x_j y_j$, and $x^2 = \sum_{j=1}^{d} x_j^2$. The operator K_V is known as the Kramers–Fokker–Planck operator and is defined as

$$K_V = p\partial_q - \partial_q V(q)\partial_p + \frac{1}{2}(-\Delta_p + p^2), \tag{3}$$

where the potential V(q) is a real-valued function defined in the whole space $\mathbb{R}^d$.

The study of Kramers–Fokker–Planck equations, arising as the partial differential counterpart of Langevin stochastic processes for Brownian motion, encountered a renewal of interest by the mathematic community in the last decade. Several advances have been made for computing analytically and accurately the lifetimes of metastable states in rather general cases. Meanwhile the general theory of associated dynamics has been understood and extended to various geometric settings by the works of Hérau–Nier [19], Bismut–Lebeau [7], Helffer–Nier [16], Eckmann–Hairer [11], and Hérau–Hitrik–Sjostrand [20]. Among those works, the question was asked and partially answered by Helffer and Nier [16], about the relationship between the spectral properties of the Kramers–Fokker–Planck operator K_V (phase-space stochastic motion with diffusion only in the velocity variable) and the ones of the Witten Laplacian $\Delta_V^{(0)} = -\Delta_q + |\nabla V(q)|^2 - \Delta V(q)$ (the partial differential counterpart of the overdamped Langevin process describing the motion of particles in the position variable with a complete diffusion). In [16], as reviewed in Sect. 2.2, the link between the two operators $\Delta_V^{(0)}$ and K_V has been precisely described and resulted in the following conjecture:

Conjecture 1 *The Fokker–Planck operator K_V has a compact resolvent if and only if the Witten Laplacian $\Delta_V^{(0)}$ has a compact resolvent.*

Here, we recall that if these dynamics admit an equilibrium state, namely a function $f \in L^2$ such that $\Delta_V^{(0)} f = 0$ (respectively, $K_V f = 0$), the compactness of the resolvent ensures an exponential return to equilibrium with a strictly positive rate given by a spectral gap. It is also worth mentioning that the necessary part of Conjecture 1 has already been proven by Helffer and Nier (see [16], Theorem 1.1). However, the reverse implication is still open up to now for a general potential, and it is indeed verified under certain conditions on the potential V (see, for example, [16, 19, 29]).

When studying the Witten Laplacian, numerous researchers have made significant contributions to its understanding, including the analysis of the resolvent compactness issue (see, e.g., [14, 16] and [8]). For instance, it was proven in [16] (see Proposition 3) that the operator $\Delta_V^{(0)}$ has a compact resolvent under the condition that

$$t\,|\nabla V(q)|^2 - \Delta V(q) \to +\infty, \qquad \text{as} \quad |q| \to +\infty,$$

for some $t \in]0, 2[$. Additionally, following the Rothschild–Stein [39] and Helffer–Nourrigat [18] nilpotent Lie groups approach, a thorough study has been carried out by Helffer–Nier in [16] which shows that if no nonzero polynomial from a certain canonical set containing V has a local minimum, then the Witten Laplacian $\Delta_V^{(0)}$ has compact resolvent. In particular, this study highlights the influence of the sign of V on the compactness of the resolvent of $\Delta_V^{(0)}$. The most well-known example in this regard is $V(q_1, q_2) = q_1^2 q_2^2$ in dimension 2. Indeed, for this example, the operator $\Delta_{-V}^{(0)}$ has a compact resolvent, whereas $\Delta_{+V}^{(0)}$ does not.

On the other side, the difficulty in studying the Kramers–Fokker–Planck operator K_V arises from the fact that it is neither self-adjoint nor elliptic. Several works have been dedicated to investigating the case of a polynomial potential V(q) with a degree less than or equal to 2 (see [1, 21, 23, 42, 43]). Nevertheless, as soon as a general potential is considered, different types of sufficient conditions on $V(q)$ have been examined by Hérau–Nier [19], Helffer–Nier [16], Villani [41], and Wei-Xi Li [29]. These first results considered only variations of elliptic situation at infinity (for a nondegenerate potential), which do not distinguish the sign $\pm V(q)$. Then, a significant improvement of these works has been made by Wei-Xi Li [30] based on some multiplier methods. In his work [30], Wei-Xi Li provided the compactness of the resolvent of the Kramers–Fokker–Planck operator, under some sufficient conditions relying primarily on the sign of the eigenvalues of the Hessian matrix $(\partial_{q_i q_j} V)_{1 \le i,j \le d}$. His contribution presents the first results reflecting the microlocal dependency property on the sign of V for the compactness of the resolvent of the Kramers–Fokker–Planck operator K_V, which plays an important role in the analysis of the compactness of the resolvent for the Witten Laplacian. These initial results are reviewed and discussed in Sect. 3.1.

In this vein, well-satisfied sufficient conditions on V so that K_V has a compact resolvent were provided in the recent works [3, 5] and [4]. The thorough analysis presented in these works bears a strong resemblance to the compactness criteria already established for the Witten Laplacian $\Delta_V^{(0)}$, thus supporting the idea that Conjecture 1 is true. These contributions are reviewed in detail in Sects. 3.2 and 3.3.

Overall, this problem contains hence many aspects from the mathematical analysis point of view: degenerate ellipticity (hypoellipticity and subellipticity); Lie group approach (with Hörmander condition for hypoellipticity) and nilpotent Lie group (Rotschild–Stein [39] and Helffer–Nourrigat maximal hypoellipticity approach [16, 18, 32]); extension to the Riemannian differential setting via Witten Laplacians and Bismut hypoelliptic Laplacian; and semiclassical spectral analysis since the low temperature limit (small temperature case for chemists, physicists, and all sciences referring to Brownian motion) is nothing but a semiclassical limit.

2 Analysis of the Kramers–Fokker–Planck Operator

2.1 *Maximal Accretivity*

Generally speaking, a linear operator A in a Hilbert space H with domain D(A) and range $\mathrm{Im}(A)$ is said to be accretive if

$$\mathrm{Re}\langle Au, u\rangle_H \geq 0, \qquad \text{for every} \quad u \in D(A),$$

or equivalently if

$$\|(A+\lambda)u\| \geq \lambda\|u\| \qquad \text{for all } u \in D(A) \text{ and } \lambda > 0.$$

If in particular $\mathrm{Im}(A+\lambda)$ is dense in H for some (and hence for every) $\lambda > 0$, we say that A is essentially maximal-accretive and its closure $\overline{A}$ is said to be maximal-accretive. Note that in this case $-\overline{A}$ is the infinitesimal generator of a contraction semigroup. Moreover, the adjoint of a linear maximal-accretive operator in a Hilbert space H is likewise maximal-accretive. For further results on the concept of maximal accretivity, we refer the reader to the books by Dautray–Lions [9] (Volume 5, Chapter XVII), Reed–Simon [37], and Davies [10]. In the context of our study, we have the following result.

Proposition 1 *Let V be a $\mathcal{C}^\infty$ potential on $\mathbb{R}^d$, then the closure of the Kramers–Fokker–Planck operator defined on $\mathcal{C}_0^\infty(\mathbb{R}^{2d})$ by*

$$K_V = p\partial_q - \partial_q V(q)\partial_p + \frac{1}{2}(-\Delta_p + p^2) \tag{4}$$

is maximally accretive.

Proposition 1 is proven in detail in [16] (see Proposition 5.5, page 45).

Remark 1 As $(K_V, \mathcal{C}_0^\infty(\mathbb{R}^{2d}))$ is essentially maximal-accretive, the domain of its closure is given by

$$D(K_V) = \left\{u \in L^2(\mathbb{R}^{2d}),\ K_V u \in L^2(\mathbb{R}^{2d})\right\}.$$

Since $(K_V, D(K_V))$ is a maximal-accretive operator, its numerical range, defined as the set $\{\langle u, K_V u\rangle,\ u \in D(K_V)\}$, is *a priori* the full closed half plane $\{z \in \mathbb{C},\ \mathrm{Re} z \geq 0\}$. Hence, the spectrum of K_V, denoted by $\sigma(K_V)$, is such that

$$\sigma(K_V) \subset \{z \in \mathbb{C}, \ \mathrm{Re}z \geq 0\}.$$

For explicit computations of the spectrum of the quadratic Kramers–Fokker–Planck operator, we recommend referring to Section 5.5.1 of the book authored by Helffer and Nier [16]. Additionally, for a detailed description of the spectrum of more general quadratic operators, we suggest consulting [40] and [21].

Within this context, one of the basic questions that occurs is about the exponential decay to equilibrium of e^{-tK_V} with respect to time $t \geq 0$. By this, we mean the existence of $C, \tau > 0$ such that

$$\|e^{-tK_V} - \Pi_0\|_{\mathcal{L}(L^2(\mathbb{R}^{2d}))} \leq Ce^{-\tau t},$$

where Π_0 denotes the spectral projection on

$$\mathrm{Ker}\, K_V = \begin{cases} \mathbb{C}m & \text{if } m \in D(K_V) \\ \{0\} & \text{otherwise,} \end{cases}$$

with

$$m(q,p) = e^{-\frac{1}{2}(\frac{p^2}{2} + V(q))}.$$

Owing to

$$(K_V - z)^{-1} = \int_0^{+\infty} e^{-t(K_V - z)}dt, \qquad \mathrm{Re}\, z < 0,$$

and

$$e^{-tK_V} = \frac{1}{2i\pi}\int_{-i\infty}^{+i\infty} e^{-tz}(z - K_V)^{-1}dz,$$

we remark that in order to study the exponential decay of e^{-tK_V}, an efficient method is to investigate the spectral gap, which is reduced to analyze the compactness of resolvent of the Kramers–Fokker–Planck operator.

2.2 *Links with Witten Laplacians*

By virtue of Simader's theorem, which states that a nonnegative Schrödinger operator $-\Delta + W(x)$ with $W \in \mathcal{C}^\infty(\mathbb{R}^d)$ is essentially self-adjoint on $\mathcal{C}_0^\infty(\mathbb{R}^d)$, it follows that for any $V \in \mathcal{C}^\infty(\mathbb{R}^d)$, the Witten Laplacian $\Delta_V^{(0)} = -\Delta_q + |\nabla V(q)|^2 - \Delta V(q)$ is essentially self-adjoint on $\mathcal{C}_0^\infty(\mathbb{R}^d)$. In other words, it has a unique self-adjoint extension, and its self-adjoint closure $\Delta_V^{(0)}$ has the domain

$$D(\Delta_V^{(0)}) = \left\{ u \in L^2(\mathbb{R}^d),\ \Delta_V^{(0)} u \in L^2(\mathbb{R}^d) \right\}.$$

In the context of the analysis of the Kramers–Fokker–Planck operator K_V, there is a strong correlation with the Witten Laplacian $\Delta_V^{(0)}$, as highlighted in [19, 32], and [16]. In particular, in [16], Helffer–Nier established the following connections between the two operators $(K_V, D(K_V))$ and $(\Delta_V^{(0)}, D(\Delta_V^{(0)}))$.

Proposition 2 *Assume that V is a $\mathcal{C}^\infty$ function.*

- *If the operator K_V has a compact resolvent, then $\Delta_V^{(0)}$ has a compact resolvent.*
- *If 0 belongs to the essential spectrum of $\Delta_V^{(0)}$, then 0 is in the essential spectrum of K_V.*

We hereby recall that $\lambda \in \mathbb{C}$ belongs to the essential spectrum of an operator $(A, D(A) \subseteq L^2(\mathbb{R}^d))$ if there exists an orthonormal sequence $(u_k)_{k\in\mathbb{N}}$ in $L^2(\mathbb{R}^d)$, with $u_k \in D(A)$, such that $\lim_{k\to+\infty} \|(A - \lambda)u_k\|_{L^2(\mathbb{R}^d)} = 0$. This condition is generally sufficient, and it is a necessary and sufficient condition if the operator A is self-adjoint. For a detailed proof of Proposition 2, we refer the reader to [16] (see Proposition 5.19, page 58).

In connection with Proposition 2, Helffer and Nier put forward Conjecture 1. While Conjecture 1 remains open in general, significant progress has been made in understanding this relationship under certain conditions on the potential $V(q)$. In the subsequent section, we review and discuss several contributions that are connected to this conjecture, providing an analysis of their findings and implications.

3 Compactness Criteria for the Resolvent of Kramers–Fokker–Planck Operator

In this section we consider the Kramers–Fokker–Planck operator

$$K_V = p\partial_q - \partial_q V(q)\partial_p + \frac{1}{2}(-\Delta_p + p^2), \qquad (q, p) \in \mathbb{R}^{2d}, \tag{5}$$

with domain

$$D(K_V) = \left\{ u \in L^2(\mathbb{R}^{2d}),\ K_V u \in L^2(\mathbb{R}^{2d}) \right\},$$

where the potential $V(q)$ is a real-valued function defined in the whole space $\mathbb{R}^d_q$.

Notation Throughout this section, we use the notation

$$\langle\cdot\rangle = \sqrt{1 + |\cdot|^2}\,.$$

3.1 Initial Known Results

The analysis of the compactness of the resolvent is performed using subelliptic estimates. At least two types of methods can be considered: (1) the Kohn method as adapted by Hérau–Nier [19], Helffer–Nier [16], and Wei-Xi Li [29, 30] or in the hypocoercive methods of Villani [41], and (2) the Helffer–Nourrigat method, as used, for example, to treat the Witten Laplacian, which involves a structure of nilpotent Lie group (see [18, 39]). The first method does not provide optimal exponents, unlike the second method, which thus provides hints for searching for counterexamples. In related problems, we also note the work of Helffer–Mohamed [15].

The Kohn method was then used by Hérau–Nier [19] to study the case of potentials $V(q)$ that behave like homogeneous functions at infinity. More precisely, they considered the case where $V(q)$ is a $\mathcal{C}^\infty$ function that satisfies the following assumption:

Assumption 5

- *There exist $n \geq 1$ and, for all $\alpha \in \mathbb{N}^d$, a positive constant C_α so that*

$$\forall q \in \mathbb{R}^d, \quad |\partial_q^\alpha V(q)| \leq C_\alpha\Big(1 + \langle q\rangle^{2n-\min\{|\alpha|,2\}}\Big).$$

- *There exist two constants $C_0 = C_0(V) > 0$ and $C_1 = C_1(V) > 0$ such that*

$$\forall q \in \mathbb{R}^d, \quad \pm V(q) \geq C_0^{-1}\langle q\rangle^{2n} - C_0 \quad \text{and} \quad |\partial_q V(q)| \geq C_1^{-1}\langle q\rangle^{2n-1} - C_1.$$

Under Assumption 5, Hérau–Nier [19] established the following isotropic subelliptic estimate using pseudo-differential calculus.

Theorem 1 *If $V(q)$ satisfies Assumption 5, then there is a constant $C_V > 0$ such that*

$$\|\Lambda^\epsilon u\|^2_{L^2(\mathbb{R}^{2d})} \leq C_V\Big(\|K_V u\|^2_{L^2(\mathbb{R}^{2d})} + \|u\|^2_{L^2(\mathbb{R}^{2d})}\Big), \qquad \forall u \in \mathcal{C}_0^\infty(\mathbb{R}^{2d}) \tag{6}$$

$$\epsilon = \min(\frac{1}{4}, \frac{1}{4n-2}), \qquad \Lambda = \Big(1 - \Delta_p - \Delta_q + |\partial_q V(q)|^2 - \Delta V(q) + \frac{1}{2}|p|^2\Big)^{\frac{1}{2}}.$$

By expanding upon the results and approach of Hérau–Nier, Helffer–Nier [16] obtained the following estimate:

$$\|\Lambda^{\frac{1}{4}} u\|^2_{L^2(\mathbb{R}^{2d})} \leq C_V\Big(\|K_V u\|^2_{L^2(\mathbb{R}^{2d})} + \|u\|^2_{L^2(\mathbb{R}^{2d})}\Big), \tag{7}$$

for more general potentials $V \in \mathcal{C}^\infty$ satisfying the following hypothesis.

Assumption 6

$$\exists C, M \geq 1, \quad \forall q \in \mathbb{R}^d, \quad C^{-1}\langle q\rangle^{\frac{1}{M}} \leq \langle \partial_q V(q)\rangle \leq C\langle q\rangle^M,$$

$$\forall \alpha \in \mathbb{N}^d, \quad |\alpha| \geq 1, \exists C_\alpha \geq 1, \quad |\partial_q^\alpha V(q)| \leq C_\alpha \langle \partial_q V(q)\rangle.$$

Here we note that the exponent $\frac{1}{4}$ in the estimate (7) is not optimal. A better exponent, which appears to be $\frac{2}{3}$ as seen in [39], can be obtained by explicit methods in the particular case where $V(q)$ is a nondegenerate quadratic form (see [16], page 62). In [16], Helffer–Nier also studied the case where $V(q)$ satisfies the following assumption:

Assumption 7

$$\forall \alpha \in \mathbb{N}^d, \ |\alpha| = 2, \quad |\partial_q^\alpha V(q)| \leq C_\alpha \langle \partial_q V(q)\rangle^{1-\varrho_0} \quad \textit{with} \quad \varrho_0 > \frac{1}{3},$$

and we obtained the following estimate:

$$\||\partial_q V(q)|^{\frac{2}{3}} u\|^2_{L^2(\mathbb{R}^{2d})} \leq C_V \Big(\|K_V u\|^2_{L^2(\mathbb{R}^{2d})} + \|u\|^2_{L^2(\mathbb{R}^{2d})} \Big). \tag{8}$$

This result generalizes in particular the case of nondegenerate quadratic forms. Moreover, the exponent $\frac{2}{3}$ in (8) is better than $\frac{1}{4}$ appearing in the estimate (7). Afterward, Wei-Xi Li proved in [29] under ellipticity assumptions the following result.

Theorem 2 *Let $V(q) \in \mathcal{C}^2(\mathbb{R}^d)$ be a real-valued function that satisfies, for an $s < \frac{4}{3}$, the property:*

$$\forall |\alpha| = 2\,, \exists C_\alpha > 0\,, \quad |\partial_q^\alpha V(q)| \leq C_\alpha \Big(1 + |\partial_q V(q)|^2\Big)^{\frac{s}{2}}. \tag{9}$$

Then there exist $C > 0$ and $\delta > 0$ such that for every $u \in \mathcal{C}_0^\infty(\mathbb{R}^{2d})$,

$$\||\partial_q V(q)|^{\frac{2}{3}} u\|_{L^2(\mathbb{R}^{2d})} \leq C\Big(\|K_V u\|_{L^2(\mathbb{R}^{2d})} + \|u\|_{L^2(\mathbb{R}^{2d})} \Big),$$

and

$$\begin{aligned} &\|(1-\Delta_q)^{\frac{\delta}{2}} u\|_{L^2(\mathbb{R}^{2d})} + \|(1-\Delta_p + p^2)^{\frac{1}{2}} u\|_{L^2(\mathbb{R}^{2d})} \\ &\leq C\Big(\|K_V u\|_{L^2(\mathbb{R}^{2d})} + \|u\|_{L^2(\mathbb{R}^{2d})} \Big), \end{aligned}$$

where δ equals $\frac{2}{3}$ if $s \leq \frac{2}{3}$, $\frac{4}{3} - s$ if $\frac{2}{3} < s \leq \frac{10}{9}$, and $\frac{2}{3} - \frac{s}{2}$ if $\frac{10}{9} < s < \frac{4}{3}$.

According to this theorem, if we take in particular $s = \frac{2}{3}$, we obtain the following optimal hypoelliptic estimate:

$$\||\partial_q V(q)|^{\frac{2}{3}} u\|_{L^2(\mathbb{R}^{2d})} + \|(1-\Delta_q)^{\frac{2}{3}} u\|_{L^2(\mathbb{R}^{2d})} \leq C(\|K_V u\|_{L^2(\mathbb{R}^{2d})} + \|u\|_{L^2(\mathbb{R}^{2d})}).$$

Recall that the criterion established by Helffer–Nier, stated in the second point of Theorem 10.16 in [16], demonstrates the significance of the sign of $V(q)$ for the compactness of the resolvent of the Witten Laplacian. As for the Kramers–Fokker–Planck operator, the Helffer–Nier Conjecture 1 strongly suggests that it should possess the same microlocal property as the Witten Laplacian. We note here that the drawback of condition (9) is that it provides no information regarding the dependence of the sign of $V(q)$.

However, Wei-Xi Li has provided in [30] (Theorems 1.2 and 1.3) certain sufficient conditions for the compactness of the resolvent of the Kramers–Fokker–Planck operator, primarily based on the sign of the eigenvalues of the Hessian matrix $(\partial_{q_i q_j} V)_{1\leq i,j\leq d}$. Thus, these advancements present the initial results reflecting this dependence property, which plays a crucial role in the analysis of the resolvent's compactness for the Witten Laplacian.

The following two results are taken from [30] (Theorems 1.2 and 1.3).

Theorem 3 *Let $V(q) \in \mathcal{C}^2(\mathbb{R}^d)$. Denote by $\lambda_l(q)$, where $1 \leq l \leq d$, the eigenvalues of the Hessian matrix*

$$(\partial_{q_i q_j} V(q))_{1\leq i,j\leq d}.$$

With each $q \in \mathbb{R}^d$, we associate a set I_q of indexes defined by

$$I_q = \{1 \leq l \leq d\ ,\ \textit{tel que}\ \ \lambda_l(q) > 0\}\,.$$

If there exists a constant $c > 0$ such that

$$\forall q \in \mathbb{R}^d, \qquad \sum_{j\in I_q} \lambda_j(q) \leq c\langle \partial_q V(q)\rangle^{\frac{4}{3}}, \tag{10}$$

then the following conclusions hold:

(i) There exists a constant $c > 0$ such that for all $u \in \mathcal{C}_0^\infty(\mathbb{R}^{2d})$,

$$\||\partial_q V(q)|^{\frac{1}{16}} u\|_{L^2(\mathbb{R}^{2d})} \leq c\Big(\|K_V u\|_{L^2(\mathbb{R}^{2d})} + \|u\|_{L^2(\mathbb{R}^{2d})}\Big). \tag{11}$$

As a result, when the limit $\lim_{|q|\to+\infty} |\partial_q V(q)| = +\infty$ is satisfied, it implies the compactness of the resolvent for the Fokker–Planck operator K_V.

(ii) If we suppose furthermore the existence of a constant $\alpha \geq 0$ such that

$$\lim_{|q|\to+\infty}(\alpha|\partial_q V(q)|^2 - \Delta_q V(q)) = +\infty, \tag{12}$$

then there exists a constant $\widetilde{c}_\alpha > 0$ *such that for all* $u \in C_0^\infty(\mathbb{R}^{2d})$,

$$\| |\alpha|\partial_q V(q)|^2 - \Delta_q V(q)|^{\frac{1}{80}} u\|_{L^2(\mathbb{R}^{2d})} \leq \widetilde{c}_\alpha \Big(\|K_V u\|_{L^2(\mathbb{R}^{2d})} + \|u\|_{L^2(\mathbb{R}^{2d})} \Big).$$

Consequently, the resolvent of the Kramers–Fokker–Planck operator K_V *is compact.*

Theorem 4 *Let* $V(q) \in C^2(\mathbb{R}^d)$*. Suppose that there exists a constant* $\tau > 0$ *such that the matrix*

$$A_\tau(q) = (a_{ij}^\tau(q))_{1\leq i,j\leq d}, \qquad a_{ij}^\tau = \tau \langle \partial_q V\rangle^{\frac{4}{5}} (\partial_{q_i} V)(\partial_{q_j} V) - \partial_{q_i q_j} V + \tau \delta_{ij}$$

is positive-definite for all $q \in \mathbb{R}^d$*, where* δ_{ij} *denotes the Kronecker Delta. Then there is a constant* $c > 0$ *such that for all* $u \in C_0^\infty(\mathbb{R}^{2d})$,

$$\begin{aligned} &\| |\partial_q V(q)|^{\frac{1}{20}} u\|_{L^2(\mathbb{R}^{2d})} + \sum_{1\leq i,j\leq d} \| |a_{ij}^\tau(q)|^{\frac{1}{80}} u\|_{L^2(\mathbb{R}^{2d})} \\ &\leq c\Big(\|K_V u\|_{L^2(\mathbb{R}^{2d})} + \|u\|_{L^2(\mathbb{R}^{2d})} \Big). \end{aligned}$$

Therefore, if $\lim_{|q|\to+\infty}\Big(|\partial_q V(q)| + \sum_{1\leq i,j\leq d} |a_{ij}^\tau(q)| \Big) = +\infty$*, the Kramers–Fokker–Planck operator has a compact resolvent.*

In the case where $V(q) = -q_1^2 q_2^2$, it is well known (see Proposition 10.19 in [16]) that the Witten Laplacian $\Delta_V^{(0)}$ has a compact resolvent, while 0 belongs to the essential spectrum of $\Delta_{-V}^{(0)}$ and its resolvent cannot be compact. Considering the same aforementioned potential, we observe that the assumptions of Theorems 3 and 4 are valid for $V(q) = -q_1^2 q_2^2$ and violated by $V(q) = q_1^2 q_2^2$. Note that the estimate (11) does not provide any information about the compactness of the resolvent of the Kramers–Fokker–Planck operator associated with $V(q_1, q_2) = -q_1^2 q_2^2$ (since $\partial_q V(q)$ vanishes along the lines $q_1 = 0$ and $q_2 = 0$). In this case, we use item (ii) of Theorem 3. Indeed, the considered potential satisfies hypothesis (ii) of Theorem 3: For all $\alpha \geq 0$,

$$\lim_{|q|\to+\infty}(\alpha|\partial_q V(q)|^2 - \Delta_q V(q)) = \lim_{|q|\to+\infty}(4\alpha q_1^2 q_2^2 |q|^2 + 2|q|^2) = +\infty$$

(in the statement of Theorem 3, it suffices to find a single constant $\alpha \geq 0$ that satisfies the hypothesis). The detailed proofs of both aforementioned theorems can be found in [30], where they heavily rely on multiplier techniques.

With the goal of developing a systematic study of degenerate Kramers–Fokker–Planck operators, the research conducted by the author and her collaborators, as documented in [3, 5], and [4], continues the line of inquiry established by the previous studies presented in this subsection.

3.2 *Case of Polynomial Potential*

In this subsection we consider the Kramers–Fokker–Planck operator (5) with a polynomial potential $V(q)$. To simplify the written expressions, we introduce notations for its real and imaginary parts as follows.

Notation 1

$$O_p = \frac{1}{2}(D_p^2 + p^2), \qquad X_V = p\partial_q - \partial_q V(q)\partial_p.$$

Degree Less than 3

The investigation of certain quadratic differential operators, which are neither self-adjoint nor elliptic, has been the subject of several scholarly works. The existing literature, encompassing references such as [21–23, 40, 42, 43], and [1], delves into various aspects of these operators and offers valuable insights into their properties and behavior. Among these operators, considerable attention has been given to the Kramers–Fokker–Planck operator, which incorporates a polynomial potential $V(q)$ of degree less than or equal to 2. In our research conducted jointly with F. Nier and J. Viola [5], we have made notable contributions by establishing global subelliptic estimates for the Kramers–Fokker–Planck operator K_V with a polynomial potential $V(q)$ of degree less than or equal to 2. What is particular about our results, which are summarized in Theorems 5 and 6, is the uniform control of the inequalities with respect to the coefficients of the polynomial V. In our proofs, we draw inspiration from the techniques developed by Hitrik, Pravda-Starov, Viola, and Aleman in [21, 42, 43], and [1].

To state our results accurately, we need to introduce the following notations. First, consider a real-valued polynomial potential $V(q) = \sum_{|\alpha|\leq 2} V_\alpha q^\alpha$ defined on $\mathbb{R}^d$ with $d^\circ V = 2$. By using an orthogonal change of variables, we can assume that the Hessian of V takes the form:

$$\text{Hess } V = \begin{pmatrix} \nu_1 & 0 & \dots & 0 \\ 0 & \nu_2 & \dots & 0 \\ \vdots & \vdots & \ddots & \vdots \\ 0 & 0 & \dots & \nu_d \end{pmatrix}.$$

Note that the constant term V_0 has no impact on the expression of the operator K_V and can thus be set to 0. Consequently, we distinguish two cases:

- If Hess V is nondegenerate, a translation in q reduces the problem to

$$V(q) = \sum_{i=1}^{d} \frac{\nu_i}{2} q_i^2 . \tag{13}$$

- If Hess V is degenerate, a suitable choice of an orthonormal basis and a translation yields

$$V(q) = \lambda_1 q_1 + \sum_{i=2}^{d} \frac{\nu_i}{2} q_i^2 , \tag{14}$$

  where λ_1 is defined by $|\lambda_1| = \min_{q \in \mathbb{R}^d} |\nabla V(q)| \geq 0$.

Now, let us introduce the following notations:

Notations 1

$$\begin{aligned}
\mathrm{Tr}_+ &= \sum_{\nu_i > 0} \nu_i \,, \\
\mathrm{Tr}_- &= -\sum_{\nu_i \leq 0} \nu_i \,, \\
A &= \max\{(1 + \mathrm{Tr}_+)^{2/3}, 1 + \mathrm{Tr}_-\} \,, \\
B &= \max\{|\lambda_1|^{4/3}, \frac{1 + \mathrm{Tr}_-}{(\log(2 + \mathrm{Tr}_-))^2}\}.
\end{aligned}$$

The subelliptic estimates presented in [5] are as follows.

Theorem 5 *Let $V(q)$ be a potential as in (13) or (14). Then there exists a constant $c > 0$ that does not depend on V such that the subelliptic estimate with a remainder term*

$$\begin{aligned}
\|K_V u\|^2_{L^2(\mathbb{R}^{2d})} + A\|u\|^2_{L^2(\mathbb{R}^{2d})} \geq c\Big(&\|O_p u\|^2_{L^2(\mathbb{R}^{2d})} + \|X_V u\|^2_{L^2(\mathbb{R}^{2d})} \\
&+ \|\langle \partial_q V(q)\rangle^{2/3} u\|^2_{L^2(\mathbb{R}^{2d})} + \|\langle D_q\rangle^{2/3} u\|^2_{L^2(\mathbb{R}^{2d})}\Big)
\end{aligned} \tag{15}$$

holds for all $u \in D(K_V)$.

It is worth noting that the compactness of the resolvent of K_V can be directly inferred from Theorem 5 when $|\partial_q V(q)|$ tends to infinity as $|q|$ goes to infinity.

Theorem 6 *Let $V(q)$ be as in (13) or (14). Then there is a constant $c > 0$ independent of the polynomial V so that the subelliptic estimate without a remainder*

$$\|K_V u\|^2_{L^2(\mathbb{R}^{2d})} \geq \frac{c}{1+\frac{A}{B}}\Big(\|O_p u\|^2_{L^2(\mathbb{R}^{2d})} + \|X_V u\|^2_{L^2(\mathbb{R}^{2d})}$$
$$+ \|\langle \partial_q V(q)\rangle^{2/3} u\|^2_{L^2(\mathbb{R}^{2d})} + \|\langle D_q\rangle^{2/3} u\|^2_{L^2(\mathbb{R}^{2d})}\Big) \quad (16)$$

holds for all $u \in D(K_V)$.

These results are demonstrated using regularization estimates of the associated semigroup. Specifically, both of the aforementioned theorems were derived as consequences of the following result.

Proposition 3 *Let $V(q)$ be as in (13) or (14). Then there exists a constant $c > 0$, independent of V, such that*

$$\sum_{i=1}^{d}\Big(\||D_{q_i}|e^{-t(K_V+\sqrt{A})}\|_{\mathcal{L}(L^2(\mathbb{R}^{2d}))} + \||\partial_{q_i}V(q)|e^{-t(K_V+\sqrt{A})}\|_{\mathcal{L}(L^2(\mathbb{R}^{2d}))}\Big) \leq \frac{c}{t^{\frac{3}{2}}}$$

for all $t > 0$.

Moreover,

$$\|K_V^{-1}\|_{\mathcal{L}(L^2(\mathbb{R}^{2d}))} \leq \int_0^{+\infty} \|e^{-tK_V}\|_{\mathcal{L}(L^2(\mathbb{R}^{2d}))}dt \leq \frac{c}{\sqrt{B}}.$$

The core idea of the proof of the previously mentioned results can be summarized as follows. Our approach relies on the utilization of two main types of arguments:

- Since the estimates in Proposition 3 are expressed in terms of the semigroup $(e^{-tK_V})_{t\geq 0}$, their proof can be accomplished by a separation of variables for a polynomial of the form (13) or (14). Indeed, e^{-tK_V} can be regarded as a composition of commutative contraction semigroups associated with each variable (q_j, p_j), and it suffices to write

$$\sum_{i=1}^{d} \|M_i e^{-t(K_V+\sqrt{A})}\|_{\mathcal{L}(L^2(\mathbb{R}^{2d}))} \leq \sum_{i=1}^{d} \|M_i e^{-t(K_{V_i(q_i)}+\alpha_i)}\|_{\mathcal{L}(L^2(\mathbb{R}^{2}))}$$
$$M_i = |D_{q_i}| \text{ or } M_i = |\partial_{q_i} V(q)|,$$

where $V_i(q_i)$ denotes the one-dimensional potential in the q_i variable, with $V_1(q_1) = \frac{\nu_1 q_1^2}{2}$ or $V_1(q_1) = \lambda_1 q_1$, $V_i(q_i) = \frac{\nu_i q_i^2}{2}$ for $i \geq 2$, $\alpha_i = |\nu_i|^{1/2}$ if $\nu_i < 0$, $\alpha_i = \nu_i^{1/3}$ if $\nu_i > 0$, and $\alpha_i = 0$ if $\partial^2_{q_i} V = 0$.

The second estimate of Proposition 3 can be demonstrated even in a simpler manner. Consequently, Proposition 3 is the outcome of a thorough analysis of

three one-dimensional potentials: $V(q) = \pm\frac{\nu q^2}{2}$ with $\nu > 0$, and $V(q) = \lambda_1 q$ for $\lambda_1 \in \mathbb{R}$. Once the problem is reduced to one dimension, our results and formulas are explicitly expressed using biquaternions (or complex quaternions).
- The second argument consists of using the interpolation results of Lunardi (see Remark 5.11, Theorem 5.12, and Corollary 5.13 in [31]). Indeed, by combining the first inequality of Proposition 3 with the fact that

$$|\mathbb{R}\mathrm{e}\langle [O_p, X_V]u, u\rangle| \le C_\epsilon(\| \, |D_q|^{\frac{2}{3}}u\|^2 + \| \, |\partial_q V(q)|^{\frac{2}{3}}u\|^2) + \epsilon\|O_p u\|^2,$$

for every $u \in D(K_V)$ (where $\epsilon > 0$ is sufficiently small), we obtain the estimate of Theorem 5. Then, the first estimate of Theorem 5 is just a rewriting of the resolvent estimate given in Proposition 3. By combining the resolvent estimate with the result of Theorem 5, we obtain the subelliptic estimate without the remainder of Theorem 6.

Once the case of polynomials of degree less than or equal to 2 has been precisely addressed, we can tackle subelliptic estimates with potentials of higher degree, but whose asymptotic models at infinity after scaling are of degree less than or equal to 2. The logarithmic corrections, which cannot be eliminated in the quadratic case, are fortunately absorbed by the Tarski–Seidenberg theorem, a very classical tool in an algebraic framework (polynomial potential). This is the subject of the following subsection.

Degree Greater than 2

This part covers the content of the article [3], which provides a study of the Kramers–Fokker–Planck operator K_V associated with a polynomial potential $V(q)$ of degree $r \ge 3$ having quadratic limiting behavior. In the following, we present the main result from [3] and outline some key elements from its proof. To begin, let us clarify some notations. We denote by $R_V^{\ge n}$, where $n \in \mathbb{N}^*$, the function defined on $\mathbb{R}^d$ by

$$R_V^{\ge n}(q) = \sum_{n \le |\alpha| \le r} |\partial_q^\alpha V(q)|^{\frac{1}{|\alpha|}}.$$

Furthermore, for all $\kappa > 0$, we denote

$$\Sigma(\kappa) = \left\{q \in \mathbb{R}^d, \ |\nabla V(q)|^{\frac{4}{3}} \ge \kappa\Big(|\mathrm{Hess}\ V(q)| + R_V^{\ge 3}(q)^4 + 1\Big)\right\}.$$

We introduce the following hypothesis.

Assumption 8 *There exist two large constants $\kappa_0, C_1 > 1$ such that for all $\kappa \ge \kappa_0$ the polynomial $V(q)$ satisfies the following properties:*

$$\mathrm{Tr}_{-,V}(q) > \frac{1}{C_1}\mathrm{Tr}_{+,V}(q), \quad \textit{for all } q \in \mathbb{R}^d \setminus \Sigma(\kappa) \textit{ with } |q| \geq C_1. \tag{17}$$

Moreover, if $\mathbb{R}^d \setminus \Sigma(\kappa)$ *is not bounded,*

$$\lim_{\substack{|q|\to+\infty \\ q\in\mathbb{R}^d\setminus\Sigma(\kappa)}} \frac{R_V^{\geq 3}(q)^4}{|\mathrm{Hess}\ V(q)|} = 0. \tag{18}$$

The main theorem from [3] can be restated as follows.

Theorem 7 *Let* $V(q)$ *be a polynomial of degree* r *greater than 2 verifying Assumption 8. Then there exists a strictly positive constant* $C_V > 1$ *(depending on* V*) such that*

$$\|K_V u\|_{L^2}^2 + C_V\|u\|_{L^2}^2 \geq \frac{1}{C_V}\Big(\|L(O_p)u\|_{L^2}^2 + \|L(\langle\nabla V(q)\rangle^{\frac{2}{3}})u\|_{L^2}^2$$
$$+ \|L(\langle \mathrm{Hess}\ V(q)\rangle^{\frac{1}{2}})u\|_{L^2}^2 + \|L(\langle D_q\rangle^{\frac{2}{3}})u\|_{L^2}^2\Big) \tag{19}$$

holds for all $u \in D(K_V)$*, where* $L(s) = \frac{s+1}{\log(s+1)}$ *for any* $s \geq 1$*.*

As a consequence of this theorem, if $V(q)$ satisfies Assumption 8, then the Kramers–Fokker–Planck operator K_V has a compact resolvent and, thereby, a discrete spectrum (see Corollary 1.2 in [3]).

It is worth mentioning that one knows that for a potential V satisfying Assumption 8, the Witten Laplacian $\Delta_V^{(0)}$ also has a compact resolvent since the asymptotic models at infinity are of degree less than or equal to 2 without a local minimum (see Theorem 10.16 in [16]).

Furthermore, it is important to highlight that despite the logarithmic loss in (19), we are very close to the optimal exponents shown in the estimate (15).

Before proceeding with the overview of the proof of Theorem 7, it is crucial to provide a deeper understanding of Assumption 8 and, in particular, the underlying concept behind the partition $\mathbb{R}^d = \Sigma(\kappa) \sqcup (\mathbb{R}^d \setminus \Sigma(\kappa))$: The region $\Sigma(\kappa)$ corresponds to the domain where the gradient dominates the Hessian and higher order derivatives. As a result, the analysis conducted in this region is essentially the same as in the various elliptic cases treated in [16, 19], and [29]. On the other hand, in the complementary region $\mathbb{R}^d \setminus \Sigma(\kappa)$, it is the Hessian of V that dominates the gradient and higher order derivatives near infinity. In this region, the accurate estimates of the quadratic model given by the second-order Taylor expansion have to be used. Distinguishing the sign of the potential arises in particular when the region $\mathbb{R}^d \setminus \Sigma(\kappa)$ is considered. Actually $\mathrm{Tr}_{+,V}$ and $\mathrm{Tr}_{-,V}$ play different roles in the accurate subelliptic estimate without remainder (16) for polynomials of degree less than 3. The Tarski–Seidenberg Theorem and some of its consequences reviewed in Appendix 3.B of [3] transform Assumption (18) into

$R_V^{\geq 3}(q)^4 = O(|\text{Hess } V(q)|^{1-\nu})$ as $|q| \to +\infty$, $q \in \mathbb{R}^d \setminus \Sigma(\kappa)$, for some $\nu > 0$ (with $|\text{Hess } V(q)| \to +\infty$ as $|q| \to +\infty$, $q \in \mathbb{R}^d \setminus \Sigma(\kappa)$). Alternatively one could simply assume from the beginning the existence of such a $\nu > 0$. This approach allowed us to absorb the logarithmic corrections that cannot be avoided in the quadratic case. Note here that if we want to extend the result of Theorem 7 to classes of functions $V(q)$ that are not polynomials, we no longer have the right to use the algebraic argument of Tarski–Seidenberg. However, this result obviously remains valid in the case where $V = V_1 + V_2$, where V_1 is a polynomial satisfying Assumption 8, and V_2 is a function in $\mathcal{S}(\mathbb{R}^d)$. Now shifting our focus to the proof of Theorem 7, its foundation primarily rests on two key techniques:

- The first technique consists of constructing a locally finite partition of unity with respect to the position variable $q \in \mathbb{R}^d$

$$\sum_{j\in\mathbb{N}} \chi_j^2(q) = \sum_{j\in\mathbb{N}} \widetilde{\chi}_j^2\Big(R_V^{\geq 3}(q_j)(q-q_j)\Big) = 1, \tag{20}$$

  where

$$\text{supp } \widetilde{\chi}_j \subset B(0,a) \text{ and } \widetilde{\chi}_j \equiv 1 \text{ in } \quad B(0,b),$$

  for some $q_j \in \mathbb{R}^d$ with $0 < b < a$ independent of $j \in \mathbb{N}$. The existence of such a partition of unity is an immediate consequence of the slowness of the metric $R_V^{\geq 3}(q)^2 dq^2$ ($\exists C > 1,\ \forall q, q' \in \mathbb{R}^d,\ \ R_V^{\geq 3}(q)|q-q'| \leq C^{-1} \Rightarrow \left(\frac{R_V^{\geq 3}(q)}{R_V^{\geq 3}(q')}\right)^{\pm 1} \leq C$). This property has been verified in Lemma A.5 of [3]. This allows, in general, to obtain global estimates from a local analysis. Using this partition of unity, we derive the following estimate (see Lemma 2.1 in [3]):

$$\|K_V u\|^2_{L^2(\mathbb{R}^{2d})} \geq \sum_{j\in\mathbb{N}} \Big(\|K_V(\chi_j u)\|^2_{L^2(\mathbb{R}^{2d})} - c_{d,r} R_V^{\geq 3}(q_j)^2 \|p\chi_j u\|^2_{L^2(\mathbb{R}^{2d})}\Big), \tag{21}$$

  for all $u \in C_0^\infty(\mathbb{R}^{2d})$, where the constant $c_{d,r} > 0$ depends only on the dimension d and the degree of V.
- The second approach is centered around decomposing the operator K_V onto two parts so that the first one be a Kramers–Fokker–Planck operator with polynomial potential of degree less than 3. Indeed, it is natural to consider a well-chosen quadratic approximation for $V(q)$ in order to extend the results of [5], which are already established for the case $d^\circ V \leq 2$. For a given $\kappa > 1$ and for any nonnegative integer j, this latter approximation is given by

$$V_j^{(2)}(q) = \sum_{0\leq|\alpha|\leq 2} \frac{\partial_q^\alpha V(q'_j)}{\alpha!}(q-q'_j)^\alpha, \tag{22}$$

where

$$\begin{cases} q_j' = q_j & \text{if } \operatorname{supp} \chi_j \subset \Sigma(\kappa) \\ q_j' \in (\operatorname{supp} \chi_j) \cap \left(\mathbb{R}^d \setminus \Sigma(\kappa)\right) & \text{otherwise.} \end{cases}$$

By associating each polynomial $V_j^{(2)}$ with the Kramers–Fokker–Planck operator $K_{V_j^{(2)}}$ and making use of the parallelogram law $2(|x|^2 + |y|^2) - |x+y|^2 = |x-y|^2 \geq 0$, we arrive at the following inequality:

$$\begin{aligned} &\sum_{j \in \mathbb{N}} \|K_V u_j\|^2_{L^2(\mathbb{R}^{2d})} \\ &\quad = \sum_{j \in \mathbb{N}} \|K_{V_j^{(2)}} u_j + (K_V - K_{V_j^{(2)}}) u_j\|^2_{L^2(\mathbb{R}^{2d})} \\ &\quad \geq \sum_{j \in \mathbb{N}} \left(\frac{1}{2} \|K_{V_j^{(2)}} u_j\|^2_{L^2(\mathbb{R}^{2d})} - \|(\nabla V(q) - \nabla V_j^{(2)}(q)) \partial_p u_j\|^2_{L^2(\mathbb{R}^{2d})}\right), \end{aligned} \tag{23}$$

where $u_j := \chi_j u$.

It turns out that when using the two aforementioned techniques, it is necessary to carefully control the errors arising from both the partition of unity and the quadratic approximation. This is ensured by adjusting the parameter κ at the end of the proof. Additionally, the transition from an estimation for $K_{V_j^{(2)}}$ to an estimation for K_V requires the use of the following result.

Lemma 1 *Assume V is a polynomial of degree r larger than 2. Consider a locally finite partition of unity described as in (20). For a multi-index $\alpha \in \mathbb{N}^d$ of length $|\alpha| \in \{1, 2\}$ and all $j \in \mathbb{N}$, one has*

$$\left|\partial_q^\alpha V(q) - \partial_q^\alpha V_j^{(2)}(q)\right| \leq c_{\alpha,d,r} \left(R_V^{\geq 3}(q_j')\right)^{|\alpha|} \tag{24}$$

for any $q \in \operatorname{supp} \chi_j = B(q_j, a R_V^{\geq 3}(q_j)^{-1})$, where $c_{\alpha,d,r} = \sum_{3 \leq |\beta| \leq r} (a\, C)^{|\beta| - |\alpha|}$.

As a consequence, if V satisfies Assumption 8, there exists a large constant $\kappa_1 \geq \kappa_0$ so that for all $\kappa \geq \kappa_1$ and every $j \in \mathbb{N}$,

$$2^{-1} \left|\partial_q V_j^{(2)}(q)\right| \leq \left|\partial_q V(q)\right| \leq 2 \left|\partial_q V_j^{(2)}(q)\right|, \tag{25}$$

for every $q \in (\operatorname{supp} \chi_j) \cap \Sigma(\kappa)$ and

$$2^{-1} \left|\text{Hess } V_j^{(2)}(q)\right| \leq |\text{Hess } V(q)| \leq 2 \left|\text{Hess } V_j^{(2)}(q)\right|, \tag{26}$$

for any $q \in (\operatorname{supp} \chi_j) \cap \left(\mathbb{R}^d \setminus \Sigma(\kappa)\right)$ *with* $|q| \geq C_2(\kappa)$*, where* $C_2(\kappa) > 0$ *is a large constant that depends on* κ.

We wrap up this subsection by summarizing some applications covered in [3]. In each of the following examples, we examine that Assumption 8 is well fulfilled. We recall here that one knows that for a potential V satisfying Assumption 8, the resolvent of the Witten Laplacian $\Delta_V^{(0)}$ is compact (see Theorem 10.16 in [16]). In the case of the Witten Laplacian, the following examples were in particular considered in [16] (cf. Proposition 10.19 and Proposition 10.21)

Example 1 As a first example of application, let us consider the case

$$V(q_1, q_2) = -q_1^2 q_2^2, \text{ with } q = (q_1, q_2) \in \mathbb{R}^2.$$

By direct computation,

$$\partial_q V(q) = \begin{pmatrix} -2q_1 q_2^2 \\ -2q_2 q_1^2 \end{pmatrix}, \ |\partial_q V(q)| = 2|q_1 q_2||q|,$$

$$\operatorname{Hess} V(q) = \begin{pmatrix} -2q_2^2 & -4q_1 q_2 \\ -4q_1 q_2 & -2q_1^2 \end{pmatrix}, \quad |\operatorname{Hess} V(q)| = 2\sqrt{|q|^4 + 6q_1^2 q_2^2} \asymp |q|^2,$$

$$R_V^{\geq 3}(q) = |4q_2|^{1/3} + |4q_1|^{1/3} + 2 \times 4^{1/4}.$$

Note that the trace of Hess $V(q)$ given by $-2|q|^2$ is negative for all $q \in \mathbb{R}^2 \setminus \{0\}$. Consequently,

$$\operatorname{Tr}_{-,V}(q) > \operatorname{Tr}_{+,V}(q) \quad \text{for all } q \in \mathbb{R}^2, |q| \geq 1.$$

Furthermore, for any $\kappa > 0$, the algebraic set $\mathbb{R}^2 \setminus \Sigma(\kappa)$ is unbounded, as $(0, q_2) \in \mathbb{R}^2 \setminus \Sigma(\kappa)$ for all $q_2 \in \mathbb{R}$. Moreover, for any $\kappa > 1$ arbitrarily chosen, we have

$$\lim_{\substack{|q| \to +\infty \\ q \in \mathbb{R}^2 \setminus \Sigma(\kappa)}} \frac{R_V^{\geq 3}(q)^4}{|\operatorname{Hess} V(q)|} = \lim_{\substack{|q| \to +\infty \\ q \in \mathbb{R}^2 \setminus \Sigma(\kappa)}} \frac{|q|^{4/3}}{|q|^2} = 0,$$

as $R_V^{\geq 3}(q)^4 \leq |q|^{4/3}$ when $|q| \geq 2^3 \times 4^{3/4}$.

Below, we present an example where we depict $\Sigma(800)$ using the color blue (Fig. 1).

Given that Assumption 8 is well satisfied, we can then deduce the compactness of the resolvent of K_V in this case from Corollary 1.2 in [3].

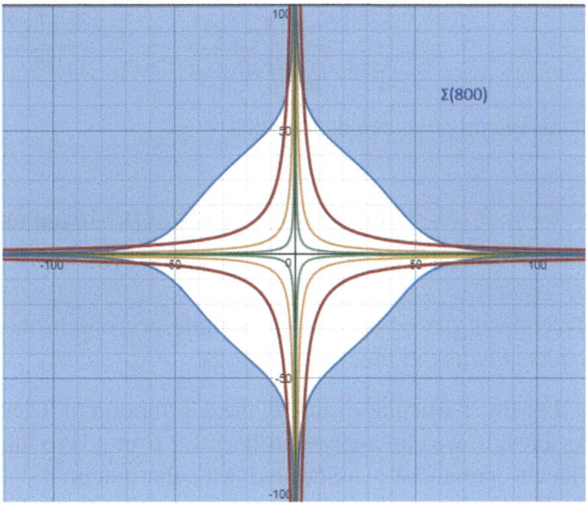

Fig. 1 Contour lines of $V(q_1, q_2) = -q_1^2 q_2^2$

Example 2 Let $n \in \mathbb{N}$. In the case of the polynomial $V(q) = -q_1^2(q_1^2 + q_2^2)^n$, Assumption 8 is only fulfilled for $n = 1$.

By a straightforward computation,

$$\partial_q V(q) = -\begin{pmatrix} 2q_1(|q|^{2n} + nq_1^2|q|^{2(n-1)}) \\ 2nq_2q_1^2|q|^{2(n-1)} \end{pmatrix},$$

$$\text{Hess } V(q) =$$

$$-2|q|^{2(n-2)} \begin{pmatrix} |q|^4 + 5nq_1^2|q|^2 + 2n(n-1)q_1^4 & 2nq_1q_2|q|^2 + 2n(n-1)q_1^3q_2 \\ 2nq_1q_2|q|^2 + 2n(n-1)q_1^3q_2 & nq_1^2|q|^2 + 2n(n-1)q_1^2q_2^2 \end{pmatrix}.$$

Observe that the trace of Hess $V(q)$ given by

$$-2|q|^{2(n-2)}\left(|q|^4 + 5nq_1^2|q|^2 + 2n(n-1)q_1^4 + nq_1^2|q|^2 + 2n(n-1)q_1^2q_2^2\right)$$

is strictly negative for all $q \in \mathbb{R}^2$, $|q| \geq 1$. Then,

$$-\operatorname{Tr}_{-,V}(q) + \operatorname{Tr}_{+,V}(q) < 0, \quad \text{for any } q \in \mathbb{R}^2, \ |q| \geq 1.$$

Moreover, for all $\kappa > 0$, the set $\mathbb{R}^2 \setminus \Sigma(\kappa)$ is unbounded since $(0, q_2) \in \mathbb{R}^2 \setminus \Sigma(\kappa)$ for all $q_2 \in \mathbb{R}$.

Notice that, for q large enough, $|\text{Hess } V(q)| \asymp |q|^{2n}$ and $|D^3V(q)| \asymp |q|^{2n-1}$. Then, for sufficiently large q, we have

$$\frac{R_V^{\geq 3}(q)^4}{|\text{Hess } V(q)|} \asymp \frac{(|q|^{2n-1})^{4/3}}{|q|^{2n}}.$$

Consequently, we can deduce that

$$\lim_{\substack{|q|\to+\infty \\ q\in\mathbb{R}^2\setminus\Sigma(\kappa)}} \frac{R_V^{\geq 3}(q)^4}{|\text{Hess } V(q)|} = 0 \quad \text{if and only if} \quad n < 2.$$

By choosing $\kappa = 800$ as an example, we obtain the shape of $\Sigma(800)$, which is depicted below in blue color (Fig. 2).

For this particular example, the assumption of Theorem 7 holds only for $n = 1$. As a result, we can apply Corollary 1.2 from [3] to infer that the Kramers–Fokker–Planck operator K_V associated with the potential $V(q) = -q_1^2(q_1^2 + q_2^2)$ has a compact resolvent.

Once the work presented in [3] was completed, almost all the examples presented by Helffer and Nier in [16] to compare the Witten Laplacian and the Kramers–Fokker–Planck operator were fully addressed, except for the simple case of homogeneous potentials, not necessarily polynomial, of degree greater than 2. This last point is partially addressed in the author's article [4] and reviewed in the

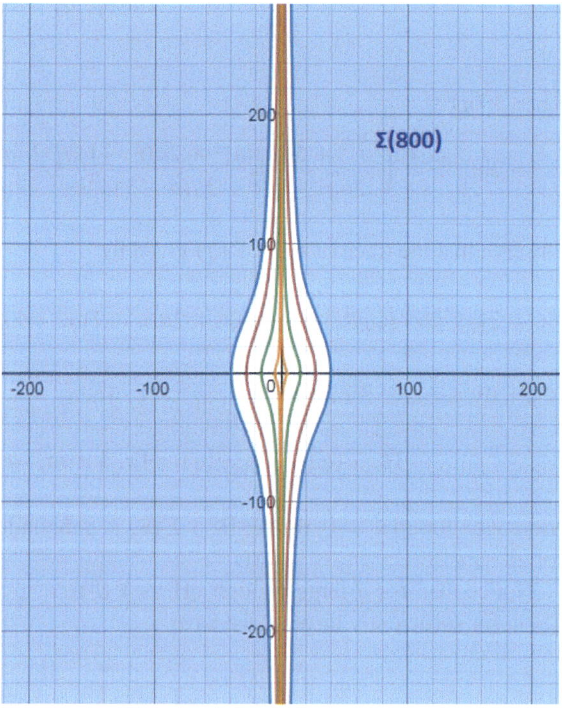

Fig. 2 Contour lines of $V(q_1, q_2) = -q_1^2(q_1^2 + q_2^2)$

following subsection, where homogeneity replaces the more implicit arguments provided by Tarski–Seidenberg and yields better results in the remaining cases.

3.3 Case of Homogeneous Potential

In this subsection we study the Kramers–Fokker–Planck operator defined in (5) where the potential $V(q)$ is a real-valued function defined in the whole space $\mathbb{R}_q^d$.

Setting

$$O_p = \frac{1}{2}(D_p^2 + p^2) \qquad \text{and} \qquad X_V = p\partial_q - \partial_q V(q)\partial_p,$$

the operator K_V reads $K_V = X_V + O_p$.

We firstly list some notations used throughout the part. We denote for an arbitrary function $V(q)$ in $\mathcal{C}^\infty(\mathbb{R}^d)$

$$\mathrm{Tr}_{+,V}(q) = \sum_{\substack{\nu \in \mathrm{Spec}(\mathrm{Hess}\, V(q)) \\ \nu > 0}} \nu(q),$$

$$\mathrm{Tr}_{-,V}(q) = -\sum_{\substack{\nu \in \mathrm{Spec}(\mathrm{Hess}\, V(q)) \\ \nu \leq 0}} \nu(q).$$

This part is dedicated to providing an overview and discussion of the key points and findings presented in the article [4]. This article provides an extension of Theorem 7 to encompass homogeneous potentials of degree $2 < r < 6$ in $\mathcal{C}^\infty(\mathbb{R}^d \setminus \{0\})$, under the condition outlined below.

Assumption 9

$$\forall \ q \in \mathcal{S} := \left\{ q \in \mathbb{R}^d, \ |q| = 1 \right\}, \qquad \partial_q V(q) = 0 \Rightarrow \mathrm{Tr}_{-,V}(q) > 0. \tag{27}$$

The main result established in [4] is stated as follows.

Theorem 8 *If the potential $V(q)$ verifies Assumption 9, then there exists a strictly positive constant $C_V > 1$ (which depends on V) such that*

$$\begin{aligned} \|K_V u\|_{L^2}^2 + C_V \|u\|_{L^2}^2 \geq \frac{1}{C_V}\Big(&\|L(O_p)u\|_{L^2}^2 + \|L(\langle \nabla V(q)\rangle^{\frac{2}{3}})u\|_{L^2}^2 \\ &+ \|L(\langle \mathrm{Hess}\ V(q)\rangle^{\frac{1}{2}})u\|_{L^2}^2 + \|L(\langle D_q\rangle^{\frac{2}{3}})u\|_{L^2}^2\Big) \end{aligned} \tag{28}$$

holds for all $u \in D(K_V)$, where $L(s) = \frac{s+1}{\log(s+1)}$ for any $s \geq 1$.

It is worth noting that the subelliptic estimate established in Theorem 8 is identical to the one obtained in the polynomial case described in Theorem 7. By virtue of Theorem 8, we derive the subsequent corollary.

Corollary 1 *The Kramers–Fokker–Planck operator K_V with a potential $V(q)$ satisfying Assumption 9 has a compact resolvent.*

It is important to highlight that the result of Corollary 1 is not valid in the case of a homogeneous polynomial of degree 2 with a degenerate Hessian. In this case, with a degenerate Hessian, the Kramers–Fokker–Planck operator is indeed invariant by translation in the direction of the kernel of the Hessian and therefore cannot have a compact resolvent.

Furthermore, it should be mentioned that Corollary 1 is in accordance with the results of Wei-Xi Li [29, 30] and those of Helffer–Nier on the Witten Laplacian with homogeneous potential [17]. For a detailed proof of Corollary 1, the readers are referred to Corollary 1.2 in [3].

To establish the validity of Theorem 8, we employ the following lines of reasoning:

- Firstly, we consider a dyadic partition with respect to the position variable $q \in \mathbb{R}^d$ (see Proposition 2.1 in [4] and page 59 in [2]).

$$\sum_{j\geq -1} \chi_j^2(q) = \tilde{\chi}_{-1}^2(2|q|) + \tilde{\chi}_0^2(2|q|) + \sum_{j\geq 1} \tilde{\chi}^2(2^{-j}|q|) = 1, \tag{29}$$

where the cutoff functions $\widetilde{\chi}_0$, $\widetilde{\chi}$ and $\widetilde{\chi}_{-1}$ belong, respectively, to $\mathcal{C}_0^\infty(\left]\frac{3}{4}, \frac{8}{3}\right[)$, $\mathcal{C}_0^\infty(\left]\frac{3}{4}, \frac{8}{3}\right[)$, and $\mathcal{C}_0^\infty(\left[0, \frac{4}{3}\right[)$. Then, by Lemma 2.2 in [4], there exists a uniform constant $c > 0$ such that

$$(1+4c)\|K_V u\|^2_{L^2(\mathbb{R}^{2d})} + c\|u\|^2_{L^2(\mathbb{R}^{2d})} \geq \sum_{j\geq -1} \|K_V u_j\|^2_{L^2(\mathbb{R}^{2d})}, \tag{30}$$

for all $u \in \mathcal{C}_0^\infty(\mathbb{R}^{2d})$ with $u_j := \chi_j u$.

By exploiting the homogeneity of the potential V, we can rewrite the inequality (30) as follows:

$$(1+4c)\|K_V u\|^2_{L^2(\mathbb{R}^{2d})} + c\|u\|^2_{L^2(\mathbb{R}^{2d})} \geq \sum_{j\geq -1} \|K_{j,V} v_j\|^2_{L^2(\mathbb{R}^{2d})}, \tag{31}$$

where

$$K_{j,V} = p(h^{\frac{1}{2(r-1)}}\partial_q) - h^{-\frac{1}{2}}\partial_q V(q)\partial_p + \frac{1}{2}(-\Delta_p + p^2), \qquad h = 2^{-2(r-1)j},$$

$$v_j(q,p) = 2^{\frac{jd}{2}} u_j(2^j q, p)\,, \qquad \operatorname{supp} v_j \subset \overline{\mathcal{C}} := \left\{ q \in \mathbb{R}^d,\ \frac{3}{4} \le |q| \le \frac{8}{3} \right\}.$$

– The idea is then to introduce the set of zeros of the gradient of $V(q)$,

$$K_0 := \left\{ q \in \overline{\mathcal{C}}\,,\quad \partial_q V(q) = 0 \right\},$$

which is compact due to the continuity of the mapping $q \mapsto \partial_q V(q)$ on the fixed shell $\overline{\mathcal{C}}$.

Notice that since $q \mapsto \frac{\mathrm{Tr}_{-,V}(q)}{1+\mathrm{Tr}_{+,V}(q)}$ is uniformly continuous on any compact neighborhood of K_0, there exists $\varepsilon_1 > 0$ such that

$$d(q, K_0) \le \epsilon_1 \Rightarrow \frac{\mathrm{Tr}_{-,V}(q)}{1 + \mathrm{Tr}_{+,V}(q)} \ge \frac{\epsilon_0}{2}, \tag{32}$$

where $\epsilon_0 := \min\limits_{q \in K_0} \frac{\mathrm{Tr}_{-,V}(q)}{1+\mathrm{Tr}_{+,V}(q)}$.

On the other hand, by the definition of K_0 and the continuity of $q \mapsto \partial_q V(q)$ on $\overline{\mathcal{C}}$, there exists a constant $\epsilon_2 > 0$ (depending on ϵ_1) such that

$$\forall\, q \in \overline{\mathcal{C}}\,,\ \ d(q, K_0) \ge \epsilon_1 \Rightarrow |\partial_q V(q)| \ge \epsilon_2. \tag{33}$$

These remarks lead us to consider a new partition along the position variable $q \in \mathbb{R}^d$, namely a locally finite partition of unity

$$\sum_{k \in K_h} \theta_{k,h}(q)^2 = 1 \tag{34}$$

in the neighborhood of $\overline{\mathcal{C}}$, where for any index $k \in K_h := \left\{ k \in \mathbb{Z}^d,\ |k| \le \frac{3}{|\ln(h)|h^{\nu}} \right\}$,

$$q_{k,h} = |\ln(h)|h^{\nu} k\,, \qquad \operatorname{supp} \theta_{k,h} \subset B(q_{k,h}, |\ln(h)|h^{\nu})\,,$$

$$\theta_{k,h} \equiv 1 \ \text{ in } \ B(q_{k,h}, \frac{1}{2}|\ln(h)|h^{\nu}),$$

with ν a fixed parameter such that

$$\frac{3}{16} + \frac{1}{16(r-1)} < \nu \le \frac{1}{8} + \frac{3}{8(r-1)}. \tag{35}$$

Next, we consider the set

$$I(\epsilon_1) = \left\{ k \in K_h\,,\ \ \operatorname{supp} \theta_{k,h} \subset \{ q \in \mathcal{C}\,,\ d(q, K_0) \ge \epsilon_1 \} \right\}.$$

The idea behind introducing the second partition of unity is to be able to analyze and control the following two cases effectively.

Case 1 $k \notin I(\epsilon_1)$. In this situation, the support of the cutoff function $\theta_{k,h}$ may intersect the set of zeros of the gradient of V.

Case 2 $k \in I(\epsilon_1)$. In this case, the gradient of V is nonzero for every q in the support of $\theta_{k,h}$.

Using the partition (34), we get once again through Lemma 2.2 in [4]

$$\|K_{j,V}v_j\|_{L^2}^2 \geq \sum_{k\in K_h} \Big(\|K_{j,V}\theta_{k,h}v_j\|_{L^2}^2 - |\ln(h)|^{-2}h^{\frac{1}{r-1}-2\nu}\|p\theta_{k,h}v_j\|_{L^2}^2\Big). \tag{36}$$

In order to reduce the written expressions, we denote in what follows

$$w_{k,j} = \theta_{k,h}v_j \ .$$

Taking into account (36),

$$\begin{aligned}\|K_{j,V}v_j\|_{L^2}^2 &\geq \sum_{k\in K_h} \Big(\|K_{j,V}w_{k,j}\|_{L^2}^2 - 2|\ln(h)|^{-2}h^{\frac{1}{r-1}-2\nu}\|w_{k,j}\|_{L^2}\|K_{j,V}w_{k,j}\|_{L^2}\Big)\\ &\geq \sum_{k\in K_h} \Big(\frac{1}{2}\|K_{j,V}w_{k,j}\|_{L^2}^2 - 2(|\ln(h)|^{-2}h^{\frac{1}{r-1}-2\nu})^2\|w_{k,j}\|_{L^2}^2\Big).\end{aligned} \tag{37}$$

Observe that in the last inequality we simply use, respectively, the fact that

$$\|pw_{k,j}\|_{L^2}^2 \leq 2\mathrm{Re}\langle w_{k,j}, K_{j,V}w_{k,j}\rangle \leq 2\|w_{k,j}\|_{L^2}\|K_{j,V}w_{k,j}\|_{L^2}$$

and the Cauchy inequality. The idea is then to use a quadratic approximating polynomial

$$V_{k,h}^{(2)}(q) = \sum_{|\alpha|\leq 2} \frac{\partial_q^\alpha V(q'_{k,h})}{\alpha!}(q - q'_{k,h})^\alpha$$

near some $q'_{k,h} \in \operatorname{supp}\theta_{k,h}$ to write

$$\sum_{k\in K_h} \|K_{j,V}w_{k,j}\|_{L^2}^2 \geq \sum_{k\in K_h} \Big(\frac{1}{2}\|K_{j,V_{k,h}^{(2)}}w_{k,j}\|_{L^2}^2 - \|(K_{j,V} - K_{j,V_{k,h}^{(2)}})w_{k,j}\|_{L^2}^2\Big),$$

or equivalently

$$\sum_{k\in K_h} \|K_{j,V} w_{k,j}\|_{L^2}^2$$
$$\geq \sum_{k\in K_h} \Big(\frac{1}{2}\|K_{j,V^{(2)}_{k,h}} w_{k,j}\|_{L^2}^2 - \|\frac{1}{\sqrt{h}}(\partial_q V(q) - \partial_q V^{(2)}_{k,h}(q))\partial_p w_{k,j}\|_{L^2}^2\Big). \tag{38}$$

Then based on the estimates written in [5], which are valid for the operator $K_{V^2_{k,h}}$, we deduce a subelliptic estimate for K_V, after a careful control of the errors which appear in (37) and (38) in the aforementioned two cases.

4 Conclusion and Perspectives

The study of Kramers–Fokker–Planck operators on various spaces has resulted in too many publications. While it is not possible to provide an exhaustive list, here are some notable contributions and references in the field that we were not covered in this chapter [6, 24–26, 33, 35, 45, 46]. Numerous questions have arisen, been resolved, or remained partially open in various geometric frameworks. Among them is the analysis of the Helffer–Nier Conjecture:

Conjecture 2 *The operator K_V has a compact resolvent if and only if $\Delta_V^{(0)}$ has a compact resolvent.*

Recalling that the necessary part of this conjecture has already been established by Helffer and Nier (cf. [16], Theorem 1.1]), the objective of the works [3, 5], and [4] was particularly to consider examples and situations for which the analysis of compactness for the Witten Laplacian is well developed and to see if the reverse implication of the conjecture still holds.

In their work, Helffer and Nier [16] were the first to propose the idea of using the criteria of maximal hypoellipticity, a concept developed by Helffer and Nourrigat [18, 34], to derive a criterion for compactness of the resolvent for the Witten Laplacian.

This criterion, based in particular on a recurrence on the degree of the polynomial potential, was the basis for the idea pursued in [3, 5]: to develop a systematic analysis for the Kramers–Fokker–Planck operator. Naturally, our starting point was the study of the case of a polynomial potential with degree less than or equal to 2. This basic study led to optimal subelliptic estimates with uniform control of the constants. This is the subject of the article [5].

The emergence of quaternionic structure in [5], which greatly simplified the study, opens the door to possible research avenues for other models of operators with quadratic symbols.

Once we have carefully dealt with the case of polynomials of degree 2 or less, we proceeded to consider polynomials of degree 3 or higher, whose asymptotic models at infinity, after rescaling, have degree 2 or less. Despite the logarithmic corrections present in the quadratic case, we were able to establish a subelliptic estimate for K_V that is very close to being optimal, primarily relying on the Tarski–Seidenberg Theorem. This work is presented in the context of the article [3].

With this recent work, we have addressed almost all the examples presented by Helffer and Nier in [16] for comparing the Witten Laplacian and the Kramers–Fokker–Planck operator, except for the case of homogeneous potentials, not necessarily polynomial, of degree higher than 2. This latter case is partially addressed in [4] and has led to improved results.

After all, the Helffer–Nier Conjecture 2 is still far from being completely resolved. It can be broken down into several unanswered questions, including:

1. In the case of a polynomial potential, is it possible to show that the Kramers–Fokker–Planck operator K_V has a compact resolvent if it satisfies the same criteria developed by Helffer–Nier in the case of the Witten Laplacian (cf. Theorem 10.16 in [16])?
2. Can we extend the result of Theorem 8 (established in the article [4]) to the case of a homogeneous potential of degree $r \geq 6$ in $\mathcal{C}^\infty(\mathbb{R}^d \setminus \{0, 1\})$? What happens in particular for the polynomial case $V(q) = -q_1^2(q_1^2 + q_2^2)^n$ with $n \in \mathbb{N} \setminus \{0, 1\}$?
3. Is the equivalence (2) true for a general potential $V \in \mathcal{C}^\infty(\mathbb{R}^d)$?

Furthermore, these questions provide opportunities for further research and potential avenues for expansion.

References

1. A. Aleman, J. Viola, On weak and strong solution operators for evolution equations coming from quadratic operators. J. Spectr. Theory **8**(1), 33–121 (2018)
2. H. Bahouri, J-Y. Chemin, R. Danchin, Fourier analysis and nonlinear partial differential equations. *Grundlehren der Mathematischen Wissenschaften [Fundamental Principles of Mathematical Sciences]*, vol. 343 (Springer, Heidelberg, 2011), xvi+523 pp.
3. M. Ben Said, Global subelliptic estimates for Kramers-Fokker-Planck operators with some class of polynomials. J. Inst. Math. Jussieu **21**(2), 675–711 (2022)
4. M. Ben Said, Kramers-Fokker-Planck operators with homogeneous potentials. Math. Methods Appl. Sci. **45**(2), 914–927 (2022)
5. M. Ben Said, F. Nier, J. Viola, Quaternionic structure and analysis of some Kramers-Fokker-Planck operators. Asymptot. Anal. **119**(1–2), 87–116 (2020)
6. J.M. Bismut, A survey of the hypoelliptic Laplacian. Géométrie différentielle, physique mathématique, mathématiques et société II. Astérisque **322**, 39–69 (2008)
7. J.M. Bismut, G. Lebeau, *The Hypoelliptic Laplacian and Ray-Singer Metrics*. Annals of Mathematics Studies, vol. 167 (Princeton University Press, Princeton, 2008), x+367 pp.
8. P. Bolley, M. Dauge, B. Helffer, Conditions suffisantes pour l'injection compacte d'espaces de Sobolev à poids (ou autour d'une question de F. Mignot). Séminaire de l'université de Nantes (1989–1990)
9. R. Dautray, J.L. Lions, *Mathematical Analysis and Numerical Methods for Science and Technology* (Springer, Berlin, 1992)

10. E.B. Davies, *One-Parameter Semigroups* (Academic Press Inc. [Har-court Brace Jovanovich Publishers], London, 1980)
11. J.P. Eckmann, M. Hairer, Non-equilibrium statistical mechanics of strongly anharmonic chains of oscillators. Comm. Math. Phys. **212**(1), 105–164 (2000)
12. A. Einstein, Über die von der molekularkinetischen Theorie der Wärme geforderte Bewegung von in ruhenden Flüssigkeiten suspendierten Teilchen. Ann. Phys. **17**(4), 549–560 (1905)
13. A.D. Fokker, Die mittlere Energie rotierender elektrischer Dipole im Strahlungsfeld. Ann. Phys. **348**(5), 810–820 (1914)
14. B. Helffer, *Semiclassical Analysis, Witten Laplacians, and Statistical Mechanics*. Series in Partial Differential Equations and Applications, vol. 1 (World Scientific Publishing Co., Inc., River Edge, 2002)
15. B. Helffer, A. Mohamed, Caractérisation du spectre essentiel de l'opérateur de Schrödinger avec un champ magnétique. Ann. Inst. Fourier (Grenoble) **38**(2), 95–112 (1988)
16. B. Helffer, F. Nier, *Hypoelliptic Estimates and Spectral Theory for Fokker-Planck Operators and Witten Laplacians*. Lecture Notes in Mathematics, vol. 1862 (Springer, Berlin, 2005), x+209 pp.
17. B. Helffer, F. Nier, Criteria to the poincare inequality associated with Dirichlet Forms in $\mathbb{R}^d$, $d \geq 2$. Int. Math. Res. Not. **22**, 1199–1223 (2003)
18. B. Helffer, J. Nourrigat: *Hypoellipticité maximale pour des opérateurs polynômes de champs de vecteurs*. Progress in Mathematics, vol. 58 (1985)
19. F. Hérau, F. Nier, Isotropic hypoellipticity and trend to equilibrium for the Fokker-Planck equation with a high-degree potential. Arch. Ration. Mech. Anal. **171**(2), 151–218 (2004)
20. F. Hérau, M. Hitrik, J. Sjöstrand, Tunnel effect for Kramers-Fokker-Planck type operators: return to equilibrium and applications. Int. Math. Res. Not. IMRN, no. 15, Art. ID rnn057, 48 (2008)
21. M. Hitrik, K. Pravda-Starov, Spectra and semigroup smoothing for non-elliptic quadratic operators. Math. Ann. **344**(4), 801–846 (2009)
22. M. Hitrik, K. Pravda-Starov, J. Viola, From semigroups to subelliptic estimates for quadratique operators. Trans. Am. Math. Soc. **370**(10), 7391–7415 (2018)
23. L. Hörmander, Symplectic classification of quadratic forms, and general Mehler formulas. Math. Z. **219**, 413–449 (1995)
24. Z. Karaki, Trend to the equilibrium for the Fokker-Planck system with an external magnetic field. Kinet. Rel. Models **13**(2), 309 (2020)
25. Z. Karaki, Maximal estimates for the Fokker-Planck operator with magnetic field. J. Spectral Theory **11**(3), 1179–1213 (2021)
26. Z. Karaki, Study of the Kramers-Fokker-Planck quadratic operator with a constant magnetic field. J. Math. Phys. **63**, 081503 (2022)
27. H.A. Kramers, Brownian motion in a field of force and the diffusion model of chemical reactions. Physica **7**(4), 284–360 (1940)
28. P. Langevin, Sur la théorie du mouvement brownien. C. R. Acad. Sci. (Paris) **146**, 530–533 (1908)
29. W.-X. Li, Global hypoellipticity and compactness of resolvent for Fokker-Planck operator. Ann. Sc. Norm. Super. Pisa Cl. Sci. (5) **11**(4), 789–815 (2012)
30. W.-X. Li, Compactness criteria for the resolvent of Fokker-Planck operator. Prepublication. ArXiv1510.01567 (2015)
31. A. Lunardi, *Interpolation Theory*, 2nd edn. Appunti (Scuola Normale Superiore di Pisa (Nuova Serie), Pisa, 2009), xiv+191 pp.
32. F. Nier, *Hypoellipticity for Fokker-Planck Operators and Witten Laplacians*. Lectures on the analysis of nonlinear partial differential equations. Morningside Lectures in Mathematics, vol. 1, Part 1 (International Press, Somerville, 2012), pp. 31–84
33. F. Nier, *Boundary Conditions and Subelliptic Estimates for Geometric Kramers-Fokker-Planck Operators on Manifolds with Boundaries*, vol. 252 (American Mathematical Society, Providence, 2018)

34. J. Nourrigat, *Subelliptic Estimates for Systems of Pseudo-Differential Operators*. Course in Recife (University of Recife, Recife, 1982)
35. R. Novak, X.P. Wang, On the Kramers-Fokker-Planck equation with decreasing potentials in dimension one. J. Spectral Theory **10**(1), 1–32 (2020)
36. V. Planck, Über einen satz der statistischen dynamik und seine erweiterung in der quantentheorie. Sitzungberichte der (1917)
37. M. Reed, B. *Simon, Methods of Modern Mathematical Physics. I. Functional Analysis*, 2nd edn. (Academic Press, Inc., Cambridge, 1980), xv+400 pp.
38. H. Risken, *The Fokker-Planck Equation. Methods of Solution and Applications*, 2nd edn. (Springer, Berlin, 1989). J. Spectral Theory **8**(1), 33–121 (2018)
39. L.P. Rothschild, E.M. Stein, Hypoelliptic differential operators and nilpotent groups. Acta Math. **137**, 248–315 (1977)
40. J. Sjöstrand, Parametrices for pseudodifferential operators with multiple characteristics. Ark. Mat. **12**, 85–130 (1974)
41. C. Villani, *Hypocoercivity* (Memoirs of the American Mathematical Society, Providence, 2009), vol. 202, no. 950, iv+141 pp.
42. J. Viola, Spectral projections and resolvent bounds for partially elliptic quadratic differential operators. J. Pseudo-Diff. Oper. Appl. **4**(2), 145–221 (2013)
43. J. Viola, The elliptic evolution of non-self-adjoint degree-2 Hamiltonians. ArXiv 1701.00801 (2017)
44. M. von Smoluchowski, Zur kinetischen Theorie der Brownschen Molekularbewegung und der Suspensionen. Ann. Phys. **326**(14), 756–780 (1906)
45. X.P. Wang, Large-time asymptotics of solutions to the Kramers-Fokker-Planck equation with a short-range potential. Comm. Math. Phys. **336**(3), 1435–1471 (2015)
46. X.P. Wang, L. Zhu, Global-in-time $L^p - L^q$ estimates for solutions of the Kramers-Fokker-Planck equation. arxiv.org/abs/2107.03246 (2021)

Exponential Decay of Solutions to Linear Evolution Equations with Time-Dependent Time Delay

Elisa Continelli and Cristina Pignotti

2010 Mathematics Subject Classification 37L15, 93D15

1 Introduction

The study of evolution equations in the presence of delay or memory terms attracted, in recent years, the interest of many researchers. The presence of a time delay makes the problems more difficult to deal with, and, of course, it is important to include in the models time delays/memory terms to take into account time lags, such as reaction times, maturation times, times needed to receive some information, etc., commonly present in real-life phenomena.

On the other hand, it is well known that a time delay can produce instability phenomena. In particular, for the damped wave equation, it has been proven that an arbitrarily small delay can make the model unstable even if it is uniformly asymptotically stable in the absence of delay effects (see, e.g., [11, 12, 21, 30]).

Nevertheless, suitable feedback laws can ensure the delayed model has the same stability properties as the undelayed one (see [21, 30]). Here, we are interested in proving exponential stability estimates for abstract linear evolution equations in the presence of a time delay feedback. In particular, we will consider time variable time delays, that is, time delays that are functions of the time variable.

Let $\tau : [0, +\infty) \to (0, +\infty)$ be the time delay function belonging to $C(0, +\infty)$. We assume that

$$0 \le \tau(t) \le \bar{\tau}, \quad \forall\, t \ge 0, \tag{1.1}$$

for a suitable positive constant $\bar{\tau}$.

E. Continelli · C. Pignotti (✉)
Dipartimento di Ingegneria e Scienze dell'Informazione e Matematica, Università degli Studi di L'Aquila, L'Aquila, Italy
e-mail: elisa.continelli@graduate.univaq.it; cristina.pignotti@univaq.it

K. Ammari et al. (eds.), *Advances in Partial Differential Equations and Control*, Trends in Mathematics, https://doi.org/10.1007/978-3-031-62265-6_8

We consider the following abstract model:

$$\begin{aligned} &U'(t) = AU(t) + k(t)BU(t - \tau(t)), \quad t \in (0, \infty), \\ &U(t) = f(t) \quad t \in [-\bar{\tau}, 0], \end{aligned} \tag{1.2}$$

where the operator A generates an exponentially stable C_0-semigroup $(S(t))_{t\geq 0}$ in a Hilbert space H, and B is a continuous linear operator of H into itself. Moreover, the delay damping coefficient $k : [-\bar{\tau}, +\infty) \to \mathbb{R}$ belongs to $\mathcal{L}^1_{loc}([-\bar{\tau}, +\infty); \mathbb{R})$, and τ is a continuous time delay function satisfying (1.1). We denote with $U_0 := f(0)$.

By the assumptions on the operator A, there exist two positive constants M and ω such that

$$\|S(t)\|_{\mathcal{L}(H)} \leq Me^{-\omega t}, \quad t \geq 0. \tag{1.3}$$

Under some mild assumptions on the involved functions and parameters, we will establish the well-posedness of the problem (1.2), and we will obtain exponential decay estimates for its solutions.

Moreover, on the delay feedback coefficient, we assume that the integrals on intervals of length $\bar{\tau}$ are uniformly bounded, namely,

$$\int_{t-\bar{\tau}}^{t} |k(s)|ds \leq K, \quad \forall t \geq 0, \tag{1.4}$$

for some $K > 0$.

Stability results for abstract evolution equations with delay have been already studied in [18, 23, 24]. In [23, 24], it is analyzed the case of a single constant delay, and also the delay damping coefficient is assumed to be constant. The analysis has then been extended in [18] by considering, as here, (multiple) time-dependent time delays. However, here we work in a more general setting. Indeed, in [18], the classical set of assumptions usually employed to deal with wave-type equations in the presence of time variable time delays (see, e.g., [9, 14, 25]) is used. In particular, in addition to (1.1), it is required that $\tau \in W^{1,\infty}(0, +\infty)$ and that $\tau'(t) \leq c < 1$. On the contrary, in the present chapter, we only assume that the time delay is a function bounded from above. Therefore, our results significantly extend and improve the ones in [18].

A concrete model that can be rewritten in the form (1.2) is, e.g., the wave equation with frictional damping and delay feedback. In the case of constant delay feedback coefficient and constant time delay, this model has been first studied in [28]. Under a suitable smallness condition on the delay term coefficient, an exponential decay estimate has been proven. This result has then been extended to linear wave equations with internal delay feedback and boundary dissipative condition in [5]. In [2, 15] it is instead analyzed the case of the wave equations with delay feedback and viscoelastic damping.

For other stability estimates in the presence of time delay effects, for specific models, mainly in the case of constant time delay and constant delay damping coefficient, we quote [1, 3, 4, 6, 10, 20, 26, 29], among others. We mention also the recent papers [17] and [8] dealing with delayed Korteweg–de Vries–Burgers and higher order dispersive equations, respectively.

The rest of the chapter is organized as follows. In Sect. 2 we prove the well-posedness of the model. We give two different proofs: the first, via a step-by-step procedure, under the additional assumption that the time delay function is bounded from below by a positive constant, and the second, in the general setting, using a fixed point theorem. In Sect. 3, we deduce the exponential stability estimate for the linear model. In Sect. 4, we extend the well-posedness and exponential stability results to a nonlinear model with a Lipschitz continuous nonlinearity. Finally, Sect. 5 illustrates some concrete examples for which the abstract theory is applicable: namely the damped wave equation with delay and the elasticity system with delay.

2 Well-Posedness

Now, we prove the following well-posedness result.

Theorem 1 *Let $f : [-\bar{\tau}, 0] \to H$ be a continuous function. Then, the problem* (1.2) *has a unique (weak) solution given by Duhamel's formula*

$$U(t) = S(t)U_0 + \int_0^t S(t-s)\,k(s)BU(s-\tau(s))\,ds, \tag{2.1}$$

for all $t \geq 0$.

Proof Let $f \in C([-\bar{\tau}, 0]; H)$. We give two different proofs, the first one only valid under the additional assumption that the time delay is bounded from below by a positive constant.

Case 1 Assume that

$$\tau(t) \geq \tau_0 > 0, \quad \forall t \geq 0, \tag{2.2}$$

for a suitable positive constant τ_0. We can argue step by step, as in the proof of Proposition 2.1 of [18], by restricting ourselves each time to time intervals of length τ_0.

First we consider $t \in [0, \tau_0]$. Then, from (2.2), $t - \tau(t) \in [-\bar{\tau}, 0]$. So, setting $F(t) = k(t)BU(t-\tau(t))$, $t \in [0, \tau_0]$, we have that $F(t) = k(t)Bf(t-\tau(t))$, $t \in [0, \tau_0]$. Then, problem (1.2) can be rewritten, in the interval $[0, \tau_0]$, as a standard inhomogeneous evolution problem:

$$\begin{aligned} &U'(t) = AU(t) + F(t) \quad \text{in } (0, \tau_0),\\ &U(0) = U_0. \end{aligned} \tag{2.3}$$

Since $k \in \mathcal{L}^1_{loc}([-\bar{\tau}, +\infty); \mathbb{R})$, B is a bounded linear operator, and $f \in C([-\bar{\tau}, 0]; H)$, we have that $F \in \mathcal{L}^1((0, \tau_0); H)$. Therefore, applying [27, Corollary 2.2], there exists a unique solution $U \in C([0, \tau_0]; H)$ of (2.3) satisfying Duhamel's formula

$$U(t) = S(t)U_0 + \int_0^t S(t-s)F(s)ds, \quad t \in [0, \tau_0].$$

Therefore,

$$U(t) = S(t)U_0 + \int_0^t S(t-s)k(s)BU(s-\tau(s))ds, \quad t \in [0, \tau_0].$$

Next, we consider the time interval $[\tau_0, 2\tau_0]$ and also define $F(t) = k(t)BU(t - \tau(t))$, for $t \in [\tau_0, 2\tau_0]$. Note that if $t \in [\tau_0, 2\tau_0]$, then $t - \tau(t) \in [-\bar{\tau}, \tau_0]$ and so $U(t - \tau(t))$ is known from the first step. Then $F_{|[\tau_0, 2\tau_0]}$ is a known function, and it belongs to $\mathcal{L}^1((\tau_0, 2\tau_0); H)$. So we can rewrite our model (1.2) in the time interval $[\tau_0, 2\tau_0]$ as the inhomogeneous evolution problem

$$\begin{aligned} &U'(t) = AU(t) + F(t) \quad \text{for } t \in (\tau_0, 2\tau_0), \\ &U(\tau) = U(\tau_0^-). \end{aligned} \tag{2.4}$$

Then, by the standard theory of abstract Cauchy problems, we have a unique continuous solution $U : [\tau_0, 2\tau_0] \to H$ satisfying

$$U(t) = S(t-\tau_0)U(\tau_0^-) + \int_{\tau_0}^t S(t-s)F(s)ds, \quad t \in [\tau_0, 2\tau_0],$$

and so

$$U(t) = S(t-\tau_0)U(\tau_0^-) + \int_{\tau_0}^t S(t-s)k(s)BU(s-\tau(s))ds, \quad t \in [\tau_0, 2\tau_0].$$

Putting together the partial solutions obtained in the first and second steps, we have a unique continuous solution $U : [0, 2\tau_0] \to \mathbb{R}$ satisfying Duhamel's formula

$$U(t) = S(t)U_0 + \int_0^t S(t-s)k(s)BU(s-\tau(s))ds, \quad t \in [0, 2\tau_0].$$

Iterating this procedure, we can find a unique solution $U \in C([0, +\infty); H)$ satisfying the representation formula (2.1).

Case 2 Let $\tau(\cdot)$ be a continuous function satisfying (1.1). In this case, since assumption (2.2) does not necessarily hold, we cannot use a step-by-step procedure in order to get the existence of a solution to (1.2) with the initial datum f, as we did in Case 1. We have rather to employ a different method, based on the use of Banach's fixed point theorem.

Now, since $k \in \mathcal{L}^1_{loc}([-\bar{\tau}, +\infty); \mathbb{R})$, there exists $T > 0$ such that

$$\|k\|_{\mathcal{L}^1([0,T];\mathbb{R})} = \int_0^T |k(s)|ds < \frac{1}{M\|B\|}. \tag{2.5}$$

We define the set

$$C_f([-\bar{\tau}, T]; H) := \{U \in C([-\bar{\tau}, T]; H) : U(s) = f(s), \ \forall s \in [-\bar{\tau}, 0]\}.$$

Let us note that $C_f([-\bar{\tau}, T]; H) \neq \emptyset$, since it suffices to take

$$U(t) = \begin{cases} U_0, & t \in (0, T], \\ f(t), & t \in [-\bar{\tau}, 0], \end{cases}$$

to have that $U \in C_f([-\bar{\tau}, T]; H)$.

It is immediate to see that $C_f([-\bar{\tau}, T]; H)$ with the norm

$$\|U\|_{C([-\bar{\tau},T];H)} = \max_{r\in[-\bar{\tau},T]} \|U(r)\|, \quad \forall U \in C([-\bar{\tau}, T]; H),$$

is a Banach space.

Now, we define the map $\Gamma : C_f([-\bar{\tau}, T]; H) \to C_f([-\bar{\tau}, T]; H)$ given by

$$\Gamma U(t) = \begin{cases} S(t)U_0 + \int_0^t S(t-s)k(s)BU(s-\tau(s))\, ds, & t \in [0, T], \\ f(t), & t \in [-\bar{\tau}, 0). \end{cases}$$

We claim that Γ is well defined. Indeed, let $U \in C_f([-\bar{\tau}, T]; H)$. Then, from the semigroup theory, $t \mapsto S(t)U_0$ is continuous. Also, since $U(\cdot)$ is continuous in $[-\bar{\tau}, T]$, $\tau(\cdot)$ is a continuous function and B is a bounded linear operator from H into itself, $[0, T] \ni t \mapsto BU(t - \tau(t))$ is continuous. Moreover, $k \in \mathcal{L}^1([0, T]; \mathbb{R})$. So $k(\cdot)BU(\cdot - \tau(\cdot)) \in \mathcal{L}^1([0, T]; H)$. Hence, the map $t \mapsto \int_0^t S(t-s)k(s)BU(s - \tau(s))ds$ is continuous in $[0, T]$. Thus, $\Gamma U \in C([0, T]; H)$. Finally, since $\Gamma U = f$ in $[\bar{\tau}, 0]$ with $f \in C([-\bar{\tau}, 0]; H)$ and $f(0) = U_0$, $\Gamma U \in C_f([\bar{\tau}, T]; H)$ and Γ is well defined.

Now, let $U, V \in C_f([\bar{\tau}, T]; H)$. For all $t \in [-\bar{\tau}, 0]$,

$$\|\Gamma U(t) - \Gamma V(t)\| = 0.$$

On the other hand, for all $t \in (0, T]$,

$$\begin{aligned}\|\Gamma U(t)-\Gamma V(t)\| &\leq \int_0^t \|S(t-s)\||k(s)|\|BU(s-\tau(s))-BV(s-\tau(s))\|ds\\ &\leq M\|B\|\int_0^t |k(s)|\|U(s-\tau(s))-V(s-\tau(s))\|ds\\ &\leq M\|B\|\|U-V\|_{C([-\bar{\tau},T];H)}\int_0^T |k(s)|ds\\ &= MB\|k\|_{\mathcal{L}^1([0,T];\mathbb{R})}\|U-V\|_{C([-\bar{\tau},T];H)}.\end{aligned}$$

Thus,

$$\|\Gamma U(t)-\Gamma V(t)\| \leq MB\|k\|_{\mathcal{L}^1([0,T];\mathbb{R})}\|U-V\|_{C([-\bar{\tau},T];H)},\quad \forall t\in[-\bar{\tau},T],$$

from which

$$\|\Gamma U-\Gamma V\|_{C([-\bar{\tau},T];H)} \leq M\|B\|\|k\|_{\mathcal{L}^1([0,T];\mathbb{R})}\|U-V\|_{C([-\bar{\tau},T];H)}.$$

Now, from (2.5) we have that $M\|B\|\|k\|_{\mathcal{L}^1([0,T];\mathbb{R})} < 1$. Hence, Γ is a contraction. Then, from Banach's theorem, Γ has a unique fixed point $U \in C_f([-\bar{\tau},T];H)$, i.e., (1.2) has a unique solution $U \in C([0,T];H)$ given by Duhamel's formula

$$U(t) = S(t)U_0 + \int_0^t S(t-s)k(s)BU(s-\tau(s))\,ds,\quad \forall t\in[0,T].$$

Now, let us note that the solution U is bounded. Indeed, for all $t\in[0,T]$,

$$\begin{aligned}\|U(t)\| &\leq M\|U_0\| + M\|B\|\int_0^t |k(s)|\|U(s-\tau(s))\|ds\\ &\leq M\|U_0\| + M\|B\|\|k\|_{\mathcal{L}^1([0,T];\mathbb{R})}\max_{r\in[-\bar{\tau},0]}\|f(r)\|\\ &\quad + M\|B\|\int_0^t |k(s)|\max_{r\in[0,s]}\|U(r)\|ds.\end{aligned}$$

Then,

$$\begin{aligned}\max_{r\in[0,t]}\|U(r)\| &\leq M\left(\|U_0\| + \|B\|\|k\|_{\mathcal{L}^1([0,T];\mathbb{R})}\max_{r\in[-\bar{\tau},0]}\|f(r)\|\right)\\ &\quad + M\|B\|\int_0^t |k(s)|\max_{r\in[0,s]}\|U(r)\|ds.\end{aligned}$$

Hence, Gronwall's estimate yields

$$\max_{r\in[0,t]}\|U(r)\| \leq M\left(\|U_0\| + \|B\|\|k\|_{\mathcal{L}^1([0,T];\mathbb{R})}\max_{r\in[-\bar{\tau},0]}\|f(r)\|\right)e^{M\|B\|\int_0^t |k(s)|ds},$$

from which, taking into account of (2.5),

$$\|U(t)\| \le e\left(M\|U_0\| + \max_{r\in[-\bar{\tau},0]}\|f(r)\|\right), \quad \forall t \in [0,T].$$

Thus, the solution U is bounded, and we can extend it up to some maximal interval $[0,\delta)$, $\delta > 0$. We claim that $\delta = +\infty$. Indeed, assume by contradiction that $\delta < +\infty$. Then, being U bounded, we can consider the following problem:

$$\begin{aligned} &V'(t) = AV(t) + k(t)BV(t-\tau(t)), \quad t \in (0,\infty),\\ &V(t) = U(t) \quad t \in [\delta-\bar{\tau},\delta),\\ &V(\delta) = U(\delta^-). \end{aligned} \tag{2.6}$$

Arguing as before, there exists $T' > 0$ such that

$$\|k\|_{\mathcal{L}^1([\delta,T'],\mathbb{R})} = \int_\delta^{T'} |k(s)|ds < \frac{1}{M\|B\|}. \tag{2.7}$$

We then set

$$\begin{aligned} &C_U([\delta-\bar{\tau},T'];H)\\ &\quad = \{V \in C([\delta-\bar{\tau},T'];H) : V(s) = U(s), \forall s \in [\delta-\bar{\tau},\delta), V(\delta) = U(\delta^-)\}, \end{aligned}$$

which is a nonempty closed subset of $C([\delta-\bar{\tau},T'];H)$.

Next, we define the map $\Gamma : C_U([-\bar{\tau},T];H) \to C_U([-\bar{\tau},T];H)$ given by

$$\Gamma V(t) = \begin{cases} S(t-\delta)U(\delta^-) + \int_\delta^t S(t-s)k(s)BV(s-\tau(s)), & t \in [\delta,T'],\\ U(t), & t \in [\delta-\bar{\tau},\delta). \end{cases}$$

We have that Γ is well defined and, using the same arguments employed at the beginning of Case 2, (2.7) implies that Γ is a contraction. So, Γ has a unique fixed point, i.e., (2.6) has a unique continuous V solution given by the Duhamel formula

$$V(t) = S(t-\delta)U(\delta^-) + \int_\delta^t S(t-s)k(s)BV(s-\tau(s)), \quad t \in [\delta,T'].$$

Thus, putting together the solutions U, V, we get the existence of a unique continuous solution to (1.2) that satisfies Duhamel's formula (2.1) and that is defined in $[0,\delta')$, with $\delta' > \delta$. This contradicts the maximality of δ. Hence, $\delta = +\infty$, i.e., (1.2) has a unique global solution $U \in C([0,+\infty);H)$ given by (2.1).

3 Exponential Stability

Under an appropriate relation between the problem's parameters, the system (1.2) is exponentially stable. In particular, we assume that

$$M\|B\|e^{\omega\bar{\tau}}\int_0^t |k(s)|ds \le \gamma + \omega' t, \quad \forall t \ge 0, \tag{3.1}$$

for suitable constants $\gamma \ge 0$ and $\omega' \in [0, \omega)$.

Theorem 2 *Assume* (3.1). *For every* $f \in C([-\bar{\tau}, 0]; H)$, *the solution* $U \in C([0, +\infty); H)$ *to* (1.2) *with the initial datum* f *satisfies the exponential decay estimate*

$$\|U(t)\| \le Me^{\gamma}\left(\|U_0\| + e^{\omega\bar{\tau}}K\|B\| \max_{s\in[-\bar{\tau},0]}\left\{\|e^{\omega s}f(s)\|\right\}\right)e^{-(\omega-\omega')t}, \tag{3.2}$$

for any $t \ge 0$.

Proof Let $f \in C([-\bar{\tau}, 0]; H)$. Let $U \in C([0, +\infty); H)$ be the solution to (1.2) with the initial condition f. From Duhamel's formula, we have that

$$\|U(t)\| \le Me^{-\omega t}\|U_0\| + Me^{-\omega t}\int_0^t e^{\omega s}|k(s)| \cdot \|BU(s-\tau(s))\|ds, \quad \forall t \ge 0.$$

Then, for all $t \ge \bar{\tau}$, we deduce that

$$\begin{aligned}\|U(t)\| &\le Me^{-\omega t}\|U_0\| + Me^{-\omega t}\int_0^{\bar{\tau}} e^{\omega s}|k(s)| \cdot \|BU(s-\tau(s))\|ds\\ &\quad + Me^{-\omega t}\int_{\bar{\tau}}^t e^{\omega s}|k(s)| \cdot \|BU(s-\tau(s))\|ds\\ &\le Me^{-\omega t}\|U_0\| + M\|B\|e^{-\omega t}e^{\omega\bar{\tau}}\int_0^{\bar{\tau}} e^{\omega(s-\tau(s))}|k(s)| \cdot \|U(s-\tau(s))\|ds\\ &\quad + M\|B\|e^{-\omega t}e^{\omega\bar{\tau}}\int_{\bar{\tau}}^t e^{\omega(s-\tau(s))}|k(s)| \cdot \|U(s-\tau(s))\|ds.\end{aligned} \tag{3.3}$$

Now, observe that

$$\begin{aligned}&\int_0^{\bar{\tau}} e^{\omega(s-\tau(s))}|k(s)| \cdot \|U(s-\tau(s))\|ds\\ &\quad \le \int_0^{\bar{\tau}} |k(s)|\left(\max_{r\in[-\bar{\tau},0]}\left\{e^{\omega r}\|f(r)\|\right\} + \max_{r\in[0,s]}\left\{e^{\omega r}\|U(r)\|\right\}\right)ds\\ &\quad \le K\max_{r\in[-\bar{\tau},0]}\left\{e^{\omega r}\|f(r)\|\right\} + \int_0^{\bar{\tau}} |k(s)|\max_{r\in[0,s]}\left\{e^{\omega r}\|U(r)\|\right\}ds.\end{aligned} \tag{3.4}$$

Then, using (3.4) in (3.3), we deduce

$$
\begin{aligned}
||U(t)|| \leq Me^{-\omega t}\left(||U_0|| + e^{\omega\bar{\tau}}K\|B\| \max_{r\in[-\bar{\tau},0]}\left\{e^{\omega r}\|f(r)\|\right\}\right)\\
+M\|B\|e^{-\omega t}e^{\omega\bar{\tau}}\int_0^t |k(s)| \max_{r\in[s-\bar{\tau},s]\cap[0,s]}\left\{e^{\omega r}||U(r)||\right\}ds,
\end{aligned}
\tag{3.5}
$$

for all $t \geq \bar{\tau}$. On the other hand, for all $t \in [0, \bar{\tau}]$, it holds that

$$
||U(t)|| \leq Me^{-\omega t}||U_0|| + M\|B\|e^{-\omega t}e^{\omega\bar{\tau}}\int_0^t e^{\omega(s-\tau(s))}|k(s)|\cdot||U(s-\tau(s))||ds.
$$

Then, arguing as before,

$$
\begin{aligned}
&\int_0^t e^{\omega(s-\tau(s))}|k(s)|\cdot\|U(s-\tau(s))\|ds\\
&\quad\leq \int_0^t |k(s)|\left(\max_{r\in[-\bar{\tau},0]}\left\{e^{\omega r}\|f(r)\|\right\} + \max_{r\in[0,s]}\left\{e^{\omega r}\|U(r)\|\right\}\right)ds\\
&\quad\leq K\max_{r\in[-\bar{\tau},0]}\left\{e^{\omega r}\|f(r)\|\right\} + \int_0^t |k(s)|\max_{r\in[0,s]}\left\{e^{\omega r}\|U(r)\|\right\}ds
\end{aligned}
$$

for all $t \in [0, \bar{\tau}]$. Hence, (3.5) holds also for $t \in [0, \bar{\tau}]$. So, we deduce

$$
\begin{aligned}
e^{\omega t}||U(t)|| \leq M\left(||U_0|| + e^{\omega\bar{\tau}}K\|B\|\max_{s\in[-\bar{\tau},0]}\left\{\|e^{\omega s}f(s)\|\right\}\right)\\
+M\|B\|e^{\omega\bar{\tau}}\int_0^t |k(s)|\max_{r\in[s-\bar{\tau},s]\cap[0,s]}\left\{e^{\omega r}||U(r)||\right\}ds, \quad \forall t\geq 0.
\end{aligned}
$$

Now, we note that it also holds

$$
\begin{aligned}
\max_{s\in[t-\bar{\tau},t]\cap[0,t]}\left\{e^{\omega s}||U(s)||\right\} \leq M\left(||U_0|| + e^{\omega\bar{\tau}}K\|B\|\max_{s\in[-\bar{\tau},0]}\left\{\|e^{\omega s}f(s)\|\right\}\right)\\
+M\|B\|e^{\omega\bar{\tau}}\int_0^t |k(s)|\max_{r\in[s-\bar{\tau},s]\cap[0,s]}\left\{e^{\omega r}||U(r)||\right\}ds, \quad \forall t\geq 0.
\end{aligned}
$$

Hence, if we denote

$$
\tilde{u}(t) := \max_{s\in[t-\bar{\tau},t]\cap[0,t]}\left\{e^{\omega s}||U(s)||\right\},
$$

Gronwall's estimate implies

$$
\tilde{u}(t) \leq \tilde{M}e^{M\|B\|e^{\omega\bar{\tau}}\int_0^t|k(s)|ds}, \quad \forall t\geq 0,
$$

where

$$
\tilde{M} := M\left(\|U_0\| + e^{\omega\bar{\tau}}K\|B\|\max_{s\in[-\bar{\tau},0]}\left\{\|e^{\omega s}f(s)\|\right\}\right).
$$

Then,

$$e^{\omega t}\|U(t)\| \leq \tilde{M} e^{M\|B\|e^{\omega\bar{\tau}}\int_0^t |k(s)|ds}, \quad \forall t \geq 0.$$

Finally, assumption (3.1) yields

$$\begin{aligned}\|U(t)\| &\leq \tilde{M} e^{M\|B\|e^{\omega\bar{\tau}}\int_0^t |k(s)|ds} e^{-\omega t} \\ &\leq \tilde{M} e^{\gamma+\omega' t} e^{-\omega t} = \tilde{M} e^{\gamma} e^{-(\omega-\omega')t},\end{aligned}$$

for all $t \geq 0$, which proves the exponential decay estimate (3.2).

4 A Nonlinear Model

As an easy generalization of previous results, we can consider the nonlinear model

$$\begin{aligned}&U'(t) = AU(t) + k(t)BU(t-\tau(t)) + G(U(t)), \quad t \in (0, \infty), \\ &U(t) = f(t) \quad t \in [-\bar{\tau}, 0],\end{aligned} \tag{4.1}$$

where A, B, $k(\cdot)$, $\tau(\cdot)$ are as before, and we denote $U_0 := f(0)$. Moreover, $G : H \to H$ is Lipschitz continuous, namely there exists $L > 0$ such that

$$\|G(U_1) - G(U_2)\| \leq L\|U_1 - U_2\|, \quad \forall\, U_1, U_2 \in H, \tag{4.2}$$

and we assume that $G(0) = 0$.

Analogously to before, one can first give a well-posedness result. See [24] for the proof in the case of constant time delay.

Theorem 3 *Let $f : [-\bar{\tau}, 0] \to H$ be a continuous function. Then, the problem* (4.1) *has a unique (weak) solution given by Duhamel's formula*

$$U(t) = S(t)U_0 + \int_0^t S(t-s)[G(U(s)) + k(s)BU(s-\tau(s))]\, ds, \tag{4.3}$$

for all $t \geq 0$.

Proof Let $f \in C([-\bar{\tau}, 0], H)$. As before, we can give two different proofs.

Case 1 Assume that (2.2) holds true. We can argue step by step, as before, by restricting ourselves each time to time intervals of length τ_0.

First we consider $t \in [0, \tau_0]$. Then, from (2.2), $t - \tau(t) \in [-\bar{\tau}, 0]$. So, setting $F(t) = k(t)BU(t-\tau(t))$, $t \in [0, \tau_0]$, we have that $F(t) = k(t)Bf(t-\tau(t))$, $t \in [0, \tau_0]$. Then, problem (4.1) can be rewritten, in the interval $[0, \tau_0]$, as a standard inhomogeneous evolution problem:

$$\begin{aligned} &U'(t) = AU(t) + G(U(t)) + F(t) \quad \text{in } (0, \tau_0), \\ &U(0) = U_0. \end{aligned} \tag{4.4}$$

Since $k \in \mathcal{L}^1_{loc}([-\bar{\tau}, +\infty); \mathbb{R})$ and $f \in C([-\bar{\tau}, 0]; H)$, then we have that $F \in \mathcal{L}^1((0, \tau_0); H)$. Therefore, applying the standard theory for nonlinear evolution equations (see, e.g., [27]), there exists a unique solution $U \in C([0, \tau_0]; H)$ of (4.4) satisfying Duhamel's formula

$$U(t) = S(t)U_0 + \int_0^t S(t-s)[G(U(s)) + F(s)]ds, \quad t \in [0, \tau_0].$$

Therefore,

$$U(t) = S(t)U_0 + \int_0^t S(t-s)[G(U(s)) + k(s)BU(s-\tau(s))]ds, \quad t \in [0, \tau_0].$$

Next, we consider the time interval $[\tau_0, 2\tau_0]$ and also define $F(t) = k(t)BU(t - \tau(t))$, for $t \in [\tau_0, 2\tau_0]$. Note that if $t \in [\tau_0, 2\tau_0]$, then $t - \tau(t) \in [-\bar{\tau}, \tau_0]$ and so $U(t - \tau(t))$ is known from the first step. Then $F_{|[\tau_0, 2\tau_0]}$ is a known function, and it belongs to $\mathcal{L}^1((\tau_0, 2\tau_0); H)$. So we can rewrite our model (4.1) in the time interval $[\tau_0, 2\tau_0]$ as the inhomogeneous evolution problem

$$\begin{aligned} &U'(t) = AU(t) + G(U(t)) + F(t) \quad \text{for } t \in (\tau_0, 2\tau_0), \\ &U(\tau) = U(\tau_0^-). \end{aligned} \tag{4.5}$$

Then, we have a unique continuous solution $U : [\tau_0, 2\tau_0] \to H$ satisfying

$$U(t) = S(t-\tau_0)U(\tau_0^-) + \int_{\tau_0}^t S(t-s)[G(U(s)) + F(s)]ds, \quad t \in [\tau_0, 2\tau_0],$$

and so

$$U(t) = S(t-\tau_0)U(\tau_0^-) + \int_{\tau_0}^t S(t-s)[G(U(s)) + k(s)BU(s-\tau(s))]ds, \ t \in [\tau_0, 2\tau_0].$$

Putting together the partial solutions obtained in the first and second steps, we have a unique continuous solution $U : [0, 2\tau_0) \to \mathbb{R}$ satisfying Duhamel's formula

$$U(t) = S(t)U_0 + \int_0^t S(t-s)[G(U(s)) + k(s)BU(s-\tau(s))]ds, \quad t \in [0, 2\tau_0].$$

Iterating this procedure, we can find a unique solution $U \in C([0, +\infty); H)$ satisfying the representation formula (4.3).

Case 2 Let $\tau(\cdot)$ be a continuous function satisfying (1.1).

Now, since $k \in \mathcal{L}^1_{loc}([-\bar{\tau}, +\infty); \mathbb{R})$, there exists $T > 0$ sufficiently small such that

$$LT + \|B\| \|k\|_{\mathcal{L}^1([0,T];\mathbb{R})} = LT + \|B\| \int_0^T |k(s)| ds < \frac{1}{M}, \tag{4.6}$$

where L is the Lipschitz constant in (4.2). We define the set

$$\tilde{C}_f([-\bar{\tau}, T]; H) := \{U \in C([-\bar{\tau}, T]; H) : U(s) = f(s), \ \forall s \in [-\bar{\tau}, 0]\}.$$

Let us note that $\tilde{C}_f([-\bar{\tau}, T]; H)$ is a nonempty and closed subset of $C([-\bar{\tau}, T]; H)$. Hence, $(\tilde{C}_f([-\bar{\tau}, T]; H), \|\cdot\|_{C([-\bar{\tau},T];H)})$ is a Banach space.

Next, we define the map $\tilde{\Gamma} : \tilde{C}_f([-\bar{\tau}, T]; H) \to \tilde{C}_f([-\bar{\tau}, T]; H)$ given by

$$\tilde{\Gamma} U(t) = \begin{cases} S(t)U_0 + \int_0^t S(t-s)[G(U(s)) + k(s)BU(s-\tau(s))]\, ds, & t \in [0, T], \\ f(t), & t \in [-\bar{\tau}, 0). \end{cases}$$

Let us note that $\tilde{\Gamma}$ is well defined.

Moreover, $\tilde{\Gamma}$ is a contraction. Indeed, let $U, V \in \tilde{C}_f([\bar{\tau}, T]; H)$. Then, for all $t \in [-\bar{\tau}, 0]$,

$$\|\tilde{\Gamma} U(t) - \tilde{\Gamma} V(t)\| = 0.$$

On the other hand, for all $t \in (0, T]$, since G is Lipschitz continuous, we get

$$\begin{aligned} \|\tilde{\Gamma} U(t) - \tilde{\Gamma} V(t)\| &\le \int_0^t \|S(t-s)\| G(U(s)) - G(V(s))\| ds \\ &\quad + \int_0^t \|S(t-s)\| |k(s)| \|BU(s-\tau(s)) - BV(s-\tau(s))\| ds \\ &\le M(LT + \|B\| \|k\|_{\mathcal{L}^1([0,T];\mathbb{R})}) \|U - V\|_{C([-\bar{\tau},T];H)}. \end{aligned}$$

Thus,

$$\|\tilde{\Gamma} U - \tilde{\Gamma} V\|_{C([-\bar{\tau},T];H)} \le M(LT + \|B\| \|k\|_{\mathcal{L}^1([0,T];\mathbb{R})}) \|U - V\|_{C([-\bar{\tau},T];H)}.$$

As a consequence, from (4.6) $\tilde{\Gamma}$ is a contraction. Thus, from Banach's theorem, $\tilde{\Gamma}$ has a unique fixed point $U \in \tilde{C}_f([-\bar{\tau}, T]; H)$, i.e., (1.2) has a unique solution $U \in C([0, T]; H)$ given by the Duhamel formula

$$U(t) = S(t)U_0 + \int_0^t S(t-s)[G(U(s)) + k(s)BU(s-\tau(s))]\, ds, \quad \forall t \in [0, T].$$

Now, note that the solution U is bounded. So, arguing as in **Case 2** of Theorem 1, we can conclude that (4.1) has a unique continuous global solution that satisfies Duhamel's formula (4.3).

Under an appropriate relation between the problem's parameters, the system (4.1) is exponentially stable.

Theorem 4 *Assume* (3.1) *and* $L < \frac{\omega-\omega'}{M}$. *Then, for every* $f \in C([-\bar{\tau}, 0]; H)$, *the solution* $U \in C([0, +\infty); H)$ *to* (4.1) *with the initial datum* f *satisfies the exponential decay estimate*

$$||U(t)|| \leq Me^{\gamma}\left(\|U_0\| + e^{\omega\bar{\tau}}K\|B\| \max_{s\in[-\bar{\tau},0]}\left\{\|e^{\omega s}f(s)\|\right\}\right)e^{-(\omega-\omega'-ML)t}, \tag{4.7}$$

for any $t \geq 0$.

Proof Let $f \in C([-\bar{\tau}, 0]; H)$, and let U be the unique global solution to (3) with initial datum f. Then, from Duhamel's formula (4.3), we have that

$$\begin{aligned} ||U(t)|| \leq Me^{-\omega t}||U_0|| + Me^{-\omega t}\int_0^t e^{\omega s}||G(U(s))||ds \\ +M\|B\|e^{-\omega t}\int_0^t e^{\omega s}|k(s)|\cdot||U(s-\tau(s))||)ds, \end{aligned}$$

for all $t \geq 0$. Now, using the same arguments employed in Theorem 2, we get

$$\begin{aligned} &\int_0^t e^{\omega s}|k(s)|\cdot||U(s-\tau(s))||\,ds \leq e^{\omega\bar{\tau}}\int_0^{\bar{\tau}} e^{\omega(s-\tau(s))}|k(s)|\cdot||U(s-\tau(s))||\,ds \\ &+e^{\omega\bar{\tau}}\int_{\bar{\tau}}^t e^{\omega(s-\tau(s))}|k(s)|\cdot||U(s-\tau(s))||\,ds \\ &\leq e^{\omega\bar{\tau}}K \max_{r\in[-\bar{\tau},0]}\{||e^{\omega r}f(r)||\} + e^{\omega\bar{\tau}}\int_0^t |k(s)|\cdot \max_{r\in[s-\bar{\tau},s]\cap[0,s]}\{e^{\omega r}||U(r)||\}ds, \end{aligned} \tag{4.8}$$

for all $t \geq \bar{\tau}$. Also, for all $t \in [0, \bar{\tau}]$,

$$\begin{aligned} &\int_0^t e^{\omega s}|k(s)|\cdot||U(s-\tau(s))||\,ds \leq e^{\omega\bar{\tau}}\int_0^t e^{\omega(s-\tau(s))}|k(s)|\cdot||U(s-\tau(s))||\,ds \\ &\leq e^{\omega\bar{\tau}}K \max_{r\in[-\bar{\tau},0]}\{||e^{\omega r}f(r)||\} + e^{\omega\bar{\tau}}\int_0^t |k(s)|\cdot \max_{r\in[s-\bar{\tau},s]\cap[0,s]}\{e^{\omega r}||U(r)||\}ds, \end{aligned}$$

i.e., (4.8) holds true for all $t \geq 0$. Hence, from (4.8), we get

$$||U(t)|| \le Me^{-\omega t}\left(||U_0|| + e^{\omega\bar{\tau}}K\|B\| \max_{r\in[-\bar{\tau},0]}\{||e^{\omega r}f(r)||\}\right)$$

$$+ Me^{-\omega t}e^{\omega\bar{\tau}}||B|| \int_0^t |k(s)| \cdot \max_{r\in[s-\bar{\tau},s]\cap[0,s]}\{e^{\omega r}||U(r)||\}ds$$

$$+ Me^{-\omega t}\int_0^t e^{\omega s}||G(U(s))||ds.$$

Now, let us note that, being $G(0) = 0$ and being G Lipschitz continuous of constant L,

$$\int_0^t e^{\omega s}||G(U(s))||ds \le L\int_0^t e^{\omega s}||U(s)|| \le L\int_0^t \max_{r\in[s-\bar{\tau},s]\cap[0,s]}\{e^{\omega r}||U(r)||\}ds.$$

As a consequence, we can write

$$\begin{aligned}||U(t)|| \le\ & Me^{-\omega t}(||U_0|| + e^{\omega\bar{\tau}}K||B|| \max_{r\in[-\bar{\tau},0]}\{||e^{\omega r}f(r)||\})\\ &+Me^{-\omega t}\int_0^t (e^{\omega\bar{\tau}}||B|||k(s)| + L)\cdot \max_{r\in[s-\bar{\tau},s]\cap[0,s]}\{e^{\omega r}||U(r)||\}ds.\end{aligned}$$

Then, we have that

$$\begin{aligned}e^{\omega t}||U(t)|| \le\ & M\left(||U_0|| + e^{\omega\bar{\tau}}K||B|| \max_{r\in[-\bar{\tau},0]}\left\{\|e^{\omega r}f(r)\|\right\}\right)\\ &+M\int_0^t (e^{\omega\bar{\tau}}||B|||k(s)| + L)\cdot \max_{r\in[s-\bar{\tau},s]\cap[0,s]}\left\{e^{\omega r}||U(r)||\right\}ds, \quad \forall t \ge 0.\end{aligned}$$

Now, we note that

$$\begin{aligned}\max_{r\in[t-\bar{\tau},t]\cap[0,t]}\left\{e^{\omega r}||U(r)||\right\} \le\ & M\left(||U_0|| + e^{\omega\bar{\tau}}K||B|| \max_{s\in[-\bar{\tau},0]}\left\{\|e^{\omega s}f(s)\|\right\}\right)\\ &+M\int_0^t (e^{\omega\bar{\tau}}||B|||k(s)| + L)\cdot \max_{r\in[s-\bar{\tau},s]\cap[0,s]}\left\{e^{\omega r}||U(r)||\right\}ds, \quad \forall t \ge 0.\end{aligned}$$

Hence, if we denote with

$$\tilde{u}(t) := \max_{r\in[t-\bar{\tau},t]\cap[0,t]}\left\{e^{\omega r}||U(r)||\right\},$$

Gronwall's estimate yields

$$\tilde{u}(t) \le \tilde{M}e^{M\|B\|e^{\omega\bar{\tau}}\int_0^t|k(s)|ds + MLt}, \quad \forall t \ge 0,$$

where

$$\tilde{M} := M\left(\|U_0\| + e^{\omega\bar{\tau}}K\|B\| \max_{r\in[-\bar{\tau},0]}\left\{\|e^{\omega r}f(r)\|\right\}\right).$$

Then,

$$e^{\omega t}\|U(t)\| \leq \tilde{M}e^{M\|B\|e^{\omega\bar{\tau}}\int_0^t |k(s)|ds + MLt}.$$

Finally, by the assumption (3.1) and the assumption on the Lipschitz constant L, we get the exponential decay estimate (4.7).

5 Examples

In this section we consider, as concrete examples for which previous abstract well-posedness and stability results hold, the wave equation with localized frictional damping and delay feedback and an elasticity system with analogous feedback laws.

5.1 The Damped Wave Equation

Let Ω be an open bounded subset of $\mathbb{R}^d$, with boundary $\partial\Omega$ of class C^2, and let $\mathcal{O} \subset \Omega$ be an open subset which satisfies the geometrical control property in [7]. For instance, $\mathcal{O} \subset \Omega$ can be a neighborhood of the whole boundary $\partial\Omega$ or, denoting by m the standard multiplier $m(x) = x - x_0$, $x_0 \in \mathbb{R}^d$, as in [19], $\mathcal{O}$ can be the intersection of Ω with an open neighborhood of the set

$$\Gamma_0 = \{x \in \Gamma \ : \ m(x)\cdot\nu(x) > 0\}.$$

Moreover, let $\tilde{\mathcal{O}} \subset \Omega$ be another open subset. Denoting by $\chi_{\mathcal{O}}$ and $\chi_{\tilde{\mathcal{O}}}$ the characteristic functions of the sets $\mathcal{O}$ and $\tilde{\mathcal{O}}$, respectively, we consider the following wave equation:

$$\begin{aligned}
&u_{tt}(x,t) - \Delta u(x,t) + a\chi_{\mathcal{O}}(x)u_t(x,t)\\
&\qquad\qquad + k(t)\chi_{\tilde{\mathcal{O}}}(x)u_t(x,t-\tau(t)) = 0, \quad (x,t) \in \Omega\times(0,+\infty),\\
&u(x,t) = 0, \quad (x,t) \in \partial\Omega\times(0,+\infty),\\
&u(x,s) = u_0(x,s), \quad u_t(x,s) = u_1(x,s), \quad (x,s) \in \Omega\times[-\bar{\tau},0],
\end{aligned} \tag{5.1}$$

where a is a positive constant, $\tau(t)$ is the time delay function satisfying $0 \leq \tau(t) \leq \bar{\tau}$, and the delayed damping coefficient $k(\cdot) : [-\bar{\tau},+\infty) \to (0,+\infty)$ is an $\mathcal{L}^1_{loc}([-\bar{\tau},+\infty))$ function satisfying (1.4). Denoting $v(t) = u_t(t)$ and $U(t) =$

$(u(t), v(t))^T$, for any $t \geq 0$, we can rewrite system (5.1) in the abstract form (1.2), with $\mathcal{H} = H_0^1(\Omega) \times L^2(\Omega)$,

$$A = \begin{pmatrix} 0 & Id \\ \Delta & -a\chi_{\mathcal{O}} \end{pmatrix}$$

and

$$B\begin{pmatrix} u \\ v \end{pmatrix} = \begin{pmatrix} 0 \\ -\chi_{\tilde{\mathcal{O}}} v \end{pmatrix}, \quad \forall\, t \geq 0.$$

We know that A generates an exponentially stable C_0-semigroup $\{S(t)\}_{t\geq 0}$ (see, e.g., [16]), namely there exist ω, $M > 0$ such that

$$||S(t)||_{\mathcal{L}(\mathcal{H})} \leq Me^{-\omega t}, \quad \forall\, t \geq 0.$$

Hence, under the assumption (3.1), the stability estimate of Theorem 2 holds for such a model. Then, we can deduce an exponential decay estimate for the energy functional

$$E(t):=\frac{1}{2}\int_\Omega |u_t(x,t)|^2dx+\frac{1}{2}\int_\Omega |\nabla u(x,t)|^2dx+\frac{1}{2}\int_{t-\bar{\tau}}^t\int_{\tilde{\mathcal{O}}} |k(s)|\cdot|u_t(x,s)|^2dxds.$$

Theorem 5 *Under assumption* (3.1), *for all initial data* $(u_0, u_1) \in C([-\bar{\tau}, 0]; H_0^1(\Omega) \times L^2(\Omega))$, *the solution to* (5.1) *satisfies the energy decay estimate*

$$E(t) \leq C_* e^{-\beta t}, \quad t \geq 0, \tag{5.2}$$

where C_* is a constant depending on the initial data and $\beta > 0$.

Proof From the energy definition,

$$\begin{aligned} E(t) &= \frac{1}{2}\|U(t)\|^2 + \frac{1}{2}\int_{t-\bar{\tau}}^t\int_{\tilde{\mathcal{O}}} |k(s)|\cdot|u_t(x,s)|^2dxds \\ &\leq \frac{1}{2}\|U(t)\|^2 + \frac{1}{2}\int_{t-\bar{\tau}}^t |k(s)|\|U(s)\|^2ds. \end{aligned} \tag{5.3}$$

Then, from Theorem 2,

$$\|U(t)\| \leq C_0 e^{-(\omega-\omega')t}, \quad \forall t \geq 0,$$

for a suitable constant C_0 depending on the initial data. So, we can estimate

$$\int_{t-\bar{\tau}}^{t} |k(s)| \|U(s)\|^2 ds \leq C_0 K e^{(\omega-\omega')\bar{\tau}} e^{-(\omega-\omega')t}, \quad \forall t \geq 0.$$

By using the last two inequalities in (5.3), we obtain the exponential decay estimate (5.2).

Remark 1 As another example, we could consider the damped plate equation (see, e.g., [22] for the model details). The analysis is analogous to the wave case above. Then, under suitable assumptions, the exponential stability result holds for that model.

5.2 A Damped Elasticity System

Let $\Omega \subset \mathbb{R}^d$, and let $\tilde{\mathcal{O}}, \mathcal{O} \subset \Omega$ be as in the previous example. We consider the following elastodynamic system:

$$\begin{aligned}
&u_{tt}(x,t)-\mu\Delta u(x,t)-(\lambda+\mu)\nabla \operatorname{div} u + a\chi_{\mathcal{O}}(x)u_t(x,t) \\
&\qquad\qquad +k(t)\chi_{\tilde{\mathcal{O}}}(x)u_t(x,t-\tau(t))=0, \quad (x,t) \in \Omega \times (0,+\infty), \\
&u(x,t) = 0, \qquad (x,t) \in \partial\Omega \times (0,+\infty), \\
&u(x,s) = u_0(x,s), \quad u_t(x,s) = u_1(x,s), \quad (x,s) \in \Omega \times [-\bar{\tau},0],
\end{aligned} \tag{5.4}$$

where a is a positive constant, $\tau(t)$ is the time delay function satisfying $0 \leq \tau(t) \leq \bar{\tau}$, and the delayed damping coefficient $k(\cdot) : [-\bar{\tau}, +\infty) \to (0, +\infty)$ is an $\mathcal{L}^1_{loc}([-\bar{\tau}, +\infty))$ function satisfying (1.4). Note that, in this case, the function u is vector-valued and takes values in $\mathbb{R}^d$, while λ and μ are positive constants usually called Lamé coefficients. Denoting $v(t) = u_t(t)$ and $U(t) = (u(t), v(t))^T$, for any $t \geq 0$, we can rewrite system (5.4) in the abstract form (1.2), with $\mathcal{H} = H_0^1(\Omega)^d \times L^2(\Omega)^d$,

$$A = \begin{pmatrix} 0 & Id \\ \mu\Delta + (\lambda+\mu)\nabla\text{div} & -a\chi_{\mathcal{O}} \end{pmatrix}$$

and

$$B\begin{pmatrix} u \\ v \end{pmatrix} = \begin{pmatrix} 0 \\ -\chi_{\tilde{\mathcal{O}}} v \end{pmatrix}, \quad \forall t \geq 0.$$

We know that A generates an exponentially stable C_0-semigroup $\{S(t)\}_{t\geq 0}$ (see, e.g., [13]), namely there exist $\omega, M > 0$ such that

$$||S(t)||_{\mathcal{L}(\mathcal{H})} \leq M e^{-\omega t}, \quad \forall t \geq 0.$$

Hence, under the assumption (3.1), the stability estimate of Theorem 2 holds for such a model. Therefore, we can deduce an exponential decay estimate for the energy functional

$$\mathcal{E}(t) := \frac{1}{2}\int_{\Omega} |u_t(x,t)|^2 dx + \frac{1}{2}\int_{\Omega}\Big[\mu \sum_{i,j=1}^{n}\Big(\frac{\partial}{\partial x_i}u_j(x,t)\Big)^2 + (\lambda+\mu)|\operatorname{div} u|\Big]dx$$
$$+\frac{1}{2}\int_{t-\bar{\tau}}^{t}\int_{\tilde{\mathcal{O}}} |k(s)|\cdot|u_t(x,s)|^2 dx ds.$$

Theorem 6 *Assume* (3.1). *Then, for all initial data* $(u_0, u_1) \in C([-\bar{\tau}, 0]; H_0^1(\Omega)^d \times L^2(\Omega)^d)$, *the solution to* (5.4) *satisfies the energy decay estimate*

$$\mathcal{E}(t) \le \bar{C}e^{-\beta^* t}, \quad t \ge 0, \tag{5.5}$$

where $\bar{C}$ *is a constant depending on the initial data and* $\beta^* > 0$.

Proof The proof comes analogously to the one of Theorem 5.

Acknowledgments We would like to thank GNAMPA and UNIVAQ for the support.

References

1. E.M. Ait Benhassi, K. Ammari, S. Boulite, L. Maniar,Feedback stabilization of a class of evolution equations with delay. J. Evol. Equ. **9**, 103–121 (2009)
2. F. Alabau-Boussouira, S. Nicaise, C. Pignotti, Exponential stability of the wave equation with memory and time delay. *New Prospects in Direct, Inverse and Control Problems for Evolution Equations*. Springer INdAM Series, vol. 10 (2014), pp. 1–22
3. K. Ammari, B. Chentouf, Asymptotic behavior of a delayed wave equation without displacement term. Z. Angew. Math. Phys. **68**(5), Paper No. 117 (2017)
4. K. Ammari, S. Gerbi, Interior feedback stabilization of wave equations with dynamic boundary delay. Z. Anal. Anwend. **36**, 297–327 (2017)
5. K. Ammari, S. Nicaise, C. Pignotti, Feedback boundary stabilization of wave equations with interior delay. Syst. Control Lett. **59**, 623–628 (2010)
6. T.A. Apalara, S.A. Messaoudi, An exponential stability result of a Timoshenko system with thermoelasticity with second sound and in the presence of delay. Appl. Math. Optim. **71**, 449–472 (2015)
7. C. Bardos, G. Lebeau, J. Rauch, Sharp sufficient conditions for the observation, control and stabilization of waves from the boundary. SIAM J. Control Optim. **30**, 1024–1065 (1992)
8. R.A. Capistrano-Filho, F. Gallego, V. Komornik, A qualitative study of the generalized dispersive systems with time-delay: the unbounded case. ArXiv:2303.06693 (2023)
9. B. Chentouf, S. Mansouri, Boundary stabilization of a flexible structure with dynamic boundary conditions via one time-dependent delayed boundary control. Discrete Contin. Dyn. Syst. - Ser. S **15**, 1127–1141 (2022)
10. Q. Dai, Z. Yang, Global existence and exponential decay of the solution for a viscoelastic wave equation with a delay. Z. Angew. Math. Phys. **65**, 885–903 (2014)

11. R. Datko, Not all feedback stabilized hyperbolic systems are robust with respect to small time delays in their feedbacks. SIAM J. Control Optim. **26**, 697–713 (1988)
12. R. Datko, J. Lagnese, M.P. Polis, An example on the effect of time delays in boundary feedback stabilization of wave equations. SIAM J. Control Optim. **24**, 152–156 (1986)
13. V.N. Domingos Cavalcanti, J.S. Prates Filho, Exact internal controllability of the elasticity system. Southwest J. Pure Appl. Math. **1**, 29–43 (1998)
14. B. Feng, C.A. Raposo, C.A. Nonato, A. Soufyane, Analysis of exponential stabilization for Rao-Nakra sandwich beam with time-varying weight and time-varying delay: multiplier method versus observability. Math. Control Relat. Fields **13**, 631–663 (2023)
15. A. Guesmia, Well-posedness and exponential stability of an abstract evolution equation with infinite memory and time delay. IMA J. Math. Control Inform. **30**, 507–526 (2013)
16. V. Komornik, *Exact Controllability and Stabilization. The Multiplier Method* (John Wiley & Sons, Masson, 1994). Chicester
17. V. Komornik, C. Pignotti, Well-posedness and exponential decay estimates for a Korteweg-de Vries-Burgers equation with time-delay. Nonlinear Anal. **191**, Art. 111646 (2020)
18. V. Komornik, C. Pignotti, Energy decay for evolution equations with delay feedbacks. Math. Nachr. **295**, 377–394 (2022)
19. J.L. Lions, Exact controllability, stabilizability, and perturbations for distributed systems. SIAM Rev. **30**, 1–68 (1988)
20. T.F. Ma, J.G. Mesquita, P.N. Seminario-Huertas, Smooth dynamics of weakly damped Lamé systems with delay. SIAM J. Math. Anal. **53**, 3759–3771 (2021)
21. S. Nicaise, C. Pignotti, Stability and instability results of the wave equation with a delay term in the boundary or internal feedbacks. SIAM J. Control Optim. **45**, 1561–1585 (2006)
22. S. Nicaise, C. Pignotti, Stabilization of second-order evolution equations with time delay. Math. Control Signals Syst. **26**, 563–588 (2014)
23. S. Nicaise, C. Pignotti, Exponential stability of abstract evolution equations with time delay. J. Evol. Equ. **15**, 107–129 (2015)
24. S. Nicaise, C. Pignotti, Well–posedness and stability results for nonlinear abstract evolution equations with time delays. J. Evol. Equ. **18**, 947–971 (2018)
25. S. Nicaise, C. Pignotti, J. Valein, Exponential stability of the wave equation with boundary time-varying delay. Discrete Contin. Dyn. Syst. - Ser. S **4**, 693–722 (2011)
26. R.L. Oliveira, H.P. Oquendo, Stability and instability results for coupled waves with delay term. J. Math. Phys. **61**, Art. 071505 (2020)
27. A. Pazy, *Semigroups of Linear Operators and Applications to Partial Differential Equations*, vol. 44. Applied Mathematical Sciences (Springer, New York, 1983)
28. C. Pignotti, A note on stabilization of locally damped wave equations with time delay. Syst. Control Lett. **61**, 92–97 (2012)
29. B. Said-Houari, A. Soufyane, Stability result of the Timoshenko system with delay and boundary feedback. IMA J. Math. Control Inform. **29**, 383–398 (2012)
30. G.Q. Xu, S.P. Yung, L.K. Li, Stabilization of wave systems with input delay in the boundary control. ESAIM Control Optim. Calc. Var. **12**, 770–785 (2006)

Study of the Numerical Method for an Inverse Problem of a Simplified Intestinal Crypt

Marie Haghebaert, Béatrice Laroche, and Mauricio A. Sepúlveda Cortés

2010 Mathematics Subject Classification 65M32

1 The Simplified Intestinal Crypt Model

1.1 The Stem-Progenitor Interaction Model

We consider a model of epithelial cells interacting with the microbiota-derived chemicals diffusing in the crypt from the gut lumen. The original model, derived from an individual-based PDMP model (piecewise deterministic Markov process model, treats five well-differentiated cell types: stem cells (SCs), progenitor cells (PCs), enterocytes (ENT), globet cells (GCs), and deep crypt secretory (DCS) cells. For a simplification of the original model in the process of studying the inverse problem, we will consider at first only the SC, PC, and DCS cells. We set $\rho_{tot} = \rho_{sc} + \rho_{dcs} + \rho_{pc}$ and solve equations on ρ_{sc} and ρ_{pc}, with $\rho_{dcs} = \rho_{dcs}(z)$ independent of t. In this reduced model, the system of equations is given by

M. Haghebaert · B. Laroche
Université Paris-Saclay, INRAE, MaIAGE, Jouy-en-Josas, France

Université Paris-Saclay, Inria Saclay-Ile-de-France, Palaiseau, France
e-mail: marie.haghebaert@inrae.fr; beatrice.laroche@inrae.fr

M. A. Sepúlveda Cortés (✉)
DIM & CI^{2}MA, Universidad de Concepción, Concepción, Chile

K. Ammari et al. (eds.), *Advances in Partial Differential Equations and Control*,
Trends in Mathematics, https://doi.org/10.1007/978-3-031-62265-6_9

$$\begin{cases} \partial_t \rho_{sc} - \mathcal{W}\partial_z \left(\phi(z)\rho_{sc}\partial_z \rho_{tot}\right) = H_1(z, \rho_{dcs}, \rho_{sc}, \rho_{pc}), & 0 < z < z_{max},\ t > 0, \\ \rho_{sc}(t,0) = \rho_{sc}^{bot}, \qquad \partial_z \rho_{sc}(t, z_{max}) = 0, & t > 0, \\ \rho_{sc}(x,0) = \rho_{sc}^{init}(x), & 0 < z < z_{max}, \end{cases} \tag{1.1}$$

where $H_1(z, \rho_{dcs}, \rho_{sc}, \rho_{pc}) = \left[q_{sc}(1 - R_n(z))(1 - R_{sc}(D\rho_{tot})) - q_{diff} R_n(z)\right] \rho_{sc}$ with

$$R_n(z) = R(z, Z_{niche}, \kappa_{niche})$$

$$R_{sc}(\varrho) = R(\varrho, K_{sc}, \kappa_{dens})$$

$$\begin{cases} \partial_t \rho_{pc} - \mathcal{W}\partial_z \left(\phi(z)\rho_{pc}\partial_z \rho_{tot}\right) = H_2(z, \rho_{dcs}, \rho_{sc}, \rho_{pc}), & 0 < z < z_{max},\ t > 0, \\ \rho_{pc}(t,0)=0, \qquad \partial_z \rho_{pc}(t, z_{max})=0, & t > 0, \\ \rho_{pc}(x,0) = \rho_{pc}^{init}(x), & 0 < z < z_{max}, \end{cases} \tag{1.2}$$

where $H_2(z, \rho_{dcs}, \rho_{sc}, \rho_{pc}) = \left[q_{pc}(1 - R_t(z))(1 - R_{pc}(D\rho_{tot})) - q_{ex} R_e(z) R_{pc}(D\rho_{tot})\right] \rho_{pc} + q_{diff} R_n(z) \rho_{sc}$ with

$$R_t(z) = R(z, Z_{tiers}, \kappa_{tiers}),$$

$$R_e(z) = R(z, Z_{ex}, \kappa_{ex}),$$

$$R_{pc}(\varrho) = R(\varrho, K_{pc}, \kappa_{dens}).$$

We use a generic regulation function $R(y, K, \kappa)$ for $y \geq 0$, $K \geq 0$, and $\kappa \geq 0$ that models the regulation of cell fate event according to variable y. The function R is a piecewise polynomial function in $\mathcal{C}^1(0, z_{max})$ defined as (Fig. 1)

$$R(y, K, \kappa) = \begin{cases} 0 & \text{if } y \geq K - \kappa, \\ \dfrac{-y^3 + 3Ky^2 - (3K^2 - 3\kappa^2)y + (K^3 + 2\kappa^3 - 3K\kappa^2)}{4\kappa^3} & \text{if } K - \kappa < y < K + \kappa, \\ 1 & \text{if } K + \kappa \geq y. \end{cases}$$

Additionally, $\phi \in \mathcal{C}^1(0, z_{max})$, such that $\phi(0) = \phi(z_{max}) = 0$, $\phi(z) = 1$ in $[r_0 - \varepsilon, z_{max} - r_0 + \varepsilon]$, and $\phi(z) \geq 0$, represents a corrective function taking into account the impact of curvature on mechanical forces modeling by Darrigade et al. [3] (see Fig. 2)

$$\phi(z) = \begin{cases} \frac{\sqrt{(\varepsilon+z)(2r_0-\varepsilon-z)} - \sqrt{\varepsilon(2r_0-\varepsilon)}}{r_0 - \sqrt{\varepsilon(2r_0-\varepsilon)}} & \text{if } z \leq r_0 - \varepsilon \\ 1 & \text{if } r_0 - \varepsilon < z < z_{max} - r_0 + \varepsilon \\ \frac{2r_0 - \sqrt{\varepsilon(2r_0-\varepsilon)}}{r_0} - \frac{2r_0 - \sqrt{(\varepsilon - z + z_{\max})(2r0 - \varepsilon + z - z_{\max})}}{r0 - \sqrt{\varepsilon(2r_0-\varepsilon)}} & \text{if } z \geq z_{max} - r_0 + \varepsilon. \end{cases}$$

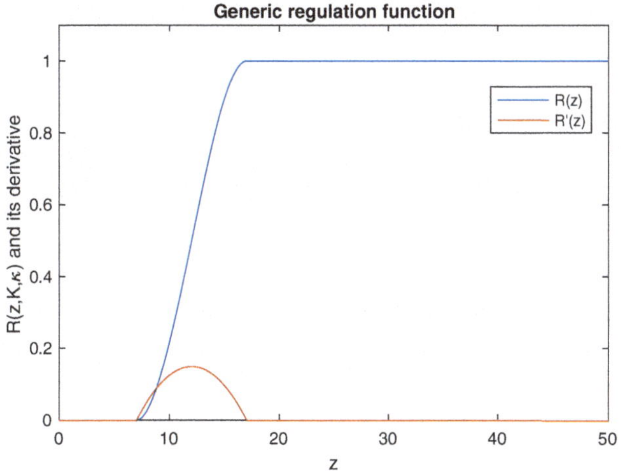

Fig. 1 Generic regulation function and its derivative, with $K = 12, \kappa = 5$

Fig. 2 Graph of the geometric curvature function and its derivative, with $z_{max} = 200, r_0 = 10, \varepsilon = 1$

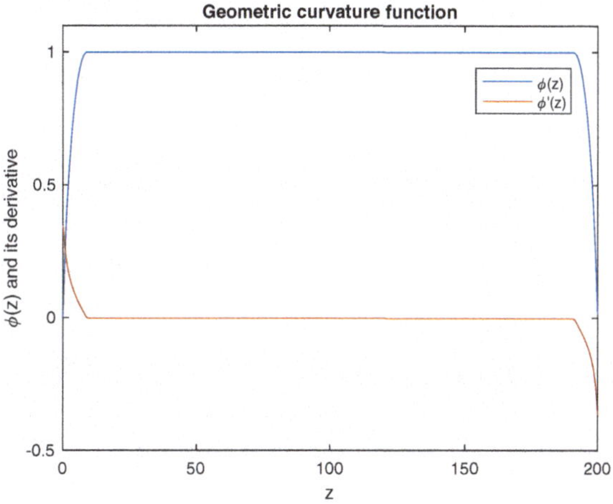

The system (1.1)–(1.2) can be rewritten as

$$\begin{cases} \partial_t \rho_{sc} + \partial_z A_{sc} = \partial_{zz} B(\rho_{sc}) + H_1, & 0 < z < z_{max}, \ t > 0, \\ \partial_t \rho_{pc} + \partial_z A_{pc} = \partial_{zz} B(\rho_{pc}) + H_2, & 0 < z < z_{max}, \ t > 0, \\ \rho_{sc}(t, 0) = \rho_{sc}^{bot}, \qquad \rho_{pc}(t, 0) = 0, & t > 0, \\ \partial_z \rho_{sc}(t, z_{max}) = \partial_z \rho_{pc}(t, z_{max}) = 0, & t > 0, \\ \rho_{sc}(x, 0) = \rho_{sc}^{init}(x), \quad \rho_{pc}(x, 0) = \rho_{pc}^{init}(x), & 0 < z < z_{max} \end{cases} \tag{1.3}$$

whose semiconservative structure (conservative in its convective and diffusive terms but with nonlinear source terms) allows us to consider a relaxation scheme as in [1].

The system (1.3) is equivalent to (1.1)–(1.2), through the relations

$$A_{sc} = A(z, \rho'_{dcs}, \partial_z \rho_{pc}, \rho_{sc})$$
$$A_{pc} = A(z, \rho'_{dcs}, \partial_z \rho_{sc}, \rho_{pc})$$
$$A(z, \xi, \eta, \rho) = -\mathcal{W}\left[\phi(\xi + \eta)\rho - \phi'\rho^2/2)\right],$$
$$B(\rho) = B(z, \rho) = \mathcal{W}\phi\rho^2/2.$$

When H_1 and H_2 does not depend on ρ_{pc} and ρ_{sc} and

$$H_1, H_2 \in L^\infty((0, z_{max} \times (0, T)),$$

the problem (1.3) is well posed, and there is a unique entropy solution in $L^\infty((0, z_{max} \times (0, T))$ such that $B(\rho) \in L^2(0, T; H^1(0, z_{max}))$ (see, for example, [2, 5]). On the other hand, the weak solution of the system (1.3) verifies

$$E(\rho_{sc}, \rho_{pc}, w_{sc}, w_{pc}; \rho_{dcs}) = 0, \qquad \text{for all } w_{sc}, w_{pc}$$

with w_{sc}, w_{pc} smooth test functions on $[0, T] \times [0, z_{max}]$, such that for all t, $w_{pc}(t, 0) = w_{pc}(t, z_{max}) = 0$ and $w_{sc}(t, 0) = w_{sc}(t, z_{max}) = 0$ and

$$\begin{aligned}
&E(\rho_{sc}, \rho_{pc}, w_{sc}, w_{pc}; \rho_{dcs}) := \\
&\int_0^T \int_0^{z_{max}} [\rho_{sc}\partial_t w_{sc} + A_{sc}\partial_z w_{sc} + B(\rho_{sc})\partial_{zz} w_{sc} + H_1 w_{sc}]\, dz\, dt \\
&+ \int_0^T \int_0^{z_{max}} \left[\rho_{pc}\partial_t w_{pc} + A_{pc}\partial_z w_{pc} + B(\rho_{pc})\partial_{zz} w_{pc} + H_2 w_{pc}\right] dz\, dt \\
&- \int_0^{z_{max}} \rho_{sc}(z, T) w_{sc}(z, T)\, dz + \int_0^{z_{max}} \rho_{sc}^{init}(z) w_{sc}(z, 0)\, dz \\
&- \int_0^{z_{max}} \rho_{pc}(z, T) w_{pc}(z, T)\, dz + \int_0^{z_{max}} \rho_{pc}^{init}(z) w_{pc}(z, 0)\, dz.
\end{aligned} \tag{1.4}$$

2 Inverse Problem

2.1 The Problem of Identifying the Function $\rho_{dcs}(z)$

(1) Given $\rho_{sc}^{obs}(z, T)$ and $\rho_{pc}^{obs}(z, T)$ (experimental, observed, or measured data), find $\rho_{dcs} \in \mathcal{C}(0, z_{max}) \cap L^\infty(0, z_{max})$, such that $\rho_{sc}(z, T) \approx \rho_{sc}^{obs}(z, T)$ and $\rho_{pc}(z, T) \approx \rho_{pc}^{obs}(z, T)$.

Equivalently, assuming a parametrization of the ρ_{dcs} density by a vector θ of N real parameters

$$\rho_{dcs}(z) = \bar{\rho}_{dcs}(z, \theta),$$

The identification problem is expressed as follows:

(2) Given $\rho_{sc}^{obs}(z, T)$ and $\rho_{pc}^{obs}(z, T)$ (experimental, observed, or measured data), find $\theta \in \mathcal{A} \subset \mathbb{R}^N$, such that $\rho_{sc}(z, T) \approx \rho_{sc}^{obs}(z, T)$ and $\rho_{pc}(z, T) \approx \rho_{pc}^{obs}(z, T)$.

We define the cost function, with $\alpha, \beta > 0$, as

$$J(\rho_{sc}, \rho_{pc}) := \frac{\alpha}{2} \int_0^{z_{max}} |\rho_{sc}(z, T) - \rho_{sc}^{obs}(z, T)|^2 dz + \frac{\beta}{2} \int_0^{z_{max}} |\rho_{pc}(z, T) - \rho_{pc}^{obs}(z, T)|^2 dz. \tag{2.1}$$

The inverse problem consists of

$$\begin{cases} \text{find } \theta \in \mathcal{A}, \text{ such that} \\ J(\{\rho_{sc}, \rho_{pc}\}(\bar{\theta})) = \min\limits_{\theta \in \mathcal{A}} J(\rho_{sc}, \rho_{pc}). \end{cases}$$

We define the Lagrangian

$$\mathcal{L}(\rho_{sc}, \rho_{pc}, w_{sc}, w_{pc}; \rho_{dcs}) := J(\rho_{sc}, \rho_{pc}) + E(\rho_{sc}, \rho_{pc}, w_{sc}, w_{pc}; \rho_{dcs}). \tag{2.2}$$

The derivatives of $\mathcal{L}$ with respect to ρ_{sc} and ρ_{pc} in the directions $\delta\rho_{sc}$ and $\delta\rho_{pc}$, respectively, are given by

$$\left\langle \frac{\partial \mathcal{L}}{\partial \rho_{sc}}, \delta\rho_{sc} \right\rangle = \alpha \int_0^{z_{max}} (\rho_{sc}(z, T) - \rho_{sc}^{obs}(z, T))\delta\rho_{sc}(z, T) dz + \left\langle \frac{\partial E}{\partial \rho_{sc}}, \delta\rho_{sc} \right\rangle = 0 \tag{2.3}$$

$$\left\langle \frac{\partial \mathcal{L}}{\partial \rho_{pc}}, \delta\rho_{pc} \right\rangle = \beta \int_0^{z_{max}} (\rho_{pc}(z, T) - \rho_{pc}^{obs}(z, T))\delta\rho_{pc}(z, T) dz + \left\langle \frac{\partial E}{\partial \rho_{pc}}, \delta\rho_{pc} \right\rangle = 0 \tag{2.4}$$

Derivating (1.4) with respect to ρ_{sc} and ρ_{pc} and replacing in (2.3)–(2.4) allow us to establish the adjoint problem given by

$$\begin{cases} \partial_t w_{sc} + \dfrac{\partial A}{\partial \rho}(z, \rho'_{dcs}, \partial \rho_{pc}, \rho_{sc})\partial_z w_{sc} - \partial_z \left[\left(\dfrac{\partial A}{\partial \eta}(z, \rho'_{dcs}, \partial \rho_{sc}, \rho_{pc}) \right) \partial_z w_{pc} \right] \\ \qquad\qquad = \dfrac{\partial B}{\partial \rho}(\rho_{sc})\partial_{zz} w_{sc} + \dfrac{\partial H_1}{\partial \rho_{sc}} w_{sc} + \dfrac{\partial H_2}{\partial \rho_{sc}} w_{pc}, \\ w_{sc}(z_{max}, t) = w_{sc}(0, t) = 0, \\ w_{sc}(z, T) = \alpha(\rho_{sc}(z, T) - \rho_{sc}^{obs}(z, T)) \\ \partial_t w_{pc} + \dfrac{\partial A}{\partial \rho}(z, \rho'_{dcs}, \partial \rho_{sc}, \rho_{pc})\partial_z w_{pc} - \partial_z \left[\left(\dfrac{\partial A}{\partial \eta}(z, \rho'_{dcs}, \partial \rho_{pc}, \rho_{sc}) \right) \partial_z w_{sc} \right] \\ \qquad\qquad = \dfrac{\partial B}{\partial \rho}(\rho_{pc})\partial_{zz} w_{pc} + \dfrac{\partial H_1}{\partial \rho_{pc}} w_{sc} + \dfrac{\partial H_2}{\partial \rho_{pc}} w_{pc}, \\ w_{pc}(z_{max}, t) = w_{pc}(0, t) = 0, \\ w_{pc}(z, T) = \beta(\rho_{pc}(z, T) - \rho_{pc}^{obs}(z, T)). \end{cases}$$

We suppose that $\rho_{dcs} = \rho_{dcs}(\theta)$ is of class $\mathcal{C}^1$ with respect to θ, the finite set of parameters to identify, and we suppose that $\rho'_{dcs} = \rho'_{dcs}(\theta)$ is also of class $\mathcal{C}^1$ with respect to θ. The derivative of the cost function with respect to θ_i is given by

$$\begin{aligned} \frac{\partial J}{\partial \theta_i} &= \frac{\partial \mathcal{L}}{\partial \theta_i} + \left\langle \frac{\partial \mathcal{L}}{\partial \rho_{sc}}, \frac{\partial \rho_{sc}}{\partial \theta_i} \right\rangle + \left\langle \frac{\partial \mathcal{L}}{\partial \rho_{pc}}, \frac{\partial \rho_{pc}}{\partial \theta_i} \right\rangle \\ &= \frac{\partial}{\partial \theta_i} \left[E(\rho_{sc}, \rho_{pc}, w_{sc}, w_{pc}; \rho_{dcs}) \right] \\ &= \left\langle \frac{\partial E}{\partial \rho_{dcs}}, \frac{\partial \rho_{dcs}}{\partial \theta_i} \right\rangle + \left\langle \frac{\partial E}{\partial \rho'_{dcs}}, \frac{\partial \rho'_{dcs}}{\partial \theta_i} \right\rangle \\ &= - \int_0^T \int_0^{z_{max}} \mathcal{W}\phi(z)(\rho_{sc}\partial_z w_{sc} + \rho_{pc}\partial_z w_{pc}) \frac{\partial \rho'_{dcs}}{\partial \theta_i} \, dz \, dt \\ &\quad + \int_0^T \int_0^{z_{max}} \left(\frac{\partial H_1}{\partial \rho_{dcs}} \frac{\partial \rho_{dcs}}{\partial \theta_i} w_{sc} + \frac{\partial H_2}{\partial \rho_{dcs}} \frac{\partial \rho_{dcs}}{\partial \theta_i} w \right) dz \, dt \\ &= - \int_0^T \int_0^{z_{max}} \mathcal{W}\phi(z)(\rho_{sc}\partial_z w_{sc} + \rho_{pc}\partial_z w_{pc}) \frac{\partial \rho'_{dcs}}{\partial \theta_i} \, dz \, dt \\ &\quad - \int_0^T \int_0^{z_{max}} D q_{sc}(1 - R_n) R'_{sc} \rho_{sc} w_{sc} \frac{\partial \rho_{dcs}}{\partial \theta_i} \, dz \, dt \\ &\quad - \int_0^T \int_0^{z_{max}} D \left[q_{pc} R_t + q_e(1 - R_e) \right] R'_{pc} \rho_{pc} w_{pc} \frac{\partial \rho_{dcs}}{\partial \theta_i} \, dz \, dt. \end{aligned}$$

This calculation is formal, and the derivative of the cost function is not guaranteed. However, it establishes a necessary condition criterion of optimality:

$$\int_0^T \int_0^{z_{max}} \mathcal{W} \partial_z \left[\phi(z)(\rho_{sc}\partial_z w_{sc} + \rho_{pc}\partial_z w_{pc}) \right] \delta \rho_{dcs} \, dz \, dt$$

$$-\int_0^T \int_0^{z_{max}} D q_{sc}(1-R_n)R'_{sc}\rho_{sc}w_{sc}\delta\rho_{dcs}\, dz\, dt$$

$$-\int_0^T \int_0^{z_{max}} D\left[q_{pc}R_t + q_e(1-R_e)\right] R'_{pc}\rho_{pc}w_{pc}\delta\rho_{dcs}\, dz\, dt \geqslant 0, \text{ for all } \delta\rho_{dcs}.$$

2.2 Example of ρ_{dcs} Shapes and $\delta\rho_{dcs}$ Directions

1. Trapezoidal shape Let

$$\rho_{trapz}(z) = (d(z-z_d)+1)\mathbb{K}_{[z_d-1/d,z_d]}(z) + \mathbb{K}_{]z_d,z_u[}(z)$$
$$+ (1+u(z-z_u))\mathbb{K}_{[z_u,z_u-1/u]}(z), \tag{2.5}$$

with its derivative

$$\rho'_{trapz}(z) = d\mathbb{K}_{[z_d-1/d,z_d]}(z) + u\mathbb{K}_{[z_u,z_u-1/u]}(z),$$

where $\mathbb{K}[a,b](z) = H(x-a) - H(x-b)$ is the characteristic function for the interval $[a,b]$. We consider the renormalization:

$$\rho_{dcs} = \frac{N_{dcs}}{\int_0^{z_{max}} \rho_{trapz}(z)\, dz}\rho_{trapz}(z), \qquad \rho'_{dcs} = \frac{N_{dcs}}{\int_0^{z_{max}} \rho_{trapz}(z)\, dz}\rho'_{trapz}(z).$$

Taking $N_{dcs} \in \mathbb{R}_+$ and $(d,u,z_d,z_u) \in \mathcal{A} := \{0 < \frac{1}{d} \leqslant z_d \leqslant z_u \leqslant z_{max} - \frac{1}{u} < z_{max}\}$, we deduce

$$\frac{\partial}{\partial d}\rho_{trapz} = (z-z_d)\mathbb{K}_{[z_d-1/d,z_d]}(z) \qquad \frac{\partial}{\partial z_d}\rho_{trapz} = -d\mathbb{K}_{[z_d-1/d,z_d]}(z)$$

$$\frac{\partial}{\partial u}\rho_{trapz} = (z-z_u)\mathbb{K}_{[z_u,z_u-1/u]}(z) \qquad \frac{\partial}{\partial z_u}\rho_{trapz} = -u\mathbb{K}_{[z_u,z_u-1/u]}(z)$$

$$\frac{\partial}{\partial d}\rho'_{trapz} = \mathbb{K}_{[z_d-1/d,z_d]}(z) - (1/d)\delta(z-z_d+1/d),$$

$$\frac{\partial}{\partial u}\rho'_{trapz} = \mathbb{K}_{[z_u,z_u-1/u]}(z) + (1/u)\delta(z-z_u+1/u),$$

$$\frac{\partial}{\partial z_d}\rho'_{trapz} = -d\left(\delta(z-z_d+1/d) - \delta(z-z_d)\right)$$

$$\frac{\partial}{\partial z_u}\rho'_{trapz} = -u\left(\delta(z-z_u) - \delta(z-z_u+1/u)\right)$$

and then

$$\frac{\partial \rho_{dcs}}{\partial N_{dcs}} = \frac{\rho_{trapz}}{\int_0^{z_{max}} \rho_{trapz}\, dz}, \qquad \frac{\partial \rho'_{dcs}}{\partial N_{dcs}} = \frac{\rho'_{trapz}}{\int_0^{z_{max}} \rho_{trapz}\, dz}$$

$$\frac{\partial \rho_{dcs}}{\partial \theta_i} = N_{dcs}\left(1 - \frac{\rho_{trapz}}{\left(\int_0^{z_{max}} \rho_{trapz}\, dz\right)^2}\right)\frac{\partial \rho_{trapz}}{\partial \theta_i},$$

$$\frac{\partial \rho'_{dcs}}{\partial \theta_i} = N_{dcs}\frac{\partial \rho'_{trapz}}{\partial \theta_i} - \frac{N_{dcs}\rho_{trapz}\dfrac{\partial \rho_{trapz}}{\partial \theta_i}}{\left(\int_0^{z_{max}} \rho_{trapz}\, dz\right)^2}$$

for $i = 2, \ldots, 5$, $\theta_1 = N_{dcs}$, $\theta_2 = d$, $\theta_3 = u$, $\theta_4 = z_d$ and $\theta_5 = z_u$.

2. Smoothing of the trapezoidal shape

   We can take $\rho^{\varepsilon}_{dcs} = \varrho_{\varepsilon} \star \rho_{dcs}$ a smooth approximation of the trapezoidal shape via convolution with mollifiers. In this case, the derivatives are applied with practically the same formulas as in the previous case, thanks to the commutativity properties of the derivative with the convolution product.

3. Simplification of the parameters

   In order to simplify the number of parameters limiting them to 3, and considering those that play a role in the main modifications of the shape of the DCS population distribution, namely height, width, and center of the distribution, let us assume a predetermined shape $\rho^0_{dcs}(z)$, which can be, for example, the trapezoidal shape (ρ_{trapz}) or the smooth form (ρ^{ε}_{dcs}) described above. We define the DCS population distribution as

$$\rho_{NLz_0}(z) := \rho_{dcs}(z; N, L, z_0) = N\rho^0_{dcs}(Lz + z_0) = N\rho^0_{dcs}(y),$$

with it derivative $\rho'_{NLz_0}(z) = NL\{\rho^0_{dcs}\}'(y)$, where $y = Lz + z_0$. For example, taking the trapezoidal shape $\rho^0_{dcs} = \rho_{trapz}$ given by (2.5), we obtain

$$\frac{\partial}{\partial N}\rho_{NLz_0}(z) = \rho_{trapz}(y), \qquad \frac{\partial}{\partial N}\rho'_{NLz_0}(z) = L\rho'_{trapz}(y)$$

$$\frac{\partial}{\partial z_0}\rho_{NLz_0}(z) = N\rho'_{trapz}(y), \qquad \frac{\partial}{\partial z_0}\rho'_{NLz_0}(z) = NL\rho''_{trapz}(y)$$

$$\frac{\partial}{\partial L}\rho_{NLz_0}(z) = Nz\rho'_{trapz}(y), \qquad \frac{\partial}{\partial L}\rho'_{NLz_0}(z) = N\rho'_{trapz}(y) + NLz\rho''_{trapz}(y),$$

where

$$\rho_{trapz}(y) \text{ is given by (2.5) with } y = Lz + z_0$$

$$\rho'_{trapz}(y) = \rho'_{trapz}(y(z)) = d\mathbb{1}_{[z_1,z_2]}(z) + u\mathbb{1}_{[z_3,z_4]}(z)$$

$$\rho''_{trapz}(y) = \rho''_{trapz}(y(z)) = d\left(\delta_{z_1} - \delta_{z_2}\right) + u\left(\delta_{z_3} - \delta_{z_4}\right)$$

with $z_1 = \frac{z_d - z_0}{L} - \frac{1}{Ld}$, $z_2 = \frac{z_d - z_0}{L}$, $z_3 = \frac{z_u - z_0}{L}$ and $z_4 = \frac{z_u - z_0}{L} - \frac{1}{Lu}$.

For instance,

$$\begin{aligned}
\frac{\partial J}{\partial L} = -\mathcal{W}N \int_0^T \Bigg[d \int_{z_1}^{z_2} & \phi(z)(\rho_{sc}\partial_z w_{sc} + \rho_{pc}\partial_z w_{pc})\, dz \\
& + u \int_{z_3}^{z_4} \phi(z)(\rho_{sc}\partial_z w_{sc} + \rho_{pc}\partial_z w_{pc})\, dz \Bigg] dt \\
& - \mathcal{W}NL \Bigg[dz_1\phi(z_1) \int_0^T (\rho_{sc}\partial_z w_{sc} + \rho_{pc}\partial_z w_{pc})\Big|_{z=z_1} dt \\
& - dz_2\phi(z_2) \int_0^T (\rho_{sc}\partial_z w_{sc} + \rho_{pc}\partial_z w_{pc})\Big|_{z=z_2} dt \\
& + uz_3\phi(z_3) \int_0^T (\rho_{sc}\partial_z w_{sc} + \rho_{pc}\partial_z w_{pc})\Big|_{z=z_3} dt \\
& - uz_4\phi(z_4) \int_0^T (\rho_{sc}\partial_z w_{sc} + \rho_{pc}\partial_z w_{pc})\Big|_{z=z_4} dt \Bigg] \\
& - DNq_{sc} \int_0^T \Bigg(d \int_{z_1}^{z_2} z(1 - R_n) R'_{sc}\rho_{sc} w_{sc}\, dz \\
& + u \int_{z_3}^{z_4} z(1 - R_n) R'_{sc}\rho_{sc} w_{sc}\, dz \Bigg) dt \\
& - DN \int_0^T \Bigg(d \int_{z_1}^{z_2} z \left[q_{pc} R_t + q_e(1 - R_e) \right] R'_{pc}\rho_{pc} w_{pc}\, dz \\
& + u \int_{z_3}^{z_4} z \left[q_{pc} R_t + q_e(1 - R_e) \right] R'_{pc}\rho_{pc} w_{pc}\, dz \Bigg) dt.
\end{aligned}$$

3 Numerical Approximation

3.1 BGK Schemes for the Direct Problem

Based on the explicit diffusive kinetic schemes introduced by Aregba-Driollet et al. [1], we obtain a general 5-point scheme in a conservative form described in detail in [4]:

$$\delta z(\rho_{sc,i}^{\varepsilon,n+1} - \rho_{sc,i}^{\varepsilon,n}) + \delta t(\mathcal{F}_{sc,i+1/2}^{\varepsilon,n} - \mathcal{F}_{sc,i-1/2}^{\varepsilon,n}) = \delta t \delta z H_1(z, \rho_{dcs}, \rho_{sc,i}^{\varepsilon,n}, \rho_{pc,i}^{\varepsilon,n}) \quad (3.1)$$

$$\rho_{sc,i}^{0} = \rho_{sc,i}^{init}, \quad (3.2)$$

$$\delta z(\rho_{pc,i}^{\varepsilon,n+1}-\rho_{pc,i}^{\varepsilon,n})+\delta t(\mathcal{F}_{pc,i+1/2}^{\varepsilon,n}-\mathcal{F}_{pc,i-1/2}^{\varepsilon,n})=\delta t\delta z H_2(z,\rho_{dcs},\rho_{sc,i}^{\varepsilon,n},\rho_{pc,i}^{\varepsilon,n}), \quad (3.3)$$

$$\rho_{pc,i}^{0}=\rho_{pc,i}^{init}, \quad (3.4)$$

where

$$\mathcal{F}_{sc,i+1/2}^{\varepsilon,n}=\sum_{\ell=1}^{4}\lambda_\ell F_\ell({}^{sc}M_{i-1,\ell}^{\varepsilon,n},{}^{sc}M_{i,\ell}^{\varepsilon,n},{}^{sc}M_{i+1,\ell}^{\varepsilon,n})$$

$$\mathcal{F}_{pc,i+1/2}^{\varepsilon,n}=\sum_{\ell=1}^{4}\lambda_\ell F_\ell({}^{pc}M_{i-1,\ell}^{\varepsilon,n},{}^{pc}M_{i,\ell}^{\varepsilon,n},{}^{pc}M_{i+1,\ell}^{\varepsilon,n})$$

with $\lambda_1=\lambda$, $\lambda_2=-\lambda$, $\lambda_3=\lambda+\frac{\theta}{\sqrt{\varepsilon}}$, $\lambda_4=-\lambda-\frac{\theta}{\sqrt{\varepsilon}}$,

$$\begin{aligned}
F_{1,2}(M_{i-1},M_i,M_{i+1}) &= F^L(M_i,M_{i+1}) \\
&\quad +\varphi^{Superbee}(M_{i-1},M_i,M_{i+1}) \\
&\quad \Big(F^H(M_i,M_{i+1})-F^L(M_i,M_{i+1})\Big) \\
F_3(M_{i-1},M_i,M_{i+1}) &= F_3(M_i,M_{i+1})=M_i+b_0(M_{i+1}-M_i) \\
F_4(M_{i-1},M_i,M_{i+1}) &= F_4(M_i,M_{i+1})=M_{i+1}+b_0(M_i-M_{i+1})
\end{aligned}$$

and

$$\begin{aligned}
{}^{sc}M_{i,1}^{\varepsilon,n} &= \frac{1}{2\lambda}\left(\lambda(\rho_{sc,i}^n-\frac{B(i\delta z,\rho_{sc,i}^n)}{\theta^2})+A_{i+\frac{1}{2}}^{sc}\right) \\
{}^{sc}M_{i,2}^{\varepsilon,n} &= \frac{1}{2\lambda}\left(\lambda(\rho_{sc,i}^n-\frac{B(i\delta z,\rho_{sc,i}^n)}{\theta^2})-A_{i+\frac{1}{2}}^{sc}\right) \\
{}^{sc}M_{i,3}^{\varepsilon,n}={}^{sc}M_{i,4}^{\varepsilon,n} &= \frac{B(i\delta z,\rho_{sc,i}^n)}{2\theta^2} \\
{}^{pc}M_{i,1}^{\varepsilon,n} &= \frac{1}{2\lambda}\left(\lambda(\rho_{pc,i}^n-\frac{B(i\delta z,\rho_{pc,i}^n)}{\theta^2})+A_{i+\frac{1}{2}}^{pc}\right) \\
{}^{pc}M_{i,2}^{\varepsilon,n} &= \frac{1}{2\lambda}\left(\lambda(\rho_{pc,i}^n-\frac{B(i\delta z,\rho_{pc,i}^n)}{\theta^2})-A_{i+\frac{1}{2}}^{pc}\right) \\
{}^{pc}M_{i,3}^{\varepsilon,n}={}^{pc}M_{i,4}^{\varepsilon,n} &= \frac{B(i\delta z,\rho_{pc,i}^n)}{2\theta^2},
\end{aligned}$$

where

$$A_{i+\frac{1}{2}} = V^{+}_{i+\frac{1}{2}}\rho_i + V^{-}_{i+\frac{1}{2}}\rho_{i+1} + \mathcal{W}(\phi'_{i+\frac{1}{2}})^{+}\rho_i^2/2 + \mathcal{W}(\phi'_{i+\frac{1}{2}})^{-}\rho_{i+1}^2/2$$

with $v^+ = \max(v, 0)$, and $v^- = \min(v, 0)$, $\phi'_{i+\frac{1}{2}} = \frac{\phi'(i\delta z)+\phi'((i+1)\delta z)}{2}$, and

$$V^{sc}_{i+\frac{1}{2}} = -\mathcal{W}\left[(\phi\rho'_{dcs})_{i+\frac{1}{2}} + \phi_{i+\frac{1}{2}}D^{+}\rho_{pc,i}\right]$$
$$V^{pc}_{i+\frac{1}{2}} = -\mathcal{W}\left[(\phi\rho'_{dcs})_{i+\frac{1}{2}} + \phi_{i+\frac{1}{2}}D^{+}\rho_{sc,i}\right]$$

with $D^+\rho_i = \dfrac{\rho_{i+1}-\rho_i}{\delta z}$. Additionally, the parameters λ and θ are chosen as

$$\theta = \sqrt{\frac{\max\{B'(\rho_l)\}}{\alpha - 1}} + \delta, \qquad \lambda = \max\left\{\frac{\partial}{\partial\rho_l}A^l_{i+\frac{1}{2}}\right\} + \delta,$$

$l = sc, pc$. Replicating these formulas independently for both ρ and ρ_{sc}, thus obtaining the parameter pairs (θ, λ) and $(\theta_{sc} = \delta, \lambda_{sc})$, respectively, and the CFL condition,

$$\delta t \leqslant \min\left\{\frac{\delta z^2}{2\theta^2}, \frac{\delta z}{\lambda_{sc}}, \frac{\delta z}{\lambda_{pc}}\right\}. \tag{3.5}$$

3.2 Adjoint Scheme Associated with BGK Scheme

Let $\rho_{sc,\delta} = (\rho^n_{sc,i})_{i,n}$, $\rho_{pc,\delta} = (\rho^n_{pc,i})_{i,n}$ Multiplying (3.1)–(3.4) by $w^n_{sc,i}$ and $w^n_{pc,i}$ and summing by parts, we obtain

$$E_\delta(\rho_{sc,\delta}, \rho_{pc,\delta}, w_{sc,\delta}, w_{pc,\delta}; \rho_{dcs}) = 0, \qquad \text{for all } w_{sc,\delta}, w_{pc,\delta}$$

with

$$E_\delta(\rho_{sc,\delta}, \rho_{pc,\delta}, w_{sc,\delta}, w_{pc,\delta}; \rho_{dcs}) = E_{1,\delta}(\rho_{sc,\delta}, \rho_{pc,\delta}, w_{sc,\delta}; \rho_{dcs}) + E_{2,\delta}(\rho_{sc,\delta}, \rho_{pc,\delta}, w_{pc,\delta}; \rho_{dcs}),$$

where

$$E_{1,\delta}(\rho_{sc,\delta}, \rho_{pc,\delta}, w_{sc,\delta}; \rho_{dcs}) := \tag{3.6}$$

$$\sum_{n=1}^{N}\sum_{i=1}^{\lfloor\frac{z_{max}}{\delta z}\rfloor-1}\left[\rho^{\varepsilon,n}_{sc,i}\left(w^n_{sc,i} - w^{n-1}_{sc,i}\right)\delta z + \mathcal{F}^{\varepsilon,n}_{sc,i+1/2}\left(w^n_{sc,i+1} - w^n_{sc,i}\right)\delta t\right.$$

$$+H^n_{1,i} w^n_{sc,i} \delta z \delta t\Big]$$

$$- \sum_{i=1}^{\lfloor \frac{zmax}{\delta z} \rfloor - 1} (\rho^N_{sc,i} w^N_{sc,i} - \rho^{init}_{sc,i} w^0_{sc,i}) \delta z$$

$$E_{2,\delta}(\rho_{sc,\delta}, \rho_{pc,\delta}, w_{pc,\delta}; \rho_{dcs}) := \tag{3.7}$$

$$\sum_{n=1}^{N} \sum_{i=1}^{\lfloor \frac{zmax}{\delta z} \rfloor - 1} \Big[\rho^{\varepsilon,n}_{pc,i} \left(w^n_{pc,i} - w^{n-1}_{pc,i}\right) \delta z + \mathcal{F}^{\varepsilon,n}_{pc,i+1/2} \left(w^n_{pc,i+1} - w^n_{pc,i}\right) \delta t$$

$$+H^n_{2,i} w^n_{pc,i} \delta z \delta t\Big]$$

$$- \sum_{i=1}^{\lfloor \frac{zmax}{\delta z} \rfloor - 1} (\rho^N_{pc,i} w^N_{pc,i} - \rho^{init}_{pc,i} w^0_{pc,i}) \delta z.$$

3.3 Discrete Inverse Problem

The discrete cost function is given by

$$J_\delta(\rho_\delta, \rho_{sc,\delta}) := \frac{\alpha}{2} \sum_{i=1}^{\lfloor \frac{zmax}{\delta z} \rfloor - 1} |\rho^N_{sc,i} - \rho^{obs}_{sc,i}|^2 \delta z + \frac{\beta}{2} \sum_{i=1}^{\lfloor \frac{zmax}{\delta z} \rfloor - 1} |\rho^N_{pc,i} - \rho^{obs}_{pc,i}|^2 \delta z \tag{3.8}$$

Thus derivating the discrete Lagrangian

$$\mathcal{L}_\delta(\rho_{sc,\delta}, \rho_{pc,\delta}, w_{sc,\delta}, w_{pc,\delta}; \rho_{dcs})$$
$$:= J_\delta(\rho_{sc,\delta}, \rho_{pc,\delta}) + E_\delta(\rho_{sc,\delta}, \rho_{pc,\delta}, w_{sc,\delta}, w_{pc,\delta}; \rho_{dcs})$$

with respect to $\rho^n_{sc,i}$ and $\rho^n_{pc,i}$, we obtain the discrete adjoint scheme

$$w^{n-1}_{sc,i} = w^n_{sc,i} + \frac{\delta t}{\delta z} \sum_{k=1}^{4} \frac{\partial}{\partial \rho_{sc,k-2}} \left[\mathcal{F}^{\varepsilon,n}_{sc,i-k+5/2}\right] (w^n_{sc,i-k+3} - w^n_{sc,i-k+2})$$

$$+ \frac{\delta t}{\delta z} \sum_{k=1}^{4} \frac{\partial}{\partial \rho_{sc,k-2}} \left[\mathcal{F}^{\varepsilon,n}_{pc,i-k+5/2}\right] \left(w^n_{pc,i-k+3} - w^n_{pc,i-k+2}\right)$$

$$+\delta t \frac{\partial H^n_{1,i}}{\partial \rho^n_{sc,i}} w^n_{sc,i} + \delta t \frac{\partial H^n_{2,i}}{\partial \rho^n_{sc,i}} w^n_{pc,i}, \tag{3.9}$$

$$w_{pc,i}^{n-1} = w_{pc,i}^{n} + \frac{\delta t}{\delta z}\sum_{k=1}^{4} \frac{\partial}{\partial \rho_{pc,k-2}} \left[\mathcal{F}_{sc,i-k+5/2}^{\varepsilon,n}\right] \left(w_{sc,i-k+3}^{n} - w_{sc,i-k+2}^{n}\right)$$

$$+\frac{\delta t}{\delta z}\sum_{k=1}^{4} \frac{\partial}{\partial \rho_{pc,k-2}} \left[\mathcal{F}_{pc,i-k+5/2}^{\varepsilon,n}\right] \left(w_{pc,i-k+3}^{n} - w_{pc,i-k+2}^{n}\right)$$

$$+\delta t \frac{\partial H_{1,i}^{n}}{\partial \rho_{pc,i}^{n}} w_{sc,i}^{n} + \delta t \frac{\partial H_{2,i}^{n}}{\partial \rho_{pc,i}^{n}} w_{pc,i}^{n}, \tag{3.10}$$

$$w_{sc,i}^{N-1} = \alpha \left(\rho_{sc,i}^{N} - \rho_{sc,i}^{obs}\right) \tag{3.11}$$

$$w_{pc,i}^{N-1} = \beta \left(\rho_{pc,i}^{N} - \rho_{pc,i}^{obs}\right). \tag{3.12}$$

Because $\left\langle \frac{\partial M_{k,\ell}^{\varepsilon,n}}{\partial \rho_{dcs}'}, \delta\rho_{dcs}' \right\rangle = -\frac{\mathcal{W}\phi_k}{2}\rho_k^n$, we deduce that the derivative of the discrete cost function is given by

$$\left\langle \frac{\partial J_\delta}{\partial \rho_{dcs}}, \delta\rho_{dcs} \right\rangle = \delta t \sum_{n=1}^{N} \sum_{i=1}^{\lfloor \frac{z_{max}}{\delta z} \rfloor - 1} \sum_{k=1}^{4} \frac{\partial}{\partial \rho_{dcs,k-2}'} \left[\mathcal{F}_{sc,i+1/2}^{\varepsilon,n}\right] \left(w_{sc,i+1}^{n} - w_{sc,i}^{n}\right)$$

$$+ \delta t \sum_{n=1}^{N} \sum_{i=1}^{\lfloor \frac{z_{max}}{\delta z} \rfloor - 1} \sum_{k=1}^{4} \frac{\partial}{\partial \rho_{dcs,k-2}'} \left[\mathcal{F}_{pc,i+1/2}^{\varepsilon,n}\right] \left(w_{pc,i+1}^{n} - w_{pc,i}^{n}\right)$$

$$+\delta t \delta z \sum_{n=1}^{N} \sum_{i=1}^{\lfloor \frac{z_{max}}{\delta z} \rfloor - 1} \left(\frac{\partial H_{1,i}^{n}}{\partial \rho_{dcs}} w_{sc,i}^{n} + \frac{\partial H_{2,i}^{n}}{\partial \rho_{dcs}} w_{pc,i}^{n}\right). \tag{3.13}$$

4 Numerical Examples

We make here a simulation for $z_{max} = 200$ and $T = 20$. We choose a reasonable discretization with $N = 250$, which gives $dz = z_{max}/(N-1) = 0.8032$. Additionally, the values proposed in [3] for the physical–biological parameters of the model are considered, that is, $\mathcal{W} = 6.01/8$, $q_{SC} = 0.15$, $Z_{niche} = 12.0$, $k_{niche} = 5.0$, $D = 12.07$, $K_{SC} = 53$, $k_{dens} = 6$, $q_{diff} = 0.2$, $q_{PC} = 0.22$, $Z_{tiers} = 40.0$, $k_{tiers} = 40.0$, $K_{PC} = 41.0$, $q_{ex} = 0.34$, $Z_{ex} = 190.0$, $k_{ex} = 15.0$. As for the geometric parameters of the crypt, these are given by $r_0 = 10$, $\varepsilon_0 = 0.1r_0$. Regarding the DCS cells, we initially consider the same trapezoidal shape of [3], with the parameters $d = 2.25$, $u = -1/8$, and $0 \leqslant N_{DCS} \leqslant 20$, $0 \leqslant z_d = z_u \leqslant 10$.

After conducting several numerical experiments, it has been confirmed that a trapezoidal shape for the DCS cells, although they generate regular solutions, they do not behave regularly in the face of small variations in the parameters of said trapezoid. Although there is a continuity of the solution with respect to these parameters, we observe through some numerical tests that the cost function (2.1) and its approximation (3.8) contain oscillations and therefore several local minima that are uncomfortable to minimize. Therefore, we choose to approximate the DCS cell distribution function by regularizing it through convolution with mollifier functions.

On the other hand, since in this instance we do not have real experimental observations, we consider a reference simulation with $N_{DCS} = 12$ and $z_d = z_u = 3$ (chosen values of [3]) at which we will call observed data. We then start from a simulation without DCS cells, that is, $N_{DCS} = 0$ and $z_d = z_u = 0$, and we try to reconstruct the shape of the DCS cells for the observed solution by solving the inverse problem 2.1.*2)* and minimizing the cost function (3.8).

Additionally, the parameters $\delta = 10^{-5}$, $\alpha = 0.9$, and $b_0 = 1$ were chosen for the description of the BGK scheme and the calculation of the values of λ_ℓ and θ. We consider a CFL condition (3.5) sufficiently strict that it covers all the range of values that $N_{DCS} \in [0; 20]$ and $z_d = z_u \in [0; 10]$ can take. This gives a value $\delta t = 3.2257 \cdot 10^{-5}$ for the set of physical, geometric, and meshing parameters described here.

Figure 3 shows the simulated cost function for various values of N_{DCS} varying from 0 to 20 and various values of $z_d = z_u$ varying from 0 to 10. The trapezoid of the DCS cell distribution function is smoothed using the smooth1DconvNE2xConv()

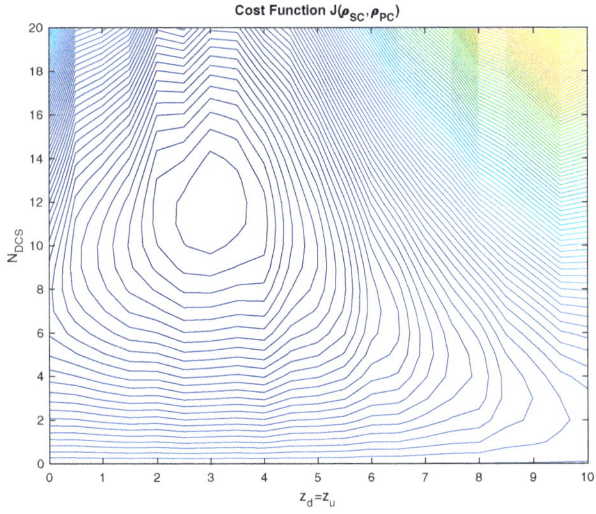

Fig. 3 Cost function $J(\rho_{sc}, \rho_{pc}) = \frac{1}{2}\int_0^{z_{max}} \left|\rho_{sc}^{obs}(z) - \rho_{sc}(z,T)\right|^2 dz + \frac{1}{2}\int_0^{z_{max}} \left|\rho_{pc}^{obs}(z) - \rho_{pc}(z,T)\right|^2 dz$

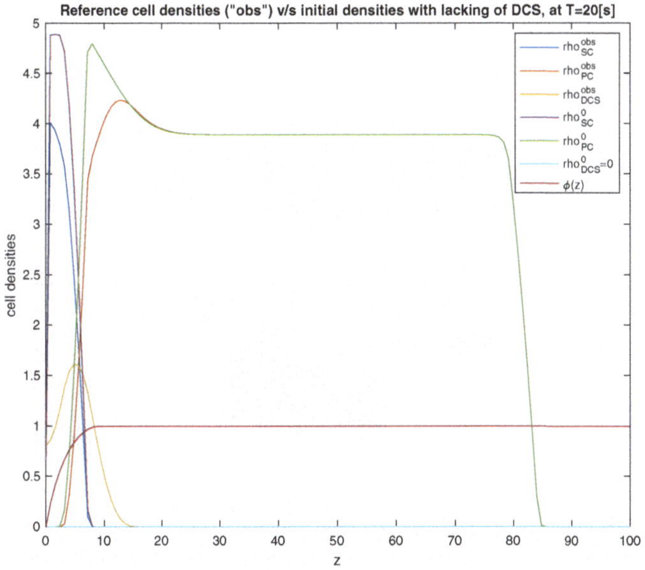

Fig. 4 Evolution of cell population: simulation of the interaction SC-PC with the presence of a smooth form for the distribution of DCS v/s simulation with lacking of DCS. Testing of the inverse problem: starting from the case with lacking of DCS, reconstruct their shape, using the numerical simulation of the case with DCS as a reference (observation) for the objective function

function of Matlab [6]. This smoothed function can be seen in Fig. 4 (yellow graph). Based on this, a cost function is obtained with a single, clearly defined minimum, locally convex, and sufficiently sensitive to data perturbation. It should be noted that the sum of two functionals $\frac{1}{2}\int_0^{z_{max}} \left|\rho_{sc}^{obs}(z) - \rho_{sc}(z,T)\right|^2 dz$ and $\frac{1}{2}\int_0^{z_{max}} \left|\rho_{pc}^{obs}(z) - \rho_{pc}(z,T)\right|^2 dz$ is being minimized in that each of them alone does not have a locally convex shape as defined as the sum of both.

4.1 *Inverse Problem Results*

In Fig. 4, the distribution of reference densities (considered as "observations") for the stem cells (cyan color) and for the progenitor cells (red color) is plotted. Additionally, the distribution of smoothed DCS cells is graphed (in yellow) that allows simulating such results and corresponds to the shape that is desired to be reconstructed through the method of solving the inverse problem. Then, for comparison, the simulation of the distributions corresponding to the starting point for the resolution using gradient descent of the minimization problem is graphed;

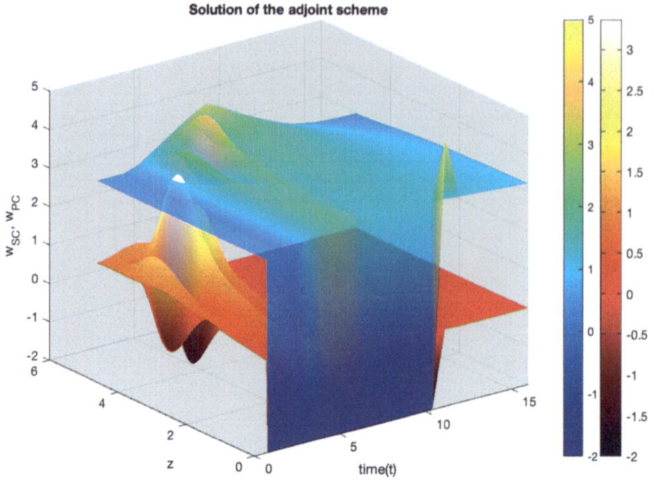

Fig. 5 Solutions of the adjoint scheme associated with the BGK scheme of the cell distributions: w_{sc} in a hot color map and $w_{pc} + 3$ (in order to be able to visualize it differentiated from w_{sc}) in a slightly transparent "parula" color map and above of the w_{sc} graph

this is a simulation without the presence of DCS cells ($N_{DCS} = z_d = z_u = 0$): stem cells (in purple) and progenitor cells (in light green). Additionally and as a reference, the representative function of the geometry of the crypt $\phi(z)$ is graphed.

We compute the adjoint state of the BGK numerical scheme, in order to obtain the critical points of the Lagrangian (2.2). The numerical simulation of the adjoint state (3.9)–(3.10) is graphed in Fig. 5. In order to simultaneously display both w_{sc} and w_{pc} in the same figure, the constant 3 is added to w_{pc} to graph it in a slightly transparent "parula" color map, in contrast to the reddish map for w_{sc}. Like Lagrange multipliers, these quantities are related to the derivatives of the constraints (in this case the equations that characterize the BGK scheme), and their stability is important for the calculation of the gradient. In this case we observe a stability of the solution except in $z = 0$ where there is a singularity in its neighborhood and for a time $t \leqslant t^*$ with $t^* \approx 12$. This could become problematic if we want to identify, for example, the boundary condition at the bottom of the crypt ($z = 0$), which is surely an ill-posed problem.

In Table 1, a summary of the identification of the parameters can be observed. Indeed, based on a situation without DCS cells, we managed to rebuild the parameters of the distribution of reference DCS cells in 17 gradient iterations and 72 evaluations of the cost function, reaching the values $N_{DCS}^{Identif} = 12$ and $z_u^{Identif} = 3$ with an error of less than $4 \cdot 10^{-6}$.

Table 1 Numerical result of the inverse problem test

Identification indicators	Values
Number of iterations	17
Evaluations of $J(\cdot)$	72
$J(\rho_{sc}^{Identif}, \rho_{pc}^{Identif})$	2.62119e−13
$(N_{DCS}^{Identif}; z_d^{Identif})$	(12; 3)
$N_{DCS}^{Identif} - N_{DCS}^{obs}$	−3.8188e−06
$z_d^{Identif} - z_d^{obs}$	4.6803e−07
$\nabla J(\rho_{sc}^{Identif}, \rho_{pc}^{Identif})^T$	(−0.1863e−06; −0.4405e−06)
Hessian of $J(\rho_{sc}^{Identif}, \rho_{pc}^{Identif})$	$\begin{pmatrix} 0.0741 & 0.1925 \\ 0.1925 & 0.6007 \end{pmatrix}$

Acknowledgments The inverse problem for the model studied in this work was discussed during the third author's long-stay visit to the MaIAGE unit of INRAE and the MUSCA team of INRIA, Université Paris-Saclay, France. The third author MS thanks these laboratories of applied mathematics and modeling for their kind support and hospitality.

Appendix: Derivative of the Flux for 5-Point Schemes

Derivative of the Flux with Respect to ρ_{sc} and ρ_{pc}

We consider the decomposition

$$\mathcal{F}^{\varepsilon,n}_{\sigma,i+1/2} = \sum_{\ell=1}^{4} \lambda_\ell F_{\sigma,\ell,i+1/2} = \sum_{\ell=1}^{4} \lambda_\ell F_\ell(M^n_{\sigma,i-1,\ell}, M^n_{\sigma,i,\ell}, M^n_{\sigma,i+1,\ell})$$

for $\sigma = sc, pc$, with $\lambda_1 = \lambda$, $\lambda_2 = -\lambda$, $\lambda_3 = \lambda + \dfrac{\theta}{\sqrt{\varepsilon}}$, $\lambda_4 = -\lambda - \dfrac{\theta}{\sqrt{\varepsilon}}$. In this sense, we have

$$\frac{\partial}{\partial \rho_{\varpi,-1}}\left[\mathcal{F}^{\varepsilon,n}_{\sigma,i+3/2}\right] = \sum_{\ell=1}^{4} \lambda_\ell \frac{\partial F_{\sigma,\ell,i+3/2}}{\partial M_{\sigma,-1,\ell}} \frac{\partial M_{\sigma,i,\ell}}{\partial \rho_{\varpi,0}}$$

$$\frac{\partial}{\partial \rho_{\varpi,0}}\left[\mathcal{F}^{\varepsilon,n}_{\sigma,i+1/2}\right] = \sum_{\ell=1}^{4} \lambda_\ell \left[\frac{\partial F_{\sigma,\ell,i+1/2}}{\partial M_{\sigma,-1,\ell}} \frac{\partial M_{\sigma,i-1,\ell}}{\partial \rho_{\varpi,,1}} + \frac{\partial F_{\sigma,\ell,i+1/2}}{\partial M_{\sigma,0,\ell}} \frac{\partial M_{\sigma,i,\ell}}{\partial \rho_{\varpi,,0}}\right]$$

$$\frac{\partial}{\partial \rho_{\varpi,1}}\left[\mathcal{F}^{\varepsilon,n}_{\sigma,i-1/2}\right]=\sum_{\ell=1}^{4}\lambda_\ell\left[\frac{\partial F_{\sigma,\ell,i-1/2}}{\partial M_{\sigma,0,\ell}}\frac{\partial M_{\sigma,i-1,\ell}}{\partial \rho_{\varpi,,1}}+\frac{\partial F_{\sigma,\ell,i-1/2}}{\partial M_{\sigma,1,\ell}}\frac{\partial M_{\sigma,i,\ell}}{\partial \rho_{\varpi,,0}}\right]$$

$$\frac{\partial}{\partial \rho_{\varpi,2}}\left[\mathcal{F}^{\varepsilon,n}_{\sigma,i-3/2}\right]=\sum_{\ell=1}^{4}\lambda_\ell\frac{\partial F_{\sigma,\ell,i-3/2}}{\partial M_{\sigma,1,\ell}}\frac{\partial M_{\sigma,i-1,\ell}}{\partial \rho_{\varpi,,1}}$$

with $\sigma = sc, pc$ and $\varpi = sc, pc$. Then, we can replace directly these expressions on (3.9) and (3.10).

Derivative of the Flux with Respect to ρ'_{dcs}

We have

$$\frac{\partial}{\partial \rho'_{dcs,-1}}\left[\mathcal{F}^{\varepsilon,n}_{\sigma,i+1/2}\right]=\sum_{\ell=1}^{4}\lambda_\ell\frac{\partial F_{\sigma,\ell,i+1/2}}{\partial M_{\sigma,-1,\ell}}\frac{\partial M_{\sigma,i-1,\ell}}{\partial \rho'_{dcs,0}}$$

$$\frac{\partial}{\partial \rho'_{dcs,0}}\left[\mathcal{F}^{\varepsilon,n}_{\sigma,i+1/2}\right]=\sum_{\ell=1}^{4}\lambda_\ell\left[\frac{\partial F_{\sigma,\ell,i+1/2}}{\partial M_{\sigma,-1,\ell}}\frac{\partial M_{\sigma,i-1,\ell}}{\partial \rho'_{dcs,,1}}+\frac{\partial F_{\sigma,\ell,i+1/2}}{\partial M_{\sigma,0,\ell}}\frac{\partial M_{\sigma,i,\ell}}{\partial \rho'_{dcs,,0}}\right]$$

$$\frac{\partial}{\partial \rho'_{dcsi,1}}\left[\mathcal{F}^{\varepsilon,n}_{\sigma,i+1/2}\right]=\sum_{\ell=1}^{4}\lambda_\ell\left[\frac{\partial F_{\sigma,\ell,i+1/2}}{\partial M_{\sigma,0,\ell}}\frac{\partial M_{\sigma,i,\ell}}{\partial \rho'_{dcs,,1}}+\frac{\partial F_{\sigma,\ell,i+1/2}}{\partial M_{\sigma,1,\ell}}\frac{\partial M_{\sigma,i+1,\ell}}{\partial \rho'_{dcs,,0}}\right]$$

$$\frac{\partial}{\partial \rho'_{dcs,2}}\left[\mathcal{F}^{\varepsilon,n}_{\sigma,i+1/2}\right]=\sum_{\ell=1}^{4}\lambda_\ell\frac{\partial F_{\sigma,\ell,i+1/2}}{\partial M_{\sigma,1,\ell}}\frac{\partial M_{\sigma,i+1,\ell}}{\partial \rho'_{dcs,,1}}$$

for $\sigma = sc, pc$. Then, we can replace these derivatives of the flux directly on (3.13).

To calculate these derivatives, it is necessary to specify the calculation of the derivative of the flux with respect to the Maxwellian ones, as well as the derivatives of the Maxwellian ones with respect to the distribution of cells. Below we detail these two levels of derivatives.

Derivative of the Flux in Respect of the Maxwellian Distributions

For $\ell = 1, 2$, we consider a second-order limiter flux with Lax–Wendroff scheme, that is,

$$\lambda_1 F_1(M_{-1}, M_0, M_1) = \lambda F_1(M_{-1}, M_0, M_1)$$

$$\begin{aligned} &= \lambda M_0 + \frac{\lambda}{2}\varphi(M_{-1}, M_0, M_1)\left(1 - \lambda\frac{\delta t}{\delta z}\right)(M_1 - M_0) \\ \lambda_2 F_2(M_{-1}, M_0, M_1) &= -\lambda F_2(M_{-1}, M_0, M_1) \\ &= -\lambda M_1 + \frac{\lambda}{2}\varphi(M_{-1}, M_0, M_1)\left(1 - \lambda\frac{\delta t}{\delta z}\right)(M_1 - M_0). \end{aligned}$$

On the other hand, the flux for $\ell = 3, 4$ is given by monotone 5-point schemes such that $F_3(M_{-1}, M_0, M_1) = F_4(M_1, M_0, M_{-1}))$, and taking the sample particular case of [1], we have

$$\begin{aligned} F_1(M_{-1}, M_0, M_1) &= M_0 + \frac{1}{2}\varphi(M_{-1}, M_0, M_1)\left(1 - \lambda\frac{\delta t}{\delta z}\right)(M_1 - M_0) \\ F_2(M_{-1}, M_0, M_1) &= M_1 - \frac{1}{2}\varphi(M_{-1}, M_0, M_1)\left(1 - \lambda\frac{\delta t}{\delta z}\right)(M_1 - M_0) \\ F_3(M_{-1}, M_0, M_1) &= M_0 + b_0(M_1 - M_0) \\ F_4(M_{-1}, M_0, M_1) &= M_1 - b_0(M_1 - M_0). \end{aligned}$$

From here the partial derivative rules that interest us are given by

$$\begin{aligned} \frac{\partial}{\partial M_{-1}} F_1(M_{-1}, M_0, M_1) &= \mathcal{S}\frac{\partial}{\partial M_{-1}} w(M_{-1}, M_0, M_1) \\ \frac{\partial}{\partial M_{-1}} F_2(M_{-1}, M_0, M_1) &= -\mathcal{S}\frac{\partial}{\partial M_{-1}} w(M_{-1}, M_0, M_1) \\ \frac{\partial}{\partial M_{-1}} F_3(M_{-1}, M_0, M_1) &= \frac{\partial}{\partial M_{-1}} F_4(M_{-1}, M_0, M_1) \;=\; 0 \\ \frac{\partial}{\partial M_0} F_1(M_{-1}, M_0, M_1) &= 1 - \mathcal{M} + \mathcal{S}\frac{\partial}{\partial M_0} w(M_{-1}, M_0, M_1) \\ \frac{\partial}{\partial M_0} F_2(M_{-1}, M_0, M_1) &= \mathcal{M} - \mathcal{S}\frac{\partial}{\partial M_0} w(M_{-1}, M_0, M_1) \\ \frac{\partial}{\partial M_1} F_1(M_{-1}, M_0, M_1) &= \mathcal{M} + \mathcal{S}\frac{\partial}{\partial M_1} w(M_{-1}, M_0, M_1) \\ \frac{\partial}{\partial M_1} F_2(M_{-1}, M_0, M_1) &= 1 - \mathcal{M} - \mathcal{S}\frac{\partial}{\partial M_1} w(M_{-1}, M_0, M_1) \\ \frac{\partial}{\partial M_0} F_3(M_{-1}, M_0, M_1) &= \frac{\partial}{\partial M_1} F_4(M_{-1}, M_0, M_1) \;=\; 1 - b_0 \\ \frac{\partial}{\partial M_1} F_3(M_{-1}, M_0, M_1) &= \frac{\partial}{\partial M_0} F_4(M_{-1}, M_0, M_1) \;=\; b_0, \end{aligned}$$

where

$$\mathcal{S} = \frac{1}{2}\left(1 - \lambda\frac{\delta t}{\delta z}\right)(M_1 - M_0)\varphi' \circ w(M_{-1}, M_0, M_1)$$

$$\mathcal{M} = \frac{1}{2}\left(1 - \lambda\frac{\delta t}{\delta z}\right)\varphi \circ w(M_{-1}, M_0, M_1),$$

$$\varphi'_{Superbee}(w) = \begin{cases} 2 & \text{if } 0 \leqslant w < 1/2, \\ 1 & \text{if } 1 \leqslant w < 2, \\ 0 & \text{otherwise} \end{cases}, \quad \varphi'_{VanLeer}(w) = \frac{2}{(1+w)^2}\text{Heaviside(w)},$$

and because $w = \dfrac{M_0 - M_{-1}}{M_1 - M_0}$

$$\frac{\partial w}{\partial M_{-1}} = -\frac{1}{M_1 - M_0}, \quad \frac{\partial w}{\partial M_0} = \frac{M_1 - M_{-1}}{(M_1 - M_0)^2}, \quad \frac{\partial w}{\partial M_1} = -\frac{M_0 - M_{-1}}{(M_1 - M_0)^2}.$$

Derivative of the Deep Crypt Secretory with Respect to Its Parameters

The deep crypt secretory (DCS) probability density is given by Darrigade et al. [3]

$$\rho_{dcs}(z) = (d(z - z_d) + 1)\mathbb{1}_{[z_d - 1/d, z_d]}(z) + \mathbb{1}_{]z_d, z_u[}(z) + (1 + u(z - z_u))\mathbb{1}_{[z_u, z_u - 1/u]}(z),$$

where $\mathbb{1}[a, b](z) = H(x - a) - H(x - b)$ is the characteristic function for the interval $[a, b]$. Thus, the derivatives of ρ_{dcs} with respect to its parameters are given by

$$\frac{\partial}{\partial d}\rho_{dcs} = (z - z_d)\mathbb{1}_{[z_d - 1/d, z_d]}(z) \qquad \frac{\partial}{\partial z_d}\rho_{dcs} = -d\mathbb{1}_{[z_d - 1/d, z_d]}(z)$$

$$\frac{\partial}{\partial u}\rho_{dcs} = (z - z_u)\mathbb{1}_{[z_u, z_u - 1/u]}(z) \qquad \frac{\partial}{\partial z_u}\rho_{dcs} = -u\mathbb{1}_{[z_d - 1/d, z_d]}(z).$$

In the derivative of the cost function, we need also to compute the derivative of ρ'_{dcs} with respect to the same parameters, which are

$$\frac{\partial}{\partial d}\rho'_{dcs} = \frac{1}{d}\delta(z - z_d - 1/d) + \mathbb{1}_{[z_d - 1/d, z_d]}(z)$$

$$\frac{\partial}{\partial z_d}\rho_{dcs} = -d(\delta(z - z_d - 1/d) - \delta(z - z_d))$$

$$\frac{\partial}{\partial u}\rho_{dcs} = -\frac{1}{u}\delta(z - z_u - 1/u) + \mathbb{1}_{[z_u-1/u, z_u]}(z)$$
$$\frac{\partial}{\partial z_u}\rho_{dcs} = -u(\delta(z - z_u) - \delta(z - z_u - 1/u)).$$

Acknowledgments This work was supported by INRIA-ANACONDA project associated to the team INRIA- MUSCA. MH's contribution was permitted in part by funding from the European Research Council (ERC) under the European Union's Horizon 2020 research and innovation program (Grant agreement ERC-2017-AdG No. 788191 - Homo.symbiosus. MS was supported by Fondecyt-ANID project 1220869, ANID-Chile through Centro de Modelamiento Matemático (FB210005), and Jean d'Alembert fellowship program, Université Paris-Saclay.

References

1. D. Aregba-Driollet, R. Natalini, S. Tang, Explicit diffusive kinetic schemes for nonlinear degenerate parabolic systems. Math. Comput. **73**(245), 63–94 (2004)
2. J. Carrillo, Entropy solutions for nonlinear degenerate problems. Arch. Ration. Mech. Anal. **147**(4), 269–361 (1999). El problema inverso
3. L. Darrigade, M. Haghebaert, C. Cherbuy, S. Labarthe, B. Laroche, A PDMP model of the epithelial cell turn-over in the intestinal crypt including microbiota-derived regulations. J. Math. Biol. **84**(7), 60 (2022)
4. M. Haghebaert, B. Laroche, Personal communication concerning the Thesis of M. Haghebaert (2023)
5. C. Mascia, A. Porretta, A. Terracina, Nonhomogeneous Dirichlet problems for degenerate parabolic-hyperbolic equations. Arch. Ration. Mech. Anal. **163**(2), 87–124 (2002)
6. P. Seibold, Smoothing 1D not equidistant curve by convolution, MATLAB Central File Exchange (2023), https://www.mathworks.com/matlabcentral/fileexchange/132528-smoothing-1d-not-equidistant-curve-by-convolution

Solving Ill-Posed Inverse Problems via the Born Approximation

Fabricio Maciá and Cristóbal J. Meroño

2010 Mathematics Subject Classification 35R30, 35P25, 65N21

1 Introduction

The Born approximation has its origin in scattering theory [7], where it has proved to be a suitable tool to study several inverse scattering problems; see [11, 16] among many others. The aim of this note is to show that the Born approximation is also meaningful and useful in order to solve inverse problems that, unlike most inverse scattering problems, are severely *ill-posed*. We will mainly focus on the inverse problem consisting in recovering a potential from the Dirichlet-to-Neumann map of the corresponding Schrödinger operator. For one-dimensional problems in the half line, this amounts essentially to the problem of recovering the potential from the spectral measure of the Schrödinger operator, whereas for higher dimensional problems this is known as the Calderón problem, which is linked to relevant applications in geological prospecting or medical imaging. The approach we present here originates from the article [3] and is still under development [4, 15, 22]. We will start by presenting it in an abstract, general setting. Next, we will show how the novel approach to inverse spectral theory for Schrödinger operators on the half line introduced by Simon in [29] fits into this framework. Finally, we will discuss in detail the Born approximation for the Calderón problem and survey some of the most important results in [3, 15, 22].

F. Maciá (✉) · C. J. Meroño
M2ASAI, Universidad Politécnica de Madrid, ETSI Caminos, Madrid, Spain
e-mail: fabricio.macia@upm.es; cj.merono@upm.es

K. Ammari et al. (eds.), *Advances in Partial Differential Equations and Control*, Trends in Mathematics, https://doi.org/10.1007/978-3-031-62265-6_10

2 An Abstract Setting for the Born Approximation to Ill-Posed Inverse Problems

Let us formulate the abstract setup of the problem we are interested in. One has a usually nonlinear, injective map

$$\Phi : X \longrightarrow \mathbb{F}$$

between an open subset X of a Banach space $\mathbb{E}$ onto some subset $\mathcal{M} := \Phi(X)$ of a topological vector space $\mathbb{F}$. The map Φ models the *forward operator* of the inverse problem we are interested in. Solving an inverse problem consists in being able to find, for any given $y \in \mathcal{M}$, the element $x \in X$ that satisfies

$$\Phi(x) = y. \tag{1}$$

This amounts to have an expression as explicit as possible of the *inverse operator*,

$$\Phi^{-1} : \mathcal{M} \longrightarrow X,$$

together with information on its stability properties.[1] Ideally, this also includes having a characterization of $\mathcal{M}$ (or at least a partial characterization), although this is in most interesting cases a very hard problem. The choice of the ambient space $\mathbb{F}$ plays a secondary role here and, as we will see, sometimes different natural choices are available.

In most inverse problems, the forward operator Φ is a Fréchet differentiable map whose inverse Φ^{-1} is possibly not continuous (this is the case in the Calderón problem). For this reason, this kind of inverse problems is said to be *ill-posed*. However, when restricted to compact sets $K \subset X$, such a map is always a homeomorphism, whose inverse $\Phi^{-1}|_{\Phi(K)}$ is uniformly continuous and therefore possess a modulus of continuity ω_K. Restricting the problem to a compact subset results in imposing *a priori* bounds on the unknowns, and estimates on ω_K give *conditional stability* results for the inverse problem:

$$\|x - y\|_{\mathbb{E}} \le \omega_K(\|\Phi(x) - \Phi(y)\|_{\mathbb{F}}), \quad \forall x, y \in K$$

(the reader may consult [20] for more insight on these issues).

Many successful numerical approaches to ill-posed problems and EIT involve at some point the use of linearization techniques, see, for example, the references in [18] for one-step linearization methods in EIT or [10] for the so-called Calderón

[1] Stability is an important issue in inverse problems, since usually one does not have access to the whole datum y, but rather to some partial data y_ϵ typically obtained through measurement, which is an approximation to y. Being able to control how the error $\|y - y_\epsilon\|_{\mathbb{F}}$ propagates to the difference of the respective solutions of $\Phi(x) = y$, and $\Phi(x_\epsilon) = y_\epsilon$ is crucial for practical purposes.

method. This is surprising given the fact that the inverse map Φ^{-1} is not even continuous. In fact, even the rigorous setting necessary to understand these linearization methods is not clear at the moment, as noted in [18].

Here we propose a formalism, based on the classical concept of Born approximation, which offers a rigorous mathematical framework that unifies many of the different linearization techniques used in EIT and in the Calderón problem. It is expected that many other ill-posed inverse problems will also fit in this scheme.

Given $x_0 \in X$, the differential of Φ at x_0 maps injectively

$$d\Phi_{x_0} : \mathbb{E} \longrightarrow T_{x_0}\mathcal{M},$$

where $T_{x_0}\mathcal{M} := d\Phi_{x_0}(\mathbb{E}) \subset \mathbb{F}$, and therefore the linearized inverse $d\Phi_{x_0}^{-1}$ is well defined. The main heuristics behind the idea of Born approximation we have in mind are the following. The space $T_{x_0}\mathcal{M}$ is expected to be a good approximation to the range $\mathcal{M}$ close to $\Phi(x_0)$. One considers a locally convex space or Fréchet space $\mathcal{E}$ in which $\mathbb{E}$ is continuously embedded and constructs an injective map $F^{\mathrm{B}}_{x_0}$ that extends $d\Phi_{x_0}$ to some neighborhood $\mathcal{X} \subseteq \mathcal{E}$ of x_0. The extension should be constructed to ensure that $F^{\mathrm{B}}_{x_0}(\mathcal{X})$ contains a neighborhood of $\Phi(x_0)$ in $\mathcal{M}$ that we denote by Y. Given $x \in \Phi^{-1}(Y)$, we can then define $x^{\mathrm{B}} := (F^{\mathrm{B}}_{x_0})^{-1}\Phi(x) \in \mathcal{X}$, an element in the larger space $\mathcal{E}$ that we call the *Born approximation* of x.

This procedure induces a factorization of Φ, and therefore of Φ^{-1}, into a linear map $F^{\mathrm{B}}_{x_0}$ and a nonlinear map

$$\Phi^{\mathrm{B}}_{x_0} := (F^{\mathrm{B}}_{x_0})^{-1} \circ \Phi|_{\Phi^{-1}(Y)},$$

as indicated in the following commuting diagram:

$$\begin{array}{ccc} \Phi^{-1}(Y) \subseteq X & \xrightarrow{\quad\Phi\quad} & Y \subseteq \mathcal{M} \\ & {\scriptstyle \Phi^{\mathrm{B}}_{x_0}} \searrow \qquad \nearrow {\scriptstyle F^{\mathrm{B}}_{x_0}} & \\ & \mathcal{X} \subseteq \mathcal{E} & \end{array}$$

One of the main advantages of the factorization $\Phi^{-1} = (\Phi^{\mathrm{B}}_{x_0})^{-1} \circ (F^{\mathrm{B}}_{x_0})^{-1}$ is that, as we shall present in some specific cases, the nonlinear map $(\Phi^{\mathrm{B}}_{x_0})^{-1}$ enjoys better stability properties than Φ^{-1}, in the sense that the instability of the problem is encoded in the ill-posed, yet linear, map $(F^{\mathrm{B}}_{x_0})^{-1}$.[2]

Moreover, this factorization is also useful for the construction of effective algorithms to solve (1). On the one hand, the linear map $F^{\mathrm{B}}_{x_0}$ can be inverted with good precision via specific algorithms that take advantage of its analytic properties,

[2] In general the linear map $(F^{\mathrm{B}}_{x_0})^{-1}$ is not continuous; if $(F^{\mathrm{B}}_{x_0})^{-1}$ were a continuous linear map, it would strongly suggest that Φ^{-1} is Fréchet differentiable at $\Phi(x_0)$, which in general is not the case, since this map might not even be continuous.

in order to minimize the undesirable effects resulting from the fact that it is ill-posed. On the other hand, since the problem of reconstructing x from its Born approximation x^{B} (the inversion of $\Phi^{\mathrm{B}}_{x_0}$) is well-posed, it can be solved efficiently via explicit, more standard algorithms (fixed-point algorithms, for instance). This approach leads to new algorithms to solve the inverse problem that cannot be obtained directly from analyzing Φ^{-1}. The paper [4] presents an application of this strategy to numerically solve the Calderón problem.

Interestingly, in all the instances presented here, this can be achieved globally: One can show that $Y = \mathcal{M}$, since the whole $\mathcal{M}$ is contained in the range of $F^{\mathrm{B}}_{x_0}$; in other words, the Born approximation x^{B} is defined for all x in X. However, when ill-posed problems are considered, it is highly nontrivial to show that the range of $F^{\mathrm{B}}_{x_0}$ contains a neighborhood of $\Phi(x_0)$ in $\mathcal{M}$ (which is essential in order to define $\Phi^{\mathrm{B}}_{x_0}$ and x^{B}). This mainly stems from the fact that one does not have a convenient characterization of the range of $F^{\mathrm{B}}_{x_0}$ (recall that $(F^{\mathrm{B}}_{x_0})^{-1}$ is not a continuous linear map). Thus, to determine whether or not the Born approximation is well defined in a given ill-posed inverse problem is in general a nontrivial result, and in fact, it can be seen as a partial characterization of $\mathcal{M}$.

3 The Born Approximation in Inverse Spectral Theory

In this section we show how the approach to inverse spectral theory for Schrödinger operators on the half line pioneered by Simon in [29] fits in the abstract framework presented in the previous section.

Consider the Schrödinger operator:

$$\widehat{H}_q := -\partial_x^2 + q, \tag{2}$$

acting on $L^2(\mathbb{R}_+)$, where $\mathbb{R}_+ := (0, \infty)$. Under certain conditions on the real-valued potential q, $\widehat{H}_q$ is known to be essentially self-adjoint on $C_c^\infty(\mathbb{R}_+)$.[3] The unique self-adjoint extension has domain $D(\widehat{H}_q) = H^2(\mathbb{R}_+) \cap H_0^1(\mathbb{R}_+)$; one then wishes to understand how the spectral properties of $\widehat{H}_q$ depend on q. To do so realize that for every $z \in \mathbb{C} \setminus [\beta, \infty)$, with $\beta \le 0$ large enough to ensure that no such z are in the spectrum of $\widehat{H}_q$, there exists a unique solution to

$$\begin{cases} \widehat{H}_q u(x; z) = z u(x; z), & x \in \mathbb{R}_+, \\ u(0; z) = 1, \quad u(\cdot; z) \in L^2(\mathbb{R}_+). \end{cases} \tag{3}$$

The *Weyl–Titchmarsh* function m_q evaluated at z is defined as

[3] The potential q is then said to be in the *limit point case* (this terminology goes back to the work of Weyl [34, 35]). A sufficient condition for a measurable potential q to be in the limit point case is that $\sup_{y>0} \int_y^{y+1} q_-(x)dx < \infty$, see, for instance, [27, Section X.1].

$$m_q(z) := \partial_x u(0; z).$$

In other words, multiplication by $-m_q(z)$ on $\mathbb{C} \cong L^2(\partial\mathbb{R}_+)$ is nothing but the Dirichlet-to-Neumann map associated with the operator $\widehat{H}_q - z$ on $\mathbb{R}_+$. The Weyl–Titchmarsch function is a holomorphic function from $\mathbb{C}_+ = \{z \in \mathbb{C} : \Im z > 0\}$ into itself that encodes the spectral data of $\widehat{H}_q$: If $\rho_q \in \mathcal{M}_+(\mathbb{R})$ denotes the spectral measure of $\widehat{H}_q$, then one has, for the weak-$*$ topology on $\mathcal{M}_+(\mathbb{R})$, the cone of positive Radon measures on $\mathbb{R}$

$$\frac{1}{\pi}\Im m_q(\lambda + i\epsilon)d\lambda \overset{*}{\rightharpoonup} \rho_q(d\lambda), \quad \epsilon \to 0^+.$$

And conversely, the spectral measure ρ_q completely determines m_q (see, for instance, the discussion in the introduction to [17]).

A direct computation (see, for instance, [32, p. 237]) shows that when $q = 0$, one has

$$m_0(z) = -\sqrt{-z},$$

where the branch of the square root is chosen in order to have $\Im\sqrt{z} > 0$ when $z \in \mathbb{C}\backslash(-\infty, 0)$. It is natural to ask whether or not the knowledge of m_q completely determines the potential q. This is an inverse problem; one can take as the forward map

$$\Phi : X \subset \mathbb{E} \longrightarrow \mathbb{F} : q \longmapsto m_q - m_0,$$

where here, for the sake of simplicity, we will set $X = \mathbb{E} = L^1(\mathbb{R}_+)$ or $L^\infty(\mathbb{R}_+)$ (in this section these spaces will be assumed to consist of real-valued functions only), and $\mathbb{F}$ is any Fréchet space that contains the functions that are holomorphic on $\mathbb{C}_+$.

The injectivity of this map was first established by Borg [6] and Marchenko [24, 25] (see also Tikhonov [33]). Moreover, the map Φ is Fréchet differentiable, and its differential at some $q_0 \in L^1(\mathbb{R}_+)$ is given by

$$M_{q_0}[q](z) := [d\Phi_{q_0}q](z) = -\partial_x R_{q_0}(z)(qu(\cdot; z)),$$

where u is given by (3) and $R_{q_0}(z) := (\widehat{H}_{q_0} - z)^{-1}$ for $z \in \mathbb{C}_+$. Integration by parts further gives that M_{q_0} is in fact an integral transform:

$$M_{q_0}[q](z) = -\int_0^\infty q(x)u(x; z)^2 dx.$$

This expression is particularly simple when $q_0 = 0$, since in that case $u(x; z) = e^{-\sqrt{-z}x}$ (see again [32, p. 237]) and

$$M_0[q](z) = -\int_0^\infty q(x)e^{-2\sqrt{-z}x}dx = -\mathcal{L}[q](2\sqrt{-z}), \tag{4}$$

where $\mathcal{L}[q]$ is nothing but the Laplace transform of q:

$$\mathcal{L}[q](\kappa) := \int_0^\infty q(x)e^{-\kappa x}dx, \qquad \kappa > 0.$$

In the rest of this section, we will focus on this case.

In [29] Simon proved that for any $q \in L^1(\mathbb{R}_+)$, $\Phi(q)$ lies in the range of some extension of M_0. The precise result is the following.

Theorem 1 ([29, Theorem 2.1] and [17, Theorem 10.3]) *Let $\mathbb{E}$ denote $L^1(\mathbb{R}_+)$ or $L^\infty(\mathbb{R}_+)$. For every $q \in \mathbb{E}$, there exist a function $A_q \in L^1_{\mathrm{loc}}(\mathbb{R}_+)$ and $C_q > 0$ such that*

$$m_q(z) - m_0(z) = -\int_0^\infty A_q(x)e^{-2\sqrt{-z}x}dx,$$

where the integral is absolutely convergent for $2\Im\sqrt{-z} > C_q$.[4]

This was extended to potentials that are locally uniformly integrable,[5] by Avdonin, Mikhaylov, and Rybkin in [1, 2].

Theorem 1 proves that $(m_q - m_0)(-\kappa^2)$ is a function that belongs to the range of the Laplace transform $\mathcal{L}$ acting on $\mathcal{E}$, the subspace of functions $f \in L^1_{\mathrm{loc}}(\mathbb{R}_+)$ such that $f(x)e^{-xD}$ belongs to $L^1(\mathbb{R}_+)$ for some $D > 0$. Therefore, because of (4), F_0^{B} can be defined as the extension of M_0 to $\mathcal{E}$, so that

$$(F_0^{\mathrm{B}})^{-1}(m_q - m_0) = A_q = q^{\mathrm{B}}.$$

In other words, Simon's *A-amplitude* A_q of q is the Born approximation of the potential q as introduced in the abstract framework of the previous section. The factorization of Φ presented there is global in this case, since one can take $X = L^\infty(\mathbb{R}_+)$:

$$\begin{array}{ccc} L^\infty(\mathbb{R}_+) \ni q & \xrightarrow{\quad \Phi \quad} & m_q - m_0 \in \mathcal{M} \\ & \searrow^{\Phi_0^{\mathrm{B}}} \qquad \nearrow_{F_0^{\mathrm{B}}} & \\ & A_q \in \mathcal{E} & \end{array}$$

One could also consider Φ defined on the space of locally uniformly integrable functions and still get an analogous factorization.

[4] One can take $C_q = \|q\|_{L^1(\mathbb{R}_+)}$, see [29], or $C_q = 2\sqrt{\|q\|_{L^\infty(\mathbb{R}_+)}}$ see [17].

[5] These are the potentials q that satisfy $\sup_{y>0}\int_y^{y+1}|q(x)|\,dx < \infty$.

This A-amplitude or Born approximation of q enjoys a series of interesting properties. First of all, it can be shown that $A_q - q \in \mathcal{C}(\mathbb{R}_+)$, and in addition, if q is of class $\mathcal{C}^m$, $m \geq 0$ in $(0, a)$ for some $a > 0$, then $A_q - q \in \mathcal{C}^{m+2}((0, a))$. We refer the reader to [17, 29] for proof and additional properties of the difference $A_q - q$.

This type of property can be viewed as a recovery of singularities result, which shows in particular that A_q contains relevant qualitative information on the potential, as its discontinuity points. We now focus on the function Φ_0^{B}, which maps a potential q into the A-amplitude/Born approximation A_q.

Theorem 2 ([29, Theorem 1.5]) *The map Φ_0^{B} is local, in the sense that for every $a > 0$ one has, for q_1, q_2 in $\mathbb{E}$,*

$$q_1(x) = q_2(x) \;\; a.e. \; on \;\; [0, a] \iff A_{q_1}(x) = A_{q_2}(x) \;\; a.e. \; on \;\; [0, a].$$

As a consequence, one can reconstruct q from the A-amplitude. In fact, the mapping $(\Phi_0^{\mathrm{B}})^{-1}$ is indeed locally Hölder continuous (in a sense made precise below), as it has been recently proved in [15]. Let us state a particular case of the actual estimate proved in [15].

Theorem 3 ([15]) *For every $R, a \geq 1$, there exists $C(a, R) > 0$ such that, for every $q_1, q_2 \in L^\infty(\mathbb{R}_+)$ of norm less than or equal to R and such that $\|A_{q_1} - A_{q_2}\|_{L^1((0,a))} < 1$, the following holds:*

$$\int_0^a |q_1(x) - q_2(x)| \, dx \leq C(a, R) \left(\int_0^a \left| A_{q_1}(x) - A_{q_2}(x) \right| \, dx \right)^{1/2}.$$

This estimate is not global though, since the constant $C(a, R)$ may blow-up as $a \to \infty$ (see Section 6 of [15] for more details). Explicit algorithms to reconstruct q form A_q can also be provided, see [29, Section 7], [2] and [15, Section 6].

Therefore, we see that for the inverse spectral problem we have a factorization that fits in the framework presented in the introduction: an unstable, ill-posed linear part and a nonlinear yet stable part. Thus, the concept of Born approximation presented here serves to introduce a rigorous notion of linearization of the inverse map for an ill-posed inverse problem. Moreover, we have shown that the Born approximation is an interesting object in itself, which leads to new uniqueness, stability, and reconstruction results for the inverse problem.

4 The Born Approximation for the Calderón Problem

We now turn to the Calderón problem on a bounded smooth domain $\Omega \subset \mathbb{R}^d$ with $d \geq 2$.

Consider a potential $q \in \mathbb{E} := L^\infty(\Omega)$. The Calderón problem asks[6] whether or not one can recover the potential q from the knowledge of the Dirichlet-to-Neumann map Λ_q defined as follows. Consider $f \in \mathcal{C}^\infty(\partial\Omega)$, and find $u \in H^2(\Omega) \cap H_0^1(\Omega)$ solving

$$\begin{cases} -\Delta u + qu = 0 & \text{in } \Omega, \\ \qquad\qquad u = f, & \text{on } \partial\Omega. \end{cases} \tag{5}$$

Then, the Dirichlet-to-Neumann map (in what follows DtN map for short) acting on f is defined as the normal derivative of u at the boundary:

$$\Lambda_q(f) := \partial_\nu u|_{\partial\Omega}.$$

This operator is well defined as soon as $q \in X$, where

$$X := \{q \in L^\infty(\Omega;\mathbb{R}) \,:\, 0 \text{ is not a Dirichlet eigenvalue of } -\Delta + q\}.$$

The DtN map corresponding to the zero potential Λ_0 is known to be a pseudo-differential operator of order 1 on $\partial\Omega$ (see, for instance, [31]), and one can prove that $\Lambda_q - \Lambda_0$ is a compact self-adjoint operator on $L^2(\partial\Omega)$. Compactness follows from integration by parts, which produces the following identity:[7] For every $f, g \in L^2(\Omega)$, one has

$$(g, (\Lambda_q - \Lambda_0)f)_{L^2(\partial\Omega)} = \int_\Omega q P_q(f)\overline{P_0(g)}dx, \tag{6}$$

where $P_q : L^2(\partial\Omega) \longrightarrow H^2(\Omega)$ is the *q-harmonic extension* operator: Given an $f \in L^2(\partial\Omega)$, the function $v := P_q(f)$ is the unique solution to

$$\begin{cases} -\Delta v + qv = 0 & \text{in } \Omega, \\ \qquad\qquad v = f, & \text{on } \partial\Omega. \end{cases}$$

Denoting by $\mathbb{F} := \mathcal{K}(L^2(\partial\Omega))$ the space of compact, self-adjoint operators on $L^2(\partial\Omega)$, one has that Φ equals in this case

$$\Phi : X \longrightarrow \mathbb{F} : q \longmapsto \Lambda_q - \Lambda_0.$$

[6] We should point out that Calderón's original problem [8, 9] was stated for *conductivities* rather than potentials. One is interested in recovering a conductivity γ from the knowledge of the Dirichlet-to-Neumann map associated with the elliptic operator $-\text{div}(\gamma\nabla\cdot)$. If $\gamma \in \mathcal{C}^2(\Omega)$, then this problem can be reduced to the one we study here.

[7] This is also known as Alessandrini's identity.

This map is known to be injective since the works [21, 26, 30], but its range $\mathcal{M} := \Phi(X)$ still remains an elusive object (we refer to [19, 28] for partial results in related specific cases).

Using (6) is easy to prove that the map Φ is Fréchet differentiable. Its differential at some $q_0 \in X$ acting on a function $q \in L^\infty(\Omega)$ is the operator

$$d\Phi_{q_0}(q) := -\partial_\nu R_{q_0}(q P_{q_0}) \in \mathcal{K}(L^2(\partial\Omega)),$$

where $R_{q_0} : L^2(\Omega) \longrightarrow H^2(\Omega) \cap H_0^1(\Omega)$ is the Dirichlet resolvent $(-\Delta + q_0)^{-1}$ (which is well defined, since $q_0 \in X$). Applying integration by parts or computing the differential using (6) gives the following alternative description: For every $q \in L^\infty(\Omega)$ and $f, g \in L^2(\partial\Omega)$,

$$(g, d\Phi_{q_0}(q) f)_{L^2(\partial\Omega)} = \int_\Omega q P_{q_0}(f)\overline{P_{q_0}(g)} dx. \tag{7}$$

In order to gain further insight on the Born approximation, we will make a series of simplifications. From now on, we will assume that Ω equals $\mathbb{B}^d$, the unit open ball in $\mathbb{R}^d$ centered at the origin, and $q_0 = 0$. In that case the integral kernel of P_0 is nothing but the *Poisson kernel*. It acts on $Y^k \in \mathfrak{h}_k$, a spherical harmonic[8] of degree k, by

$$P_0(Y^k)(x) = |x|^k Y^k(\widehat{x}), \qquad \widehat{x} := x/|x|,\ x \in \mathbb{B}^d.$$

Therefore, by virtue of (7), for $Y^k \in \mathfrak{h}_k$ and $Y^m \in \mathfrak{h}_m$,

$$(Y^m, d\Phi_0(q) Y^k)_{L^2(\mathbb{S}^{d-1})} = \int_{\mathbb{B}^d} q(x)|x|^{m+k} Y^k(\widehat{x})\overline{Y^m(\widehat{x})} dx. \tag{8}$$

Using this identity, we have a characterization of the matrix elements of the operator $d\Phi_0(q)$ in any basis of spherical harmonics. Therefore,

$$d\Phi_0 : L^\infty(\mathbb{B}^d) \longrightarrow \mathcal{K}(L^2(\mathbb{S}^{d-1}))$$

maps a function q into the operator whose matrix elements on an orthonormal basis of $L^2(\mathbb{S}^{d-1})$ consisting of spherical harmonics are given by the generalized *moments* (8).

A further simplification comes from restricting Φ to the class of *radial* potentials:

$$X_{\mathrm{rad}} := X \cap L^\infty_{\mathrm{rad}}(\mathbb{B}^d),$$

[8] Spherical harmonics are restrictions to the sphere $\mathbb{S}^{d-1}$ of the complex homogeneous polynomials of d variables that are harmonic; we denote the subspace of spherical harmonics of degree k by $\mathfrak{h}_k$. The spaces $\mathfrak{h}_k$ are mutually orthogonal in $L^2(\mathbb{S}^{d-1})$.

where

$$L^\infty_{\rm rad}(\mathbb{B}^d) := \{q \in L^\infty(\mathbb{B}^d; \mathbb{R}) \,:\, q = q_0(|\cdot|) \text{ for some } q_0 \in L^\infty((0,1))\}.$$

In that case, one has for every $q \in X_{\rm rad}$ and every $k \in \mathbb{N} \cup \{0\}$

$$\Phi(q)|_{\mathfrak{h}_k} = (\Lambda_q - \Lambda_0)|_{\mathfrak{h}_k} = (\lambda_k[q] - k)\mathrm{Id}_{\mathfrak{h}_k}.$$

The spectrum of Λ_q consists of the eigenvalues $\lambda_k[q]$ whose corresponding eigenfunctions are the spherical harmonics $\mathfrak{h}_k$; the spectrum of Λ_0 is $\mathbb{N} \cup \{0\}$. We refer the reader to [3, Theorem 2] for additional details on the structure of these eigenvalues.

Formula (8) then allows us to infer that, for every $q \in L^\infty_{\rm rad}(\mathbb{B}^d)$ and every $k \in \mathbb{N} \cup \{0\}$,

$$d\Phi_0(q)|_{\mathfrak{h}_k} = \sigma_k[q]\mathrm{Id}_{\mathfrak{h}_k},$$

where

$$\sigma_k[q] := \frac{1}{|\mathbb{S}^{d-1}|}\int_{\mathbb{B}^d} q(x)|x|^{2k}dx, \qquad k \in \mathbb{N} \cup \{0\}, \tag{9}$$

are classical Hausdorff moments of q. Therefore, $d\Phi_0$ maps in this case potentials in $L^\infty_{\rm rad}(\mathbb{B}^d)$ to operators that are multiples of the identity in each subspace $\mathfrak{h}_k$ and whose eigenvalues are the moments (9). The range of $d\Phi_0$ can be identified to the space of sequences of real numbers that are moments of some function in $L^\infty_{\rm rad}(\mathbb{B}^d)$. On this range, one has the following inversion formula (see [3, Section 3]):

$$\widehat{q}(\xi) = 2\pi^{d/2}\sum_{k=0}^{\infty}\frac{(-1)^k}{k!\Gamma(k+d/2)}\left(\frac{|\xi|}{2}\right)^{2k}\sigma_k[q], \tag{10}$$

where $\widehat{q}$ stands for the Fourier transform:

$$\widehat{q}(\xi) := \int_{\mathbb{B}^d} q(x)e^{-i\xi\cdot x}dx.$$

Thus, (10) gives an explicit and absolute convergent formula for the inverse of $d\Phi_0$. The next result states that $\Phi(q)$ lies in the range of some extension of $d\Phi_0$ (compare with Theorem 1).

Theorem 4 *For every $q \in X_{\rm rad}$, there exist a compactly supported radial distribution*[9] *$q^{\rm B} \in \mathcal{E}'(\mathbb{B}^d)$ and a constant $C_q \geq 0$ such that*

[9] A distribution $T \in \mathcal{D}'(\mathbb{R}^d)$ that is compactly supported in $\mathbb{B}^d$ (i.e., $T \in \mathcal{E}'(\mathbb{B}^d)$) is *radial* provided that $\langle T, \phi \circ R\rangle = \langle T, \phi\rangle$ for every $R \in \mathrm{SO}(d)$ and $\phi \in C^\infty_c(\mathbb{R}^d)$.

$$(\Lambda_q - \Lambda_0)|_{\mathfrak{h}_k} = \sigma_k[q^{\mathrm{B}}]\mathrm{Id}_{\mathfrak{h}_k},$$

for every $k \geq C_q$. Moreover, the Fourier transform of q^{B} is given by the absolute convergent series:

$$\widehat{q^{\mathrm{B}}}(\xi) = 2\pi^{d/2} \sum_{k=0}^{\infty} \frac{(-1)^k}{k!\Gamma(k+d/2)} \left(\frac{|\xi|}{2}\right)^{2k} (\lambda_k[q] - k). \tag{11}$$

This result is proved in [15], under much more general hypothesis on the radial potential q. Formula (11) was introduced in [3] and subsequently used in [4] to design algorithms for the reconstruction of the potential from the DtN map.[10] Theorem 4 states that the spectrum of $\Lambda_q - \Lambda_0$ is the sequence of moments of some distribution. The extension F_0^{B} in this case is defined on $\mathcal{E} = \mathcal{E}'(\mathbb{B}^d)$; its injectivity is a consequence of uniqueness in the Hausdorff moment problem on $\mathcal{E}'(\mathbb{B}^d)$ (see [15] for details). The distribution

$$(F_0^{\mathrm{B}})^{-1}(\Lambda_q - \Lambda_0) = q^{\mathrm{B}}$$

is the corresponding Born approximation, and the inversion formula (10) for the moment problem can be extended to this setting to yield (11), which can be interpreted as an explicit formula for $(F_0^{\mathrm{B}})^{-1}$.

As in the previous section, the factorization of Φ is global in this case, since it holds on the whole $X = X_{\mathrm{rad}}$:

$$\begin{array}{ccc} X_{\mathrm{rad}} \ni q & \xrightarrow{\quad\Phi\quad} & \Lambda_q - \Lambda_0 \in \mathcal{M} \\ & \searrow^{\Phi_0^{\mathrm{B}}} \qquad {}^{F_0^{\mathrm{B}}}\nearrow & \\ & q^{\mathrm{B}} \in \mathcal{E}'(\mathbb{B}^d) & \end{array} \tag{12}$$

The Born approximation captures all the discontinuities of the potential and approximates it well close to the boundary of its support.

Theorem 5 *Let $q \in X_{\mathrm{rad}}$ such that* $\operatorname{ess\,supp} q \subset B(0;\rho)$ *for some* $0 < \rho \leq 1$. *Then $q^{\mathrm{B}} - q \in \mathcal{C}(\mathbb{B}^d \setminus \{0\})$, and there exists $\alpha_q \geq 0$ such that*

$$\left|(q^{\mathrm{B}} - q)(x)\right| \leq \left(\frac{\|q\|_{L^\infty(\mathbb{B}^d)}}{2} \frac{(\rho^2 - |x|^2)_+}{|x|^{\alpha_q+1}}\right)^2. \tag{13}$$

In addition, if $q \in \mathcal{C}^m(\mathbb{B}^d)$ with $m \in \mathbb{N}$, then $q^{\mathrm{B}} - q \in \mathcal{C}^{m+2}(\mathbb{B}^d \setminus \{0\})$.

[10] In these references, q^{B} is denoted q_{exp} following the convention used in the numerical analysis literature for the Calderón problem. The Born approximation has been used in this context as a formal object in the numerical recovery of potential/conductivities, see, for instance, [5].

Again, this is a consequence of a more general and precise result proved in [15]. Theorem 5 has two main consequences: On the one hand, q^{B} is a good approximation of the potential q close to the boundary of its support. On the other hand, q^{B} has exactly the same discontinuities and higher order singularities as q, except perhaps at the origin. This makes q^{B} a tool well suited to the problem of *recovery of singularities* for the potential.

We turn now our attention to the map

$$\Phi_0^{\mathrm{B}} : X_{\mathrm{rad}} \longrightarrow \mathcal{E}'(\mathbb{B}^d) : q \longmapsto q^{\mathrm{B}}.$$

Analogously to the one-dimensional case, it is proved in [15] that this map is local and injective.

Theorem 6 *Let $q_1, q_2 \in X_{\mathrm{rad}}$ and $0 < b < 1$. Then*

$$q_1^{\mathrm{B}}(x) = q_2^{\mathrm{B}}(x) \ \ a.e. \ for \ b < |x| < 1 \iff q_1(x) = q_2(x) \ \ a.e. \ for \ b < |x| < 1.$$

Note that combining Theorems 6 and 4 and taking into account that the factorization (12) holds give a proof that Φ is injective; in other words, it provides a proof of uniqueness for the Calderón problem for radial potentials. Remarkably, this proof does not involve the use of exponentially growing solutions or CGOs which is the main technical feature used to tackle uniqueness results for this type of inverse problem since [30].

We can turn out to study its stability properties, which is one of the main advantages of the Born approximation approach to ill-posed inverse problems. As a consequence of Theorem 3, it is also shown in [15] that the following holds.

Theorem 7 *For every $R \geq 1$ and $0 < b < 1$, there exists $C(b, R) > 0$ such that, for every $q_1, q_2 \in X_{\mathrm{rad}}$ of norm less than or equal to R and such that*

$$\int_{b<|x|<1} \left| q_1^{\mathrm{B}}(x) - q_2^{\mathrm{B}}(x) \right| |x|^{2-d} \, dx < 1,$$

the following holds:

$$\int_{b<|x|<1} \frac{|q_1(x) - q_2(x)|}{|x|^{d-2}} \, dx \leq C(b, R) \left(\int_{b<|x|<1} \frac{\left|q_1^{\mathrm{B}}(x) - q_2^{\mathrm{B}}(x)\right|}{|x|^{d-2}} \, dx \right)^{1/2}.$$

The Calderón problem is known to exhibit exponential instability since the work of Mandache [23] (see also [20]). Combining Theorem 4 with Theorem 7 clarifies how this instability mechanism works, at least in the radial case. The exponential instability arises from the linear part of our factorization (12), which amounts to solving the moment problem: Find $q^{\mathrm{B}} \in \mathcal{E}'(\mathbb{B}^d)$ such that

$$\langle q^{\mathrm{B}}, \psi_k \rangle = \lambda_k[q] - k, \qquad k \in \mathbb{N} \cup \{0\},$$

with $\psi_k(x) := |\mathbb{S}^{d-1}|^{-1}|x|^{2k}$. This is a notoriously ill-posed problem, which exhibits exponential instability as well. The nonlinear part of the factorization, namely, finding q for a given q^{B}, is much better conditioned as it exhibits Hölder stability (note however that the quality of that estimate is expected to degenerate as one approaches the center of the ball $\mathbb{B}^d$). For practical purposes, computing q from q^{B} can be achieved efficiently using suitable fixed-point schemes or other methods. We refer the reader to [4] for this and for several numerical illustrations of properties of the Born approximation. In that reference, it is also shown how to define the Born approximation in the case of conductivities (see also [18]). In particular, the approximation quality of the Born approximation tends to worsen as the norm of the potential increases. This is consistent with the estimate shown in Theorem 5 and is to be expected, since the Born approximation has been defined via the linearization of the Calderón problem around the zero potential. We also remark that the notion of Born approximation is somewhat implicit in the works of Daudé, Kamran, and Nicoleau, which have applied the techniques of inverse spectral theory of [29] in the context of the Steklov problem for warped product manifolds, see [12–14]. In particular, the results in [13] imply stability and uniqueness results for the radial Calderón problem for both the conductivity and Schrödinger cases.

5 Toward a General Born Approximation for the Calderón Problem

Two questions arise naturally at this point. First, and most importantly, one would like to know whether or not the factorization implemented here holds for the general, nonradial case. Second, it is interesting to understand the behavior of Born approximations defined through linearization around potentials $q_0 \neq 0$. This could be useful, in particular, when it comes to reconstructing numerical potentials that are not close to zero. This is also a relevant intermediate step toward the problem of characterizing the range $\mathcal{M}$ (the set of all DtN maps).

To get some insight on the first of these two questions, note that an operator as $\Lambda_q - \Lambda_0$ can be identified to its integral kernel, which is a function defined on $\partial\Omega \times \partial\Omega$; in particular, it is a function of $2d - 2$ variables. One wants then to reconstruct from it the potential, or the Born approximation, a function/distribution of d variables. This problem is overdetermined as soon as $d \geq 3$, so one should not expect to obtain a meaningful factorization unless $d = 2$.

In [3, Theorem 3], a notion of Born approximation is defined for $d = 3$. In order to overcome the fact that the problem is overdetermined (four variables for the kernel of the DtN map vs. three for the potential), an averaging procedure is performed. Let us give some details on how this is done.

Recall that applying $d\Phi_0$ to a potential $q \in L^\infty(\mathbb{B}^3)$ results in an operator $L_q \in \mathcal{K}(L^2(\mathbb{B}^3))$ that is characterized by (8). Identity (8) can also be written as

$$(Y^m, L_q Y^k)_{L^2(\mathbb{S}^2)} = (\tilde{Y}^m, q\tilde{Y}^k)_{L^2(\mathbb{B}^3)} \tag{14}$$

for every spherical harmonics Y^k, Y^m on $\mathbb{S}^2$, with the convention:

$$\tilde{Y}^k(x) := |x|^k Y^k(\widehat{x}), \quad Y^k \in \mathfrak{h}_k.$$

The right-hand side of (14) is the $m+k$-order moment of $qY^k\overline{Y^m}$, and $Y^k\overline{Y^m}$ only depends on the angular variable $\widehat{x}$. It is shown in [3] that if a convenient basis of spherical harmonics is chosen, then an inversion formula similar to (10) exists. In order to state it, we must introduce some notation. Let L_3 denote the angular momentum operator with respect to the direction $e_3 := (0, 0, 1)$:

$$L_3 := \frac{1}{i}(x_1\partial_{x_2} - x_2\partial_{x_1}).$$

Then, for every $k \in \mathbb{N} \cup \{0\}$, there exists an orthonormal basis $(Y_m^k)_{|m|\le k}$ of $\mathfrak{h}_k$ (which has a dimension equal to $2k+1$) consisting of eigenfunctions of L_3:

$$L_3 Y_m^k = m Y_m^k, \qquad |m| \le k.$$

Given $\omega \in \mathbb{S}^2$ denote by $R_\omega \in \mathrm{SO}(3)$ a rotation such that $R_\omega\omega = e_3$ (there are exactly two of those when $\omega \neq e_3$), and we set $R_{e_3} = \mathrm{Id}_{\mathbb{R}^3}$. Define for $\omega \in \mathbb{S}^2$, $m \in \mathbb{N} \cup \{0\}$, and $k \ge m$ the generalized moments:

$$\mathfrak{m}_m^k[\omega, q] := (\tilde{Y}_m^k, (q \circ R_\omega^{-1})\tilde{Y}_m^m)_{L^2(\mathbb{B}^3)}.$$

These quantities are independent of the choice of the rotation R_ω, and moreover, the following holds [3, Theorem 4]: for every $q \in L^1(\mathbb{B}^3)$:

$$\widehat{q}(\xi) = \sum_{m=0}^{\infty}\sum_{k=m}^{\infty} \mu_{m,k}(-i|\xi|)^{k+m}\mathfrak{m}_m^k[\widehat{\xi}, q], \tag{15}$$

where the series converges absolutely and

$$\mu_{m,k} := \frac{4\pi}{\sqrt{(2m+1)(2k+1)}}\frac{1}{\sqrt{(2m)!(k-m)!(k+m)!}}.$$

The reader can check that when q is a radial formula, (15) coincides with (10).

If now one defines the matrix elements:[11]

$$\lambda_k^m[\omega, q] := (Y_m^k \circ R_\omega, \Lambda_q Y_m^m \circ R_\omega)_{L^2(\mathbb{S}^2)}, \qquad k \ge m,$$

[11] These are actual matrix elements with respect to a basis of spherical harmonics, since $Y^k \in \mathfrak{h}_k$ implies $Y^k \circ R \in \mathfrak{h}_k$ if $R \in \mathrm{SO}(3)$.

then the Born approximation of q should be (at this level, by analogy with (11)

$$\widehat{q^{\mathrm{B}}}(\xi) = \sum_{m=0}^{\infty} \sum_{k=m}^{\infty} \mu_{m,k}(-i|\xi|)^{k+m}(\lambda_k^m[\widehat{\xi}, q] - \delta_{m,k}k). \tag{16}$$

Further evidences that support this claim are given in [3]. Note, however, that the notation $\widehat{q^{\mathrm{B}}}$ is merely formal, since *a priori* there is no guarantee that (16) has polynomial growth as $|\xi| \to \infty$, so that one can apply the inverse Fourier transform to obtain a distribution $q^{\mathrm{B}} \in \mathcal{E}'(\mathbb{B}^3)$.

Concerning the second question, several results have been obtained in [22] for linearization around a constant potential different from zero. This is a problem of independent interest, since it corresponds to studying the Calderón problem at fixed, nonzero energy, as it is done, for instance, in [26]. In fact, in [22] analogs of (10) and (11) have been obtained in the radial case. Nonradial versions of this formulas have also been found in the two-dimensional case, and in three dimensions (15) and (16) have also been generalized for the fixed energy case. As in the case of (16), this is not yet enough to properly define q^{B}, since it is necessary to prove that $\widehat{q^{\mathrm{B}}}(\xi)$ has at most polynomial growth as $|\xi| \to \infty$.

Acknowledgments We would like to thank our collaborators Juan Antonio Barceló, Carlos Castro, Thierry Daudé, François Nicoleau, and Daniel Sánchez-Mendoza for many insightful discussions we have had on these topics. This research has been supported by grant PID2021-124195NB-C31 of Agencia Estatal de Investigación (Spain).

References

1. S. Avdonin, V. Mikhaylov, A. Rybkin, The boundary control approach to the Titchmarsh-Weyl m-function. I. The response operator and the A-amplitude. Comm. Math. Phys. **275**, 791–803 (2007)
2. S. Avdonin, V. Mikhaylov, The boundary control approach to inverse spectral theory. Inverse Probl. **26**, 045009, 19 (2010)
3. J.A. Barceló, C. Castro, F. Macià, C.J. Meroño, The Born approximation in the three-dimensional Calderón problem. J. Funct. Anal. **283**, Paper No. 109681 (2022)
4. J.A. Barceló, C. Castro, F. Macià, C.J. Meroño, The Born approximation in the three-dimensional Calderón problem II: numerical reconstruction in the radial case. Inverse Probl. Imag. **18**, 183–207 (2024)
5. J. Bikowski, K. Knudsen, J.L. Mueller, Direct numerical reconstruction of conductivities in three dimensions using scattering transforms. Inverse Probl. **27**, 015002 (2011)
6. G. Borg, Uniqueness theorems in the spectral theory of $y'' + (\lambda - q(x))y = 0$, in *Den 11te Skandinaviske Matematikerkongress, Trondheim, 1949, Johan Grundt Tanums Forlag. Oslo* (1952), pp. 276–287
7. M. Born, Quantenmechanik der Stoßvorgänge. Zeitschrift für Physik **37**, 863–868 (1926)
8. A.-P. Calderón, On an inverse boundary value problem, in *Seminar on Numerical Analysis and its Applications to Continuum Physics (Rio de Janeiro, 1980) Soc. Brasil. Mat., Rio de Janeiro* (1980), pp. 65–73

9. A.P. Calderón, On an inverse boundary value problem. Comput. Appl. Math. **25**, 133–138 (2006)
10. S. Cen, B. Jin, K. Shin, Z. Zhou, Electrical impedance tomography with deep Calderón method. J. Comput. Phys. **493**, Paper No. 112427, 14 (2023)
11. D. Colton, R. Kress, *Inverse Acoustic and Electromagnetic Scattering Theory*. Applied Mathematical Sciences, vol. 93, 3rd edn. (Springer, New York, 2013)
12. T. Daudé, B. Helffer, F. Nicoleau, Exponential localization of Steklov eigenfunctions on warped product manifolds: the flea on the elephant phenomenon Ann. Math. Qué. **47**, 295–330 (2023)
13. T. Daudé, N. Kamran, F. Nicoleau, Stability in the inverse Steklov problem on warped product Riemannian manifolds. J. Geom. Anal. **31**, 1821–1854 (2021)
14. T. Daudé, N. Kamran, F. Nicoleau, Local hölder stability in the inverse steklov and calderón problems for radial Schrödinger operators and quantified resonances in *Annales Henri Poincaré* (2023)
15. T. Daudé, F. Macià, C.J. Meroño, F. Nicoleau, Stable factorization for the Calderón problem via the Born approximation (2024). arXiv:2402.06321
16. G. Eskin, *Lectures on Linear Partial Differential Equations*. Graduate Studies in Mathematics, vol. 123 (American Mathematical Society, Providence, 2011)
17. F. Gesztesy, B. Simon, A new approach to inverse spectral theory. II. General real potentials and the connection to the spectral measure. Ann. Math. **152**(2), 593–643 (2000)
18. B. Harrach, J.K. Seo, Exact shape-reconstruction by one-step linearization in electrical impedance tomography. SIAM J. Math. Anal. **42**, 1505–1518 (2010)
19. D.V. Ingerman, Discrete and continuous Dirichlet-to-Neumann maps in the layered case. SIAM J. Math. Anal. **31**, 1214–1234 (2000)
20. H. Koch, A. Rüland, M. Salo, On instability mechanisms for inverse problems, Ars Inveniendi Analytica, (2021)
21. R. Kohn, M. Vogelius, Determining conductivity by boundary measurements. Comm. Pure Appl. Math. **37**, 289–298 (1984)
22. F. Macià, C.J. Meroño, D. Sánchez-Mendoza, The Born approximation in the three-dimensional Calderón problem III: reconstruction at fixed energy (2024). In Preparation
23. N. Mandache, Exponential instability in an inverse problem for the Schrödinger equation. Inverse Probl. **17**, 1435–1444 (2001)
24. V.A. Marčenko, Concerning the theory of a differential operator of the second order. Doklady Akad. Nauk SSSR **72**, 457–460 (1950)
25. V.A. Marčenko, Some questions of the theory of one-dimensional linear differential operators of the second order. I, Trudy Moskov. Mat. Obšč. **1**, 327–420 (1952)
26. A. Nachman, J. Sylvester, G. Uhlmann, An n-dimensional Borg-Levinson theorem. Comm. Math. Phys. **115**, 595–605 (1988)
27. M. Reed, B. Simon, *Methods of Modern Mathematical Physics. II. Fourier Analysis, Self-adjointness* (Academic [Harcourt Brace Jovanovich, Publishers], New York, 1975)
28. V.A. Sharafutdinov, The geometric problem of electrical impedance tomography in the disk. Sibirsk. Mat. Zh. **52**, 223–238 (2011)
29. B. Simon, A new approach to inverse spectral theory. I. Fundamental formalism. Ann. Math. **150**(2), 1029–1057 (1999)
30. J. Sylvester, G. Uhlmann, A global uniqueness theorem for an inverse boundary value problem. Ann. Math. **125**(2), 153–169 (1987)
31. J. Sylvester, G. Uhlmann, Inverse boundary value problems at the boundary-continuous dependence. Commun. Pure Appl. Math. **41**, 188–219 (1988)
32. G. Teschl, *Mathematical Methods in Quantum Mechanics*. Graduate Studies in Mathematics, , vol. 157, 2nd edn. (American Mathematical Society, Providence, 2014). With Applications to Schrödinger Operators
33. A.N. Tihonov, On the uniqueness of the solution of the problem of electric prospecting. Doklady Akad. Nauk SSSR **69**, 797–800 (1949)

34. H. Weyl, Über gewöhnliche differentialgleichungen mit singularitäten und die zugehörigen entwicklungen willkürlicher funktionen. Mathematische Annalen **68**, 220–269 (1910)
35. H. Weyl, Ueber gewöhnliche lineare differentialgleichungen mit singulären stellen und ihre eigenfunktionen. (2. note), in *Nachrichten von der Gesellschaft der Wissenschaften zu Göttingen, Mathematisch-Physikalische Klasse, 1910* (1910), pp. 442–467

Central Nervous System Action on Rolling Balance Board Robust Stabilization: Computer Algebra and MID-Based Feedback Design

Karim L. Trabelsi, Islam Boussaada, Amina Benarab, Csenge Molnar, Silviu-Iulian Niculescu, and Tamas Insperger

2010 Mathematics Subject Classification 34K20, 39B82, 70Q05, 47N70

1 Introduction

The intrinsic brain activity is spatio-temporally structured, and human balancing is among the few activities for which the time heterogeneity is essential. As a matter of fact, age-related decline in balance is a principal contributor to falls among older adults; these are the most common causes of accidental deaths, and in nonfatal

K. L. Trabelsi · A. Benarab
IPSA Paris Campus, Ivry sur Seine, France
e-mail: Karim.Trabelsi@ipsa.fr; Amina.Benarab@ipsa.fr

I. Boussaada (✉)
IPSA Paris Campus, Ivry sur Seine, France

Université Paris-Saclay, CNRS, CentraleSupelec, INRIA, Laboratoire des Signaux et systèmes, Gif-sur-Yvette cedex, France
e-mail: Islam.Boussaada@centralesupelec.fr

C. Molnar
Department of Mechatronics, Optics and Mechanical Engineering Informatics, Faculty of Mechanical Engineering, Budapest University of Technology and Economics, Budapest, Hungary
e-mail: csenge.molnar@mm.bme.hu

S.-I. Niculescu
Université Paris-Saclay, CNRS, CentraleSupelec, INRIA, Laboratoire des Signaux et systèmes, Gif-sur-Yvette cedex, France
e-mail: Silviu.Niculescu@l2s.centralesupelec.fr

T. Insperger
Department of Applied Mechanics, Faculty of Mechanical Engineering, Budapest University of Technology and Economics, Budapest, Hungary

HUN-REN-BME Dynamics of Machines Research Group, Budapest, Hungary
e-mail: insperger@mm.bme.hu

K. Ammari et al. (eds.), *Advances in Partial Differential Equations and Control*, Trends in Mathematics, https://doi.org/10.1007/978-3-031-62265-6_11

cases the costs related to the treatment of fall-related sequelae are on the rise as a consequence of the growing age of populations. As such, human balancing is a subject of growing interest.

The passive biomechanics of human balance are unstable as they consist of a network of interconnected inverted pendulums that are each unstable. The human balance system is the sensorimotor system that enables us to perform ordinary reflex tasks such as standing upright, walking, etc. Balance may be defined as the ability to maintain equilibrium in a gravitational field by keeping or returning the center of body mass over its base of support [30]. It is the system that depends on vestibular function, vision, and proprioception to maintain posture, navigate in one's surroundings, coordinate motion of body parts, modulate fine motor control, and initiate the vestibule-oculomotor reflexes. As such, success of balance may be assessed by the ability to perform quiet stance, compensatory postural reactions, and anticipatory postural responses. This ability is also appraised by the reaction time (RT), or the controller of the central nervous system (CNS), i.e., the length of time between a stimulus and our response to that stimulus. That said, elementary balancing tasks may be investigated by a mechanical analysis. Stick balancing on a fingertip [31], quiet stance [25, 32, 40, 51, 52], ball and beam balancing [18], and standing on a balance board [21, 22, 24] have been thoroughly investigated; see also [46] for further details. In fact, enquiries are established on the mechanical model of a single inverted pendulum, and more complex tasks require multi-degree-of-freedom models.

In this chapter, we consider the mathematical model of balancing on a uniaxial rolling balance board in the sagittal plane, see Fig. 1. The balance board has a configurable geometry in as much as the radius R of the (half) wheels, and the distance h between the straight edge of the half wheel and the board can be adjusted as in Fig. 1. The latter parameters highly affect the ability to stabilize the associated mathematical model, as preliminary computations and experiments performed by human subjects show. Namely, standing on the balance board gets more involved as the wheel radius and the board elevation decrease. Indeed, hip strategy is dominant for smaller radii, which indicates a double-inverted pendulum model for the human body. However, greater radii ($R > 100$ mm) spur the use of musculature at the ankle to maintain the equilibrium, so that the human body can be considered as a single inverted pendulum. In this chapter, greater wheel radii are considered, and therefore a 2-degree-of-freedom mechanical model is analyzed involving the balance board and the human body.

Balance deficits and disorders are the consequence of a multitude of age-associated declines in sensorimotor function, including somatosensation, vestibular function, vision, cognition, and strength. Stabilization of the human body and the balance board is performed by a control process governed by the CNS. Visual, vestibular, and somatosensory systems obtain information about the spatial orientation and surroundings of the human body. The information is delivered to the brain; CNS determines the necessary interaction to maintain the balance after processing the signals and sends an instruction to the musculature. The process described above requires a definite time lapse called RT. Consequently, the mathematical

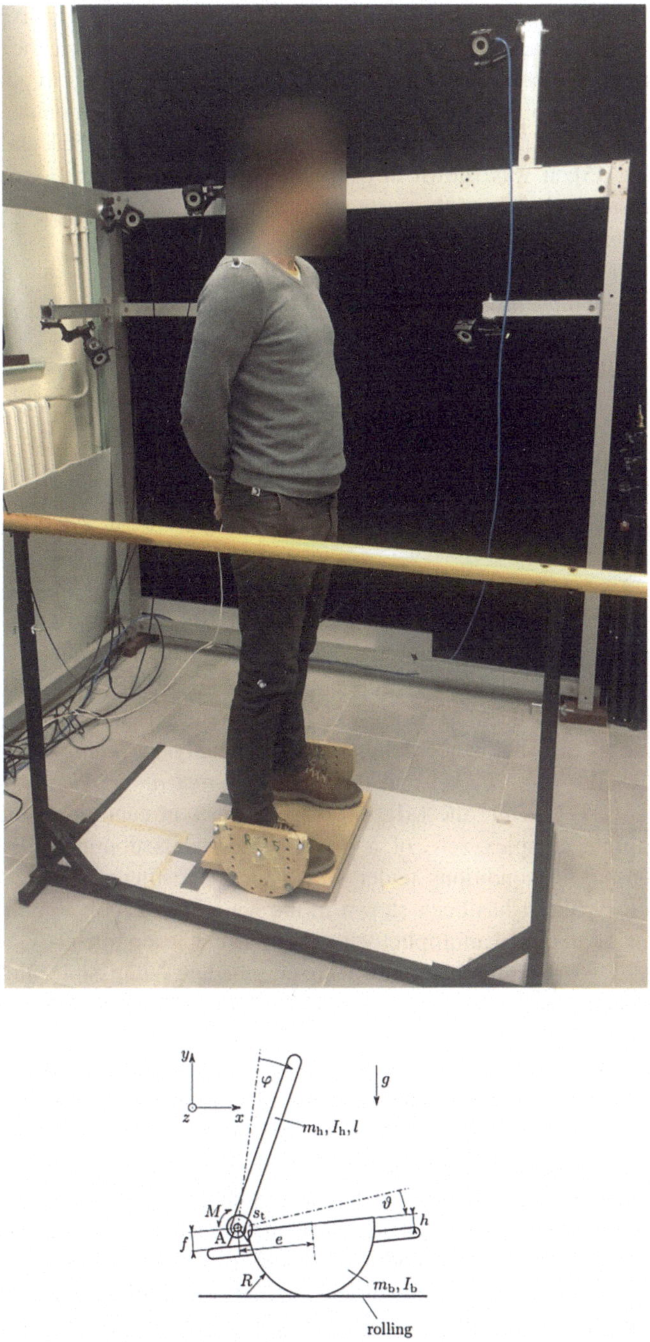

Fig. 1 (Left) Two-degree-of-freedom mechanical model of human balancing on a uniaxial rolling balance board in the sagittal plane. (Right) Human balancing on a uniaxial rolling balance board in the sagittal plane. Figures are taken from [46]

model of balancing tasks involves a delayed control law, here a proportional–derivative (PD) feedback with constant feedback delay [33, 58], since the visual and vestibular system perceive position and velocity, respectively. Other frequently used neuromuscular control models are, for instance, delayed proportional-derivative–acceleration feedback [31, 60], intermittent control [3, 19, 27, 44, 51], and predictor feedback [31, 45]. It was shown in [37] and [36] that an integral term does not enhance the stabilizability of the system in the presence of feedback delay, that is why, here, only PD feedback is considered.

Note that RT of different sensory systems differs, yet they scale alike [59]. Consequently, in most investigations related to human balancing, delays associated with different sensory organs are assumed to be similar in magnitude [3, 21, 58]. In case of quiet stance, feedback delay is estimated at 100–200 ms [3, 28]. Balancing on an unstable moving surface such as the balance board or skateboard is a trickier task; hence, a sensibly higher RT is gauged at 150–300 ms [21, 48, 58].

For the exponential behavior of solutions of the ensuing delay system, we are interested in the *spectral abscissa* of the corresponding characteristic function. Indeed, the spectral abscissa is related to the notion of decay rate of time-delay system solutions.

A series of recent works proposed a safe control methodology called *partial pole placement* (PPP) [13–15, 42]. This PPP methodology is based on the assignment of the closed-loop dominant solution's decay rate. Indeed, it has been shown that under appropriate conditions a multiple spectral value is the rightmost; see, for instance, [4, 10, 12, 14, 20]. The proposed PPP strategy relies on two properties of the spectrum distribution of delay systems, namely the *multiplicity-induced-dominancy* (MID) [14, 16, 41] and the *coexisting-real-roots-induced-dominancy* (CRRID) [2, 6, 57]. While the MID property consists in conditions under which a given multiple complex zero of a quasipolynomial is dominant, the CRRID property consists in conditions under which the coexisting real spectral values are dominant. While it has been shown in the generic quasipolynomial case that the real root of maximal multiplicity is necessarily the dominant (GMID), see, for instance, [16, 41], multiple roots with intermediate admissible multiplicities may be dominant but under appropriate conditions. Thanks to this property, an ensued control strategy is proposed in [4, 14, 39], which consists in assigning a root with an admissible multiplicity once appropriate conditions guaranteeing its dominancy are determined. Furthermore, the MID property may be used to tune standard controllers. For instance, in [39], it is applied to the systematic tuning of the stabilizing PID controller of a first-order plant. Here, we aim at assigning dominant multiple real roots with admissible codimensions, and the MID property is utilized for the mechanical model of human balancing on rolling balance board in the sagittal plane. Delayed state feedback is adopted, which gives the governing equation of the form of a delay differential equation (DDE). The rightmost characteristic root (called dominant root) is minimized by tuning the control gains according to the MID concept [43].

The main contribution of the chapter is to explore the use of the MID property for the rolling balance board stabilization. It should be mentioned that the particular structure of the system's dynamics does not allow the use of any of the existing MID results straightforwardly. Thus, for instance, due to the sparsity of the open-loop transfer function, the GMID cannot be reached, and the characterization of the generic MID proposed in [42] is not valid. Moreover, the corresponding plant is not real-rooted, but its roots are located on real and imaginary axes, and the ideas and the approach proposed in [4] cannot apply.

The chapter is organized as follows. The next section builds the mechanical model of human stance on a rolling balance board in the sagittal plane from which our mathematical model is then derived. Section 3 reminds the relevant background and some prerequisites pertaining to quasipolynomials. The main results and their proofs are presented in Sect. 4. Finally, a conclusion ends the chapter.

2 Mechanical Model

The rolling balance board, see Fig. 1, is modeled as a double-inverted pendulum, see also [47], which is an important model in human balancing research [8, 49, 53]. Indeed, the double-inverted pendulum is often referred to as the simplest nonlinear multi-body system as it features all the properties of higher degree-of-freedom nonlinear systems, such as complexity or chaos. The corresponding full state feedback involves the angular position and angular velocity of both pendulum segments, which implies that the number of control gains is 4. Hence, optimization and stabilization shall be performed in the four-dimensional space of the control gains.

Human stance on a rolling balance board in the sagittal plane is described by a 2-degree-of-freedom mechanical model as shown in Fig. 1. The generalized coordinates are the angle φ of the human body and the angle ϑ of the balance board measured from the upper (unstable) equilibrium. The human body was modeled as a homogeneous rigid bar considering that the musculature at the ankle is used during balancing only. The mass and the height of the human body are denoted by m_{h} and l; therefore, the mass moment of inertia of the human body becomes $I_{\mathrm{h}} = 1/12 m_{\mathrm{h}} l$ for the center of gravity.

The feet are considered to be fixed to the board; therefore, the ankles are modeled as a pin between the human body and the balance board. As shown in Fig. 1, the height of the ankle is denoted by f. The ankle joint is located on the left side of the symmetry axis of the balance board expressed by parameter e, since the feet are able to move the balance board in both $+\vartheta$ and $-\vartheta$ this way. The passive stiffness of the ankle musculature was considered as a torsional spring

$$s_{\mathrm{t}} = 0.91 m_{\mathrm{h}} g \frac{l}{2}, \tag{1}$$

referring to [38], while the passive damping of the muscles was neglected.

The elements of the balance board were made from plywood of 21 mm thickness and a density of 700 kg/m^3. The center of gravity l_b, mass m_b, and mass moment of inertia I_b of the balance board were calculated based on the actual set of the adjustable parameters R and h. The mass of the feet was neglected.

As it was explained in the Introduction section, the sensory organs obtain information about the angular position and angular velocity of the human body and the balance board; therefore, a delayed full state feedback

$$M(t) = P_\varphi \varphi(t-\tau) + D_\varphi \dot{\varphi}(t-\tau) + P_\vartheta \vartheta(t-\tau) + D_\vartheta \dot{\vartheta}(t-\tau) \tag{2}$$

is applied as the control torque produced by the muscles at the ankle. The feedback delay τ corresponds to the reaction time of human beings and estimated by a constant value. For the sake of simplicity, the sensory organs are involved with the same value in the control model.

The governing equation of the motion can be written by Lagrange's equation of the second kind

$$\frac{\mathrm{d}}{\mathrm{d}t}\frac{\mathrm{d}T}{\mathrm{d}\dot{q}_k} - \frac{\mathrm{d}T}{\mathrm{d}q_k} + \frac{\mathrm{d}U}{\mathrm{d}q_k} = Q_k(t) \quad k = 1, \ldots, n, \tag{3}$$

where the kinetic energy is

$$T = \frac{1}{2} m_\mathrm{b} v_{\mathrm{S_b}}^2 + \frac{1}{2} I_\mathrm{b} \dot{\vartheta}^2 + \frac{1}{2} m_\mathrm{h} v_{\mathrm{S_h}}^2 + \frac{1}{2} I_\mathrm{h} \dot{\varphi}^2 \tag{4}$$

and the potential energy is

$$U = m_\mathrm{b} g r_{\mathrm{S_{b,y}}} + m_\mathrm{h} g r_{\mathrm{S_{h,y}}} + \frac{1}{2} s_\mathrm{t} (\vartheta - \varphi)^2, \tag{5}$$

where $v_{\mathrm{S_b}}$ and $v_{\mathrm{S_h}}$ refer to the velocity of the center of gravity of the balance board and the human body, respectively. $r_{\mathrm{S_{b,y}}}$ and $r_{\mathrm{S_{h,y}}}$ stand for the vertical position of the center of gravity of the balance board and the human body, respectively, g is the gravitational acceleration, and $Q_k(t)$ denotes the generalized force. The center of gravity of the balance board

$$\mathbf{r}_{\mathrm{S_b}} = \begin{bmatrix} -e\cos\vartheta + (h-f)\sin\vartheta \\ R + (f-h)\cos\vartheta - e\sin\vartheta \\ 0 \end{bmatrix} \tag{6}$$

and that of the human body

$$\mathbf{r}_{\mathrm{S_h}} = \begin{bmatrix} -e\cos\vartheta + (h-f)\sin\vartheta - l\sin\varphi \\ R + (f-h)\cos\vartheta - e\sin\vartheta + l\cos\varphi \\ 0 \end{bmatrix} \tag{7}$$

can be obtained from the geometrical properties of the system. The velocity of the center of gravity of the balance board

$$\mathbf{v}_{\mathrm{S_b}} = \begin{bmatrix} \big(-R + (R - l_{\mathrm{b}})\cos\vartheta\big)\dot{\vartheta} \\ (R - l_{\mathrm{b}})\sin\vartheta\dot{\vartheta} \\ 0 \end{bmatrix} \tag{8}$$

and that of the human body

$$\mathbf{v}_{\mathrm{S_h}} = \begin{bmatrix} \big(-R + (h - f)\cos\vartheta + e\sin\vartheta\big)\dot{\vartheta} - l\cos\varphi\dot{\varphi} \\ -\big(e\cos\vartheta + (-h + f)\sin\vartheta\big)\dot{\vartheta} - l\sin\varphi\dot{\varphi} \\ 0 \end{bmatrix} \tag{9}$$

are the first derivatives of (6) and (7), respectively. The distance between the ground and the center of gravity of the balance board is denoted by l_{b} and can be computed using the center of gravity of the elements of the balance board. For the sake of simplicity, time dependence of the generalized coordinates and their derivatives was not indicated. Finally, the linearized equation of motion reads

$$\mathbf{M}\ddot{\mathbf{q}}(t) + \mathbf{S}\mathbf{q}(t) = \mathbf{Q}(t), \tag{10}$$

where the mass (**M**) and the stiffness matrix (**S**) are determined by the mechanical properties of the system (Table 1).

Table 1 Physical system parameters

Parameter	Explanation	Value
R	Wheel radius	0.125 m
h	Board distance from the wheel center	0.05 m
m_b	Mass of balance board	3.0988 kg
l_b	center of gravity of balance board	0.0651 m
I_b	Mass moment of inertia of the board	0.0201 kgm^2
m_h	Mass of balancing subject	57 kg
l	Height of balancing subject	1.6 m
I_h	Mass moment of inertia of the subject	12.16 kgm^2
e	Horizontal distance of ankle	0.12 m
f	Vertical distance of ankle	0.07 m
s	Passive stiffness at ankle joint	407 Nm
g	Gravitational acceleration	9.81 m/s^2
τ	Reaction time	150 ms
P_φ^*	Proportional gain of φ	8222 Nm/rad
D_φ^*	Derivative gain of φ	2667 Nms/rad
P_ϑ^*	Proportional gain of ϑ	1140 Nm/rad
D_ϑ^*	Derivative gain of ϑ	390 Nms/rad

Table 2 Dimensionless model parameters admissible range

Parameter	MIN	MAX
a_0	−5000	0
a_1	−200	0
a_2	−1000	2000
a_3	0	100

From (2) and (3), it follows that

$$\mathbf{Q}(t) = \mathbf{P}\mathbf{q}(t-\tau) + \mathbf{D}\dot{\mathbf{q}}(t-\tau) \tag{11}$$

with

$$\mathbf{P} = \begin{bmatrix} P_\varphi & P_\vartheta \\ -P_\varphi & -P_\vartheta \end{bmatrix}, \quad \mathbf{D} = \begin{bmatrix} D_\varphi & D_\vartheta \\ -D_\varphi & -D_\vartheta \end{bmatrix}, \tag{12}$$

being the matrices of the proportional and the derivative gains, respectively.

The corresponding characteristic polynomial is

$$\begin{cases} \Delta(s) = P_0(s) + P_\tau(s)e^{-s\tau} \quad \text{with} \\ P_0(s) = s^4 + a_2 s^2 + a_0 \quad \text{and} \\ P_\tau(s) = b_3 s^3 + b_2 s^2 + b_1 s + b_0. \end{cases} \tag{13}$$

Coefficients of the polynomial $P_0(s)$ are determined by the mechanical properties (mass, mass moment of inertia, geometry, passive stiffness, and damping) of the analyzed system, and the control parameters P_φ, D_φ, P_ϑ, D_ϑ are involved in the coefficients of polynomial $P_\tau(s)$. Since there is no damping in the system, $a_1 = 0 = a_3 = 0$ where the physically admissible range of parameters is indicated in Table 2.

3 Background and Prerequisites

This section provides the main methods, results, and tools used in the sequel for the investigation of the modeling of the CNS action on the human stance. First, we recall some complex analysis results delimiting the location of some analytic functions' zeros. Next, we provide a basic primer on the well-known Gröbner basis concept, as well as the powerful *Cell Decomposition* algorithm which solves systems with semi-algebraic parameters and is implemented in the Computer Algebra System `Maple`. A review of the MID-based partial poles placement strategy concludes the section.

3.1 Cauchy's Argument Principle

Cauchy's argument principle is a basic complex analysis property widely used in the stability analysis of linear time-invariant dynamical systems. Roughly speaking, the argument principle establishes a correspondence between the number of zeros minus the number of poles of a meromorphic function f in a simply connected domain $D \subset \mathbb{C}$ and a contour integral, on the boundary ∂D, of the function's logarithmic derivative, which is also the winding number of the curve ∂D. Several stability methods such as the Nyquist criterion and the Mikhaylov curve derive from the argument principle.

Theorem 1 *Let U be a simply connected region with boundary Γ (piecewise smooth and oriented anticlockwise). Let f be a meromorphic function in an open set containing the closure $\bar{U}$ with poles $p_1, \ldots, p_l$ and zeros $s_1, \ldots, s_m$ counted according to their multiplicity, none of which belonging to the closed curve Γ. Then*

$$\frac{1}{2i\pi} \oint_{\Gamma} \frac{f'(s)}{f(s)} ds = \mathcal{Z} - \mathcal{P}, \tag{14}$$

where $\mathcal{Z}$ and $\mathcal{P}$ designate, respectively, the number of zeros and the number of poles of f enclosed by Γ.

3.2 The Stepan–Hassard Approach

The main theorem from [29] underscores the link between the number of unstable spectral values $\mathbf{card}(\chi_+)$ and the number of critical spectral values $\mathbf{card}(\chi_0)$, both counting multiplicity.

Theorem 2 (Hassard, [29], pp. 223) *Consider the quasipolynomial function Δ defined by* (13). *Let $\rho_1, \ldots, \rho_r$ be the positive roots of $\mathcal{R}(y) = \Re(i^n \Delta(i\,y))$, counted by their multiplicities and ordered increasingly. For each $j = 1, \ldots, r$ such that $\Delta(i\,\rho_j) = 0$, assume that the multiplicity of $i\rho_j$ as a zero of $\Delta(\lambda, \tau)$ is the same as the multiplicity of ρ_j as a root of $\mathcal{R}(y)$. Then* $\mathbf{card}(\chi_+)$ *is given by the formula:*

$$\mathbf{card}(\chi_+) = \frac{n - \mathbf{card}(\chi_0)}{2} + \frac{(-1)^r}{2} sgn\mathcal{I}^{(\mu)}(0) + \sum_{j=1}^{r} sgn\mathcal{I}(\rho_j), \tag{15}$$

where μ designate the multiplicity of the zero spectral value of $\Delta(\lambda, \tau) = 0$ and $\mathcal{I}(y) = \Im(i^{-n}\Delta(iy))$. Furthermore, $\mathbf{card}(\chi_+)$ *is odd (respectively, even) if $\Delta^{(\mu)}(0) < 0$ ($\Delta^{(\mu)}(0) > 0$). If $\mathcal{R}(y) = 0$ has no positive zeros, set $r = 0$, and omit the summation term in the expression of* $\mathbf{card}(\chi_+)$. *If $\lambda = 0$ is not a root of the characteristic equation, set $\mu = 0$, and interpret $\mathcal{I}^{(0)}(0)$ as $\mathcal{I}(0)$ and $\Delta^{(0)}(0)$ as $\Delta(0)$.*

3.3 Background on Gröbner Bases

A wide range of problems related to the qualitative analysis of dynamical systems lead to systems of polynomial equations, see, for instance, [5, 11, 17, 55],

$$f_1 = 0, \dots, f_m = 0 \tag{16}$$

with $f_i \in \mathbb{R}[x_1, \dots, x_n]$. To solve this system, we consider the ideal $\langle f_1, \dots, f_m \rangle \subset \mathbb{R}[x_1, \dots, x_n]$. For this aim, we use Gröbner Bases computations. In this section, we recall the basic facts about Gröbner bases and refer the reader to [23] for details.

A *monomial ordering* is a total order on monomials that is compatible with the product and such that every nonempty set has the smallest element for the order. The leading term of a polynomial is the greatest monomial appearing in this polynomial.

A *Gröbner basis* of an ideal $\mathcal{I}$ for a given monomial ordering is a set G of generators of $\mathcal{I}$ such that the leading terms of G generate the ideal of leading terms of polynomials in $\mathcal{I}$. A polynomial is *reduced* with respect to the Gröbner basis G when its leading term is not a multiple of those of G. The basis is *reduced* if each element $g \in G$ is reduced with respect to $G \setminus \{g\}$. For a given monomial ordering, the reduced Gröbner basis of a given set of polynomials exists and is unique and can be computed using one's favorite general computer algebra system, like Maple, Magma, or Singular. Several efficient implementations of the Gröbner basis algorithm exist; here, we use the FGb implementation of F_4 available in Maple [26]. The complexity of a Gröbner basis computation is well known to be generically exponential in the number of variables and in the worst case doubly exponential in the number of variables. Moreover, a good choice of the monomial ordering reduces the computational cost.

The *graded reverse lexicographic order* or `grevlex` for short also denoted `tdeg` ordering is the most suited ordering for the computation of the (reduced) Gröbner basis. The monomials are first ordered by degree, and the order between two monomials of the same degree $x_\alpha = x_1^{\alpha_1} \cdots x_n^{\alpha_n}$ and $x_\beta = x_1^{\beta_1} \cdots x_n^{\beta_n}$ is given by $x_\alpha \succ x_\beta$ when the last nonzero element of $(\alpha_1 - \beta_1, \dots, \alpha_n - \beta_n)$ is negative. Thus, among the monomials of degree d, the order is

$$x_1^d \succ x_1^{d-1}x_2 \succ x_1^{d-2}x_2^2 \succ \cdots \succ x_2^d \succ x_1^{d-1}x_3 \succ x_1^{d-2}x_2x_3 \succ x_1^{d-2}x_3^2 \succ \cdots \succ x_n^d.$$

However, a Gröbner basis for the `grevlex` ordering may not be appropriate for the computation of the solutions of the system (16). The most suited ordering for this computation is the *lexicographical* ordering (or `lex` ordering for short). The monomials are ordered by comparing the exponents of the variables in lexicographical order. Thus, any monomial containing x_1 is greater than any monomial containing only variables $x_2, \dots, x_n$.

Under some hypotheses (radical ideal with a finite number of solutions and up to a linear change of coordinates), the Gröbner basis of an ideal $\langle f_1, \dots, f_m \rangle$ for the `lex` order $x_1 > \dots > x_n$ has the shape

$$\{x_1 - g_1(x_n), x_2 - g_2(x_n), \ldots, x_{n-1} - g_{n-1}(x_{n-1}), g_n(x_n)\}, \tag{17}$$

where the g_i are univariate polynomials. In this case, the computation of the solutions of the system follows easily. In the general case, the shape of the Gröbner basis for the `lex` ordering is more complicated, but it is equivalent to several triangular systems for which the computation of the solutions is straightforward.

An important point is that a Gröbner basis for the `lex` order is in general hard to compute directly. It is much faster to compute first a Gröbner basis for the `grevlex` order and then to make a change of ordering to the `lex` order.

Furthermore, if one wants to compute all the polynomials in $\mathcal{I}$, which do not depend on the variables $x_i, \ldots, x_n$, i.e., $\mathcal{I} \cap \mathbb{R}[x_1, \ldots, x_{i-1}]$, an elimination procedure has to be carried out. Geometrically speaking, this elimination of indeterminates corresponds to the projection of the associated variety into $[x_1, \ldots, x_1]$. This projection can be characterized using a Gröbner basis of $\mathcal{I}$ for `lex` ordering often designating an *elimination ordering*.

The precise ordering we use to compute the Gröbner bases of the polynomial systems occurring in this chapter is `grevlex` or `lexdeg` ordering.

Finally, we use repeatedly the *Radical Membership Theorem*:

Theorem 3 ([23]) *Let $\mathcal{I} = \langle f_1, \ldots, f_s \rangle$ be an ideal of $k[x_1, \ldots, x_n]$, then f belongs to $\sqrt{\mathcal{I}}$ if, and only if, $\langle f_1, \ldots, f_s, 1 - yf \rangle = \langle 1 \rangle = k[x_1, \ldots, x_n, y]$.*

A finite set of polynomial equations $\langle f_1, \ldots, f_m \rangle \subset \mathbb{R}[x_1, \ldots, x_n]$ is said to be *inconsistent* or *unfeasible* if, and only if, $\langle 1 \rangle$ is the corresponding reduced Gröbner basis.

3.4 Cell Decomposition in CAD Routine

The main algebraic tool we use is the so-called *discriminant variety* associated with an algebraic set, which was introduced in [35].

Let us consider the following parametric system of algebraic equations and inequations:

$$\begin{cases} f_1(T, X) = 0 \\ \vdots \\ f_n(T, X) = 0 \end{cases}, \quad \begin{cases} g_1(T, X) \neq 0 \\ \vdots \\ g_m(T, X) \neq 0, \end{cases} \tag{18}$$

where f_i and g_j are parametric multivariate polynomials in the ring $\mathbb{Q}[t_1, \ldots, t_r]$ $[x_1, \ldots, x_q]$ with the parameter vector $T = (t_1, \ldots, t_r)$ and variable vector $X = (x_1, \ldots, x_q)$. The corresponding *discriminant variety* is defined as the Zariski closure of the associated parameter space [35]. The smallest discriminant variety, which is called the minimal discriminant variety, corresponds to the intersection of all admissible discriminant varieties and characterizes such a parametric system by providing a fundamental tool for classifying regions in the parameter space with respect to the admissible number of real solutions, see, for instance, [50]. The complementary of this algebraic variety can be partitioned, using the standard the *cylindrical algebraic decomposition*, into a set of connected components where the sign vector of polynomial equations remains constant. This strategy is implemented in the certified Maple package `RootFinding[Parametric]`. More precisely, `CellDecomposition` decomposes the parameter space of a parametric system into cells in which the original system has a constant number of solutions [1, 35, 50].

3.5 The MID Paradigm: A Partial Pole-Placement Strategy

In the study of linear time-invariant (LTI) dynamical systems represented by DDEs, the characteristic function is expressed as a *quasipolynomial*, which is formally defined as follows.

Definition 1 A quasipolynomial is a particular entire function $\Delta : \mathbb{C} \to \mathbb{C}$ which may be written as follows:

$$\Delta(s) = \sum_{i=0}^{\ell} P_i(s)\, \mathrm{e}^{-\tau_i\, s}, \tag{19}$$

where ℓ is a positive integer, $\tau_i \in \mathbb{R}_+$ $(i = 0 \cdots l)$ are pairwise distinct numbers, and P_i $(i = 0 \cdots l)$ are polynomials of degree $d_i \geq 0$. Its degree $\deg(\Delta)$ is given by

$$\deg(\Delta) = \ell + \sum_{i=0}^{\ell} \deg(P_i).$$

An important result in the literature, known as *Pólya–Szegő bound* and denoted PS_B (see [54, Problem 206.2, page 144 and page 347], [9]), plays an important role in characterizing the maximal allowable multiplicity of a characteristic root of a quasipolynomial. More precisely, we have the following.

Proposition 1 *Let Δ be a quasipolynomial of degree $D = \deg(\Delta)$. Then, any characteristic root $s_0 \in \mathbb{C}$ of Δ exhibits a multiplicity at most equal to D.*

As briefly explained in the Introduction section, a characteristic root λ_0 (of Δ) satisfies the *MID property* if:

(i) Its *algebraic multiplicity* (denoted $M = M(\lambda_0)$) is *larger than one*.
(ii) It is *dominant*, meaning that the remaining characteristic roots λ_σ of the spectrum satisfy the condition $\Re(\lambda_\sigma) \leq \Re(\lambda_0)$.

Since the maximal admissible multiplicity is defined by the degree of the quasipolynomial Δ (see, e.g., [42]), it is clear that $2 \leq M \leq \deg(\Delta)$. The case $M = \deg(\Delta)$ is called *generic multiplicity*, and any multiplicity larger than one and smaller than $\deg(\Delta)$ denotes an *intermediate multiplicity*. In what follows, we give a precise definition of the *dominant root*.

It turns out that, for the characteristic equations corresponding to delay systems, the real roots of maximal multiplicity are necessarily dominant. This property is known as Generic Multiplicity-Induced-Dominancy (GMID) and consists in conditions under which a given real root of maximal multiplicity is necessarily dominant. However, multiple roots with intermediate admissible multiplicities may or may not be dominant. As for the case of a root of strictly intermediate multiplicity, one must look for conditions on the free parameters of the system for which the former is dominant, this property is called Multiplicity-Induced-Dominancy (MID).

It should be mentioned that the generic case, i.e., $M(\lambda_0) = \deg(\Delta)$, was already treated and explicit characterizations exist: The retarded case $m = n - 1$ was characterized in [42], and a unified treatment of the retarded and neutral cases, where $m \leq n$, was provided in [16].

To the best of the authors' knowledge, in the open literature, the MID with over-order multiplicities has been addressed in two particular configurations $M(\lambda_0) = n + 1$ and $M(\lambda_0) = n + m$. More precisely, in the "limit" case $M(\lambda_0) = n + 1$, sufficient conditions for their validity have been proposed in [4], where the authors exploited the particular spectrum location of the open-loop plant.

3.6 Algorithmic Investigation of the MID Property

In this subsection, we provide an effective algorithm we are explicitly using in our framework. More precisely, we first establish conditions on the system's parameters which guarantee the existence of a multiple root. Second, we perform an affine change of variable in the characteristic equation $\Delta(\lambda) = P_0(\lambda) + P_\tau(\lambda)\,\mathrm{e}^{-\lambda\tau}$ in order to reduce the corresponding quasipolynomial to a normalized form: $\tilde{\Delta}(z) = \tilde{P}_0(z) + \tilde{P}_\tau(z)\,\mathrm{e}^{-z}$. Next, we derive a bound on the imaginary part of roots of the normalized characteristic function in the complex right half plane. Lastly, a certification of the dominance of the multiple root is demonstrated. In what follows, Algorithm 2 indicates the steps to be followed to reach a suitable frequency bound (see [7, 39]).

4 Main Results

This section is dedicated to the main results. After a normalization of the characteristic equation, the maximal admissible multiplicity for a root of the latter is obtained as a root, s_0, of a fourth-order polynomial named the elimination-produced

Algorithm 2 Estimation of the MID frequency bound in time-delay differential equations with single delay

Require: $\tilde{\Delta}(z) = \tilde{P}_0(z) + \tilde{P}_\tau(z)\,\mathrm{e}^{-z}$ {Normalized quasipolynomial}
{Initialization}
ord = 0
{ord: order of truncation of the Taylor expansion of $\mathrm{e}^{2x} = \underbrace{\underbrace{1}_{\text{ord}=0} + 2x}_{\text{ord}=1} + 2x^2 + \frac{4x^3}{3} + \cdots$}
$dominance = false$
$\exists\, z_0 = x + i\omega \in \mathbb{R}_+^* + i\mathbb{R}_+^*$ s.t. $\tilde{\Delta}(z_0) = 0$
$|\tilde{P}_0(x + i\omega)|^2\,\mathrm{e}^{2x} = |\tilde{P}_\tau(x + i\omega)|^2$
while $\sim$ dominance **do**
  ord = ord+1
  $F(x, \omega) = |\tilde{P}_\tau(x + i\omega)|^2 - |\tilde{P}_0(x + i\omega)|^2\,\mathrm{T}_{ord}(\mathrm{e}^{2x}) > 0$
  {$\mathrm{T}_{ord}(\mathrm{e}^{2x})$: Taylor expansion of e^{2x} of order $= ord$}
  $\omega^2 = \Omega$
  $H(x, \Omega)$ {The polynomial characterizing the real roots of F}
end while
$\Omega_k(x)$ {k^{th} real root of H, depend on free parameters}
if $\max_x(\max_\mathrm{k}(\Omega_k(x))) < \pi^2$ **then**
  $dominance = true$
end if
return Frequency bound

polynomial. We also show that the gains are real if, and only if, s_0 is real. Then, owing to the `Maple` routine `CellDecomposition`, we investigate regions in the parameter space where we can assign such a multiple negative real root as the spectral abscissa. The proof of the dominancy is based on a well-proven algorithm described in Algorithm 2.

4.1 A Normalized Fourth-Order Friction-Free Model

We investigate a stabilizing delayed controller for the rolling balance board. In the closed loop, the corresponding characteristic function reads

$$\begin{cases} \Delta(s) = P_0(s) + P_\tau(s)e^{-s\tau} \quad \text{with} \\ P_0(s) = s^4 + a_2 s^2 + a_0 \quad \text{and} \\ P_\tau(s) = b_3 s^3 + b_2 s^2 + b_1 s + b_0. \end{cases} \tag{20}$$

Since a_0 is typically a negative parameter as shown in Table 2, the change of variables $\left(s \rightarrow \sqrt[4]{-a_0}\, s\right)$ reduces the analysis to the normalized characteristic function:

$$\begin{cases} \tilde{\Delta}(s) = \tilde{P}_0(s) + \tilde{P}_\tau(s)e^{-s\tilde{\tau}} \quad \text{with} \quad \tilde{P}_0(s) = s^4 + \tilde{a}_2 s^2 - 1 \quad \text{and} \quad \tilde{P}_\tau(s) = \tilde{b}_3 s^3 + \tilde{b}_2 s^2 + \tilde{b}_1 s + \tilde{b}_0, \\ \tilde{\tau} = \sqrt[4]{-a_0}\,\tau, \\ \tilde{a}_2 = a_2/\sqrt{-a_0}, \\ \tilde{b}_k = b_k/(-a_0)^{\frac{4-k}{4}} \quad \text{for} \quad k \in \{0, \ldots, 3\}. \end{cases} \tag{21}$$

For the sake of simplicity, the $\tilde{.}$ symbol is omitted, so the normalized quasipolynomial function studied in the sequel is the following:

$$\Delta(s) = s^4 + a_2 s^2 - 1 + \left(b_3 s^3 + b_2 s^2 + b_1 s + b_0\right) e^{-s\tau}. \tag{22}$$

At the end of our investigations, the transformation (21) shall enable the reconstruction of the appropriate stabilizing conditions for (20).

The present study is threefold. First, we address the conditions on the quasipolynomial parameters providing the maximal admissible multiplicity for a given spectral value. Second, under the maximal multiplicity conditions, we investigate additional assumptions on the parameters in order to guarantee the dominancy of the multiple spectral value. Third, how such a dominancy result can be exploited in the design of a delayed stabilizing controller?

4.2 Forcing Multiplicity

The MID property introduced in the previous section, which consists in forcing a given quasipolynomial to have a root of a prescribed multiplicity, permits under appropriate conditions to characterize the rightmost root. As a matter of fact, this multiplicity constraint defines a manifold in the parameter space enabling the tuning of the gains b_k when the delay τ is left free and guaranteeing the exponential stability of the closed-loop system solution. Hereafter, the next theorem, which is based on the MID property, provides the explicit conditions on the parameters' values guaranteeing a targeted multiplicity. Recall that the multiplicity of a given root of the generic quasipolynomial (22) is bounded by its degree; that is, 8 is the bound for multiplicity. However, forcing the multiplicity must not constrain the physical model parameters.

Theorem 4 *The following assertions hold:*

1. *If the parameters a_2 and τ are left free, then the maximal multiplicity of a given root of* (22) *is* 5. *Furthermore, such a multiplicity is reached if, and only if, $s = s_0$, a root of the elimination-produced polynomial*

$$\mathbb{P}(s) = \tau^4 s^4 + 16\,\tau^3 s^3 + \left(\tau^2 a_2 + 72\right)\tau^2 s^2 + \left(8\,\tau^2 a_2 + 96\right)\tau\, s \\ - \tau^4 + 12\,\tau^2 a_2 + 24. \tag{23}$$

2. *The root s_0 of (22) has multiplicity 5 if, and only if, the system's parameters satisfy*

$$\left\{\begin{aligned}
\frac{3\,\tau^4 b_0}{e^{s_0\tau}} &= \left(-{a_2}^2{s_0}^2 - 2\,{s_0}^2 + a_2\right)\tau^6 + \left(-23\,a_2{s_0}^3 - 8\,s_0{a_2}^2 + 26\,s_0\right)\tau^5 \\
&\quad + \left(-5\,a_2{s_0}^2 - 12\,{a_2}^2 - 254\right)\tau^4 \\
&\quad + \left(2348\,{s_0}^3 + 1612\,a_2 s_0\right)\tau^3 + \left(15{,}804\,{s_0}^2 + 3060\,a_2\right)\tau^2 \\
&\quad + 23{,}976\,s_0\tau + 6168, \\
\frac{\tau^3 b_1}{e^{s_0\tau}} &= \left(-a_2{s_0}^3 + 2\,s_0\right)\tau^5 + \left(5\,a_2{s_0}^2 - 22\right)\tau^4 \\
&\quad + \left(196\,{s_0}^3 + 148\,a_2 s_0\right)\tau^3 + \left(1380\,{s_0}^2 + 276\,a_2\right)\tau^2 \\
&\quad + 2136\,s_0\tau + 552, \\
\frac{\tau^3 b_2}{e^{s_0\tau}} &= \left(a_2{s_0}^2 - 2\right)\tau^5 + \left(18\,{s_0}^3 + 15\,a_2 s_0\right)\tau^4 + \left(138\,{s_0}^2 + 29\,a_2\right)\tau^3 \\
&\quad + 228\,s_0\tau^2 + 60\,\tau, \\
\frac{3\,\tau\, b_3}{e^{s_0\tau}} &= \left(2\,{s_0}^3 + a_2 s_0\right)\tau^3 + \left(18\,{s_0}^2 + 3\,a_2\right)\tau^2 + 36\,s_0\tau + 12.
\end{aligned}\right. \tag{24}$$

3. *The gains b_k in (22) are real if, and only if, s_0 is real.*

Remark 1 Theorem 4 gives conditions for a root of (22) to reach multiplicity 5. These conditions can be easily exploited using (21) to establish those for a root of multiplicity 5 for (20) where a_0 is a left-free parameter. A given complex number is a root of multiplicity 5 for (20) if, and only if, it is a root of the corresponding elimination-produced polynomial $\tilde{\mathbb{P}}(s) = s^4 + 16\,s^3 + \left(\tau^2 a_2 + 72\right)s^2 + \left(8\,\tau^2 a_2 + 96\right)s + \tau^4 a_0 + 12\,\tau^2 a_2 + 24$. Obviously, the nature (real/complex) of the zeros of $\tilde{\mathbb{P}}$ is the same as those of $\mathbb{P}$.

Proof of Theorem 4 First, the vanishing of the quasipolynomial Δ given in (22) yields the elimination of the exponential term as a rational function in s:

$$e^{-\tau s} = -\frac{\tilde{P}_0(s)}{\tilde{P}_\tau(s)}. \tag{25}$$

Next, we investigate potential roots with algebraic multiplicity 5 by substituting the obtained equality (25) in the ideal $\mathcal{I}_5$ generated by the first four derivatives of Δ that is $\mathcal{I}_5 = <\partial_s\Delta,\ \partial_s^2\Delta, \ldots,\ \partial_s^4\Delta>$. This enables us to investigate the following manifold of four algebraic equations in seven unknowns $b_0,\ b_1,\ b_2,\ b_3,\ a_2,\ s,\ \tau$:

$$
\begin{aligned}
0 =& \left(\tau\, \tilde{P}_0(s) + 4\,s^3 + 2\,a_2 s\right) b_0 + \left(s\,\tilde{P}_0(s)\tau + 3\,s^4 + a_2 s^2 + 1\right) b_1 \\
&+ \left(s^2\,\tilde{P}_0(s)\tau + 2\,s^5 + 2\,s\right) b_2 + \left(s^3\,\tilde{P}_0(s)\tau + s^6 - a_2 s^4 + 3\,s^2\right) b_3 \\
0 =& \left(-\tau^2\,\tilde{P}_0(s) + 12\,s^2 + 2\,a_2\right) b_0 + \left(-s\,\tilde{P}_0(s)\tau^2 + 2\,\tau\,\tilde{P}_0(s) + 12\,s^3 + 2\,a_2 s\right) b_1 \\
&+ \left(-s^2\,\tilde{P}_0(s)\tau^2 + 4\,s\,\tilde{P}_0(s)\tau + 10\,s^4 + 2\right) b_2 \\
&+ \left(-s^3\,\tilde{P}_0(s)\tau^2 + 6\,s^2\,\tilde{P}_0(s)\tau + 6\,s^5 - 4\,s^3 a_2 + 6\,s\right) b_3 \\
0 =& \left(\tau^3\,\tilde{P}_0(s) + 24\,s\right) b_0 + \left(s\,\tilde{P}_0(s)\tau^3 - 3\,\tau^2\,\tilde{P}_0(s) + 24\,s^2\right) b_1 \\
&+ \left(s^2\,\tilde{P}_0(s)\tau^3 - 6\,s\,\tilde{P}_0(s)\tau^2 + 6\,\tau\,\tilde{P}_0(s) + 24\,s^3\right) b_2 \\
&+ \left(s^3\,\tilde{P}_0(s)\tau^3 - 9\,s^2\,\tilde{P}_0(s)\tau^2 + 18\,s\,\tilde{P}_0(s)\tau + 18\,s^4 - 6\,a_2 s^2 + 6\right) b_3 \\
0 =& \left(24 - \tau^4\,\tilde{P}_0(s)\right) b_0 + \left(-s\,\tilde{P}_0(s)\tau^4 + 4\,\tau^3\,\tilde{P}_0(s) + 24\,s\right) b_1 \\
&+ \left(-s^2\,\tilde{P}_0(s)\tau^4 + 8\,s\,\tilde{P}_0(s)\tau^3 - 12\,\tau^2\,\tilde{P}_0(s) + 24\,s^2\right) b_2 \\
&+ \left(-s^3\,\tilde{P}_0(s)\tau^4 + 12\,s^2\,\tilde{P}_0(s)\tau^3 - 36\,s\,\tilde{P}_0(s)\tau^2 + 24\,\tau\,\tilde{P}_0(s) + 24\,s^3\right) b_3.
\end{aligned}
\tag{26}
$$

The above system is a linear system in the unknowns $(b_k)_{0\leq k\leq 3}$. Using standard elimination techniques, we obtain a set of three solutions: The first one asserts that $b_k = 0$ for $k \in [\![0, 3]\!]$, the second one corresponds to s as a root of the open-loop polynomial ($\tilde{P}_0(s) = 0$) with $b_2 = -s^3 b_1 - s^2 b_0 - s a_2 b_1 - s b_3 - a_2 b_0$, and the last solution corresponds to $s = s_0$ as defined in (23) and b_k as rational functions in (s_0, a_2, τ).

Observe that the first solution corresponds to the open-loop system while the second solution is inconsistent with respect to the transcendental term's elimination (25). Hence, these two solutions are discarded. Next, substituting conditions pertaining to the third solution in (25) yields the explicit values of the gains b_k allowing to tune the parameters as provided in (24), which concludes items (1) and (2).

To show (3), we assume that there exists $s^* \in \mathbb{C} - \mathbb{R}$ such that $\mathbb{P}(s^*) = 0$ and the coefficients $b_k(s^*, a_2, \tau) \in \mathbb{R}$ for $k \in [\![0, 3]\!]$. Then, we substitute $s^* = \alpha + i\,\beta$ with $\alpha \neq 0$ and $\beta \neq 0$ in the expressions of b_k for $k \in [\![0, 3]\!]$ and define an ideal $\mathcal{I}_{\mathcal{R}}$ (where the index $\mathcal{R}$ stands for the realness of the gains b_k) generated by $(p_k)_{0\leq k\leq 3}$ corresponding, respectively, to the imaginary part of the gains $(b_k)_{0\leq k\leq 3}$, and the polynomials $p_4 = \Re(\mathbb{P}(s^*))$ and $p_5 = \Im(\mathbb{P}(s^*))$; that is, $\mathcal{I}_{\mathcal{R}} =< p_0, \ldots, p_5 >$.

We proceed as advanced in Sect. 3.3 by computing a Gröbner basis with respect to the elimination order `lex` with respect to the unknowns a_2 and τ. We obtain a basis of nine elements $G = \langle G_1, \ldots, G_9 \rangle$, where

$$
\begin{aligned}
G_1 =&\beta^4, \\
G_2 =&\alpha^{12}\beta^2 - 3\,\alpha^8\beta^2 + 21\,\alpha^4\beta^2 - 125\,\beta^2, \\
G_3 =&9\,\alpha^{11}\beta^2 - 32\,\alpha^7\beta^2 + 109\,\alpha^3\beta^2 + 275\,\beta^2\tau, \\
G_4 =&2\,\alpha^{10}\beta^2 - 56\,\alpha^6\beta^2 - 208\,\alpha^2\beta^2 + 275\,\beta^2 a_2, \\
G_5 =&936\,\alpha^8\beta^2 + 825\,\alpha^8\tau^2 + 8910\,\alpha^7\tau - 3190\,\alpha^5\tau\,a_2 + 22{,}110\,\alpha^6 \\
&- 7354\,\alpha^4\beta^2 - 1375\,\alpha^4\tau^2 - 14{,}465\,\alpha^4 a_2 \\
&- 22{,}110\,\alpha^3\tau + 2200\,\alpha^2{a_2}^2 + 5500\,\alpha\,\tau\,a_2 - 56{,}760\,\alpha^2 - 11{,}500\,\beta^2 \\
&- 2200\,\tau^2 + 22{,}000\,a_2, \\
G_6 =&- 5244\,\alpha^{11}\beta^2 + 46{,}292\,\alpha^7\beta^2 + 28{,}875\,\alpha^7\tau^2 + 311{,}850\,\alpha^6\tau \\
&+ 5500\,\alpha^4\tau^3 - 111{,}650\,\alpha^4\tau\,a_2 + 773{,}850\,\alpha^5 \\
&- 182{,}714\,\alpha^3\beta^2 + 116{,}875\,\alpha^3\tau^2 - 506{,}275\,\alpha^3 a_2 + 447{,}150\,\alpha^2\tau \\
&+ 77{,}000\,\alpha\,{a_2}^2 + 5500\,\tau^3 - 55{,}000\,\tau\,a_2 \\
&+ 257{,}400\,\alpha, \\
G_7 =&- 852\,\alpha^{10}\beta^2 + 6036\,\alpha^6\beta^2 + 4125\,\alpha^6\tau^2 + 44{,}550\,\alpha^5\tau - 15{,}950\,\alpha^3\tau\,a_2 \\
&+ 110{,}550\,\alpha^4 - 18{,}862\,\alpha^2\beta^2 \\
&+ 9625\,\alpha^2\tau^2 - 72{,}325\,\alpha^2 a_2 + 2750\,\tau^2 a_2 + 54{,}450\,\tau\,\alpha \\
&+ 11{,}000\,{a_2}^2 + 46{,}200 \\
G_8 =&- 556\,\alpha^{10}\beta^2 + 7868\,\alpha^6\beta^2 + 4125\,\alpha^6\tau^2 + 41{,}250\,\alpha^5\tau \\
&- 13{,}750\,\alpha^3\tau\,a_2 + 84{,}150\,\alpha^4 - 10{,}926\,\alpha^2\beta^2 \\
&+ 4125\,\alpha^2\tau^2 + 1375\,\alpha\,\tau\,{a_2}^2 - 43{,}725\,\alpha^2 a_2 + 8250\,\tau\,\alpha + 5500\,{a_2}^2 + 6600 \\
G_9 =&295{,}576\,\alpha^9\beta^2 + 210{,}375\,\alpha^9\tau^2 + 2{,}124{,}210\,\alpha^8\tau - 688{,}490\,\alpha^6\tau\,a_2 \\
&+ 4{,}534{,}530\,\alpha^7 - 1{,}264{,}134\,\alpha^5\beta^2 \\
&- 253{,}825\,\alpha^5\tau^2 - 2{,}301{,}695\,\alpha^5 a_2 - 4{,}434{,}210\,\alpha^4\tau \\
&+ 1{,}305{,}700\,\alpha^2\tau\,a_2 + 88{,}000\,\alpha\,{a_2}^3 - 9{,}859{,}080\,\alpha^3 \\
&- 3{,}371{,}860\,\alpha\,\beta^2 - 464{,}200\,\alpha\,\tau^2 - 88{,}000\,\tau\,{a_2}^2 \\
&+ 3{,}691{,}600\,\alpha\,a_2 - 105{,}600\,\tau.
\end{aligned} \tag{27}
$$

Obviously, the first element of G signifies that a solution of the multivariate system exists only if $\beta = 0$, which contradicts the fact that $s^* \in \mathbb{C} - \mathbb{R}$. Note that this can also be recovered by intentionally adding the polynomial $\beta T - 1$, where T stands for a nonzero auxiliary unknown, to the polynomials defining the ideal $\mathcal{I}_{\mathcal{R}}$ and employing the Membership Theorem. As a matter of fact, in the latter case one obtains $\langle 1 \rangle$ as a Gröbner basis.

Remark 2 From a control theory viewpoint, the controllers' gains have to be real, and assertion (3) of Theorem 4 guarantees that they are so for the considered real solution s_0. Since the degree of the elimination-produced polynomial is even and our concern is related to such polynomials with at least one real root, then two cases are of interest: an elimination-produced polynomial with either four real roots or two real roots. In addition, in the case of a real-rooted elimination-produced polynomial, the corresponding roots need to be negative to be admissible assignable roots for the initial quasipolynomial. More precisely, the only cases of interest among elimination-produced polynomials with two negative roots are those with a negative dominant zero. Lastly, note that apart from real-rooted Hurwitz polynomials no general result exists for characterizing polynomials with negative dominant roots. One finds in Fig. 2 a chart of the number of real zeros in the parameter space (τ, a_2).

Real-Rooted Elimination-Produced Polynomial

The following proposition provides conditions on the system's parameters guaranteeing the real-rootedness of the associated elimination-produced polynomial.

Proposition 2 *Consider the following polynomials:*

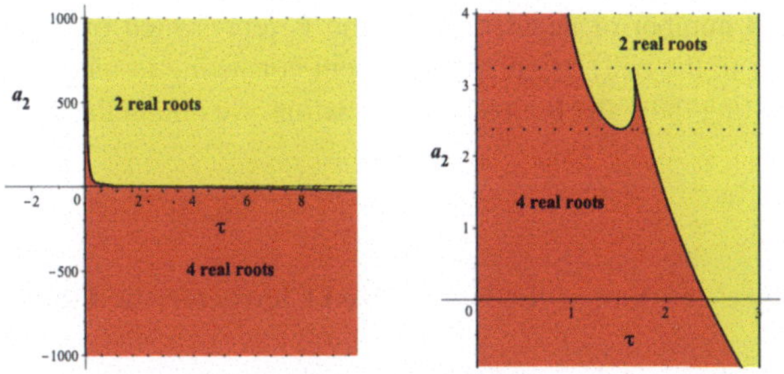

Fig. 2 Number of real roots of (23). (Left) Real roots number in the prescribed region. (Right) A zoom on the region including a cusp point at $(\tau, a_2) = \left(\sqrt{2}\sqrt[4]{3\,2^{2/3}+12\sqrt[3]{2}-18},\ \frac{\sqrt{3\,2^{2/3}+12\sqrt[3]{2}-18}}{5}\left(4\sqrt[3]{2}+3\,2^{2/3}+2\right)\right)$ corresponding to a triple root at $s = 2^{2/3} - 4$. The figures are produced using the `Maple` package `RootFinding[Parametric]` and especially the certified routine `CellDecomposition`

$$\mathring{p}(a_2) = 5\,a_2{}^6 + 36\,a_2{}^4 - 432\,a_2{}^2 - 5184, \quad \breve{p}(a_2) = 5\,a_2{}^6 + 54\,a_2{}^4 - 2592,$$

$$\bar{p}(a_2) = 3285\,a_2{}^6 + 3432\,a_2{}^4 + 25{,}216\,a_2{}^2 - 384, \quad \hat{p}(\tau) = \tau^4 - 12\,\tau^2 a_2 - 24,$$

$$\tilde{p}(\tau) = -1{,}990{,}656 + \left(a_2{}^2 + 4\right)^2 \tau^{12} + \left(4\,a_2{}^5 - 32\,a_2{}^3 - 192\,a_2\right)\tau^{10}$$
$$+ \left(-280\,a_2{}^4 + 768\,a_2{}^2 + 3456\right)\tau^8$$
$$+ 9728\,\tau^6 a_2{}^3 + \left(-152{,}064\,a_2{}^2 - 82{,}944\right)\tau^4 + 995{,}328\,\tau^2 a_2.$$

Let $\mathring{a}_2$, $\mathring{\mathring{a}}_2$, $\breve{a}_2$, $\breve{\breve{a}}_2$, $\bar{a}_2$, $\bar{\bar{a}}_2$ *be, respectively, the first and second real roots of polynomials* $\mathring{p}$, $\breve{p}$, *and* $\bar{p}$ *above. Let* $\hat{\tau}_2$ *be the second real root of* $\hat{p}$, *and let* $\tilde{\tau}_{k\in\mathbb{N}}$ *be the kth real root (ordered increasingly) of* $\tilde{p}$. *The elimination-produced polynomial* $\mathbb{P}$ *admits four negative roots if, and only if, its parameters satisfy one of the following assertions:*

1. $-1000 < a_2 < \bar{\bar{a}}_2$ *and* $0 < \tau < \hat{\tau}_2$.
2. $\breve{\breve{a}}_2 < a_2 < \mathring{\mathring{a}}_2$ *and* $\tau \in (0, \tilde{\tau}_4) \cup (\tilde{\tau}_5, \tilde{\tau}_6)$.
3. $a_2 \in (\bar{\bar{a}}_2, \breve{\breve{a}}_2) \cup (\mathring{\mathring{a}}_2, 2000)$, *and* $0 < \tau < \tilde{\tau}_2$.

Proof We use the `Maple` routine `CellDecomposition` of the `RootFinding [Parametric]` package [34, 50, 56], which decomposes the parameters' space into two manifolds: the discriminant variety and its complement. As an input for the `CellDecomposition` routine, we include the vanishing elimination-produced polynomial $\mathbb{P} = 0$, the constraints on the system's parameters $-1000 < a_2 < 2000$, a delay range $0 < \tau < 200$, and the negativity of the assigned root $s_0 < 0$. The routine computes a set of polynomials whose zeros define a discriminant variety of the parametric equation $\mathbb{P} = 0$, as well as a cylindrical algebraic decomposition of the complementary of this discriminant variety, which yields 46 cells, each of which contains a fixed number of negative roots and is represented by a sampling point (strictly interior to the cell). By a posttreatment consisting in merging neighboring cells and exploiting their algebraic characterization, we obtain the announced result (Fig. 3).

Fredholm Representation of the Normalized Characteristic Function

Let $s_0 \in \mathbb{R}_-$, and consider, as in [16], the quasipolynomial $\hat{\Delta} : \mathbb{C} \to \mathbb{C}$ obtained from Δ in (22) by the following change of variables $\hat{\Delta}(\lambda) = \tau^4 \Delta\left(\frac{\lambda}{\tau} + s_0\right)$, $\lambda \in \mathbb{C}$, then

$$\hat{\Delta}(\lambda) = \sum_{i=0}^{4} \alpha_i\,\lambda^i + \sum_{i=0}^{3} \beta i\,\lambda^i\,\mathrm{e}^{-\lambda}, \tag{28}$$

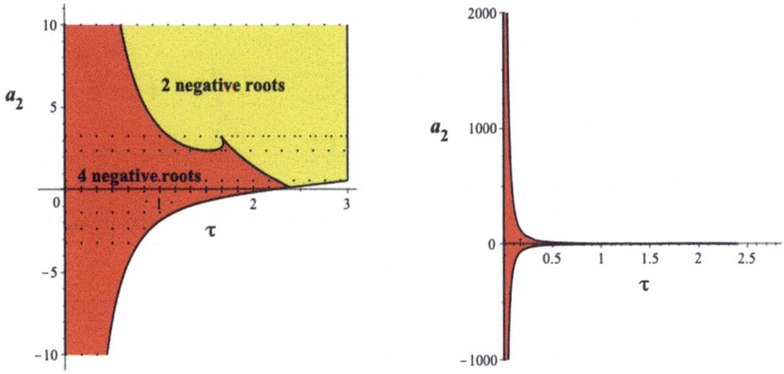

Fig. 3 (Left) Number of negative roots of (23) in a compact region. (Right) Subregion in the prescribed domain guaranteeing the real-rootedness of (23), as well as the negativity of its zeros. The figures are produced using the `Maple` package `RootFinding[Parametric]` and especially the certified routine `CellDecomposition`; see [35] for further insights on the mathematical foundation of the routine

where

$$\begin{cases} \alpha_0 = -16\,\tau^3 s_0{}^3 - 72\,\tau^2 s_0{}^2 + \dfrac{\left(-24\,\tau^2 a_2 - 288\right) s_0 \tau}{3} - 12\,\tau^2 a_2 - 24, \\ \alpha_1 = 4\,\tau^3 s_0{}^3 + 2\,\tau^3 a_2 s_0, \\ \alpha_2 = 6\,\tau^2 s_0{}^2 + \tau^2 a_2, \\ \alpha_3 = 4\,s_0 \tau, \\ \alpha_4 = 1, \\ \beta_0 = 16\,\tau^3 s_0{}^3 + 72\,\tau^2 s_0{}^2 + \dfrac{\left(24\,\tau^2 a_2 + 288\right) s_0 \tau}{3} + 12\,\tau^2 a_2 + 24, \\ \beta_1 = 12\,\tau^3 s_0{}^3 + 72\,\tau^2 s_0{}^2 + \dfrac{\left(18\,\tau^2 a_2 + 288\right) s_0 \tau}{3} + 12\,\tau^2 a_2 + 24, \\ \beta_2 = 4\,\tau^3 s_0{}^3 + 30\,\tau^2 s_0{}^2 + \dfrac{\left(6\,\tau^2 a_2 + 144\right) s_0 \tau}{3} + 5\,\tau^2 a_2 + 12, \\ \beta_3 = \dfrac{2\,\tau^3 s_0{}^3}{3} + 6\,\tau^2 s_0{}^2 + \dfrac{\left(\tau^2 a_2 + 36\right) s_0 \tau}{3} + \tau^2 a_2 + 4. \end{cases}$$

The normalized quasipolynomial $\hat{\Delta}$ defined in (28) can be factorized as

$$\hat{\Delta}(\lambda) = \lambda^5 \int_0^1 q(t)\, \mathrm{e}^{-t\,\lambda}\, \mathrm{d}t, \tag{29}$$

where

$$
\begin{aligned}
q(t) = t^4 \left(-\frac{2\,\tau^3 s_0{}^3}{3} - \frac{\tau^3 a_2 s_0}{3} - 3\,\tau^2 s_0{}^2 - \frac{\tau^2 a_2}{2} - 4\, s_0 \tau - 1\right) \\
+ t^3 \left(\frac{2\,\tau^3 s_0{}^3}{3} + \frac{\tau^3 a_2 s_0}{3}\right) \\
+ t^2 \left(3\,\tau^2 s_0{}^2 + \frac{\tau^2 a_2}{2}\right) + 4\, t s_0 \tau + 1.
\end{aligned}
\tag{30}
$$

Let us adopt the following parameterization

$$
\upsilon = \tau\, s_0 \quad \text{and} \quad \sigma = \tau^2\, a_2.
$$

Notice that the parameter υ is necessarily negative. Under this new parameterization, the integrand in (30) becomes

$$
q_{\upsilon,\sigma}(t) = (1 - t)\; \tilde{q}_{\upsilon,\sigma}, \tag{31}
$$

where

$$
\begin{aligned}
\tilde{q}_{\upsilon,\sigma}(t) = \left(4\,\upsilon^3 + 2\,\sigma\,\upsilon + 18\,\upsilon^2 + 3\,\sigma + 24\,\upsilon + 6\right) t^3 \\
+ \left(18\,\upsilon^2 + 3\,\sigma + 24\,\upsilon + 6\right) t^2 + (24\,\upsilon + 6)\, t + 6.
\end{aligned}
\tag{32}
$$

Constancy Sign of the Integrand $q_{\upsilon,\sigma}$

In our approach, the sign constancy of the polynomial $q_{\upsilon,\sigma}$ defined previously in (30) for $t \in (0, 1)$ is necessary. Therefore, the following proposition gives regions in the parameter space guaranteeing the sign constancy of $q_{\upsilon,\sigma}$ for $t \in (0, 1)$.

Proposition 3 *Let $q_{\upsilon,\sigma}$ be the parametric polynomial defined by* (31)*, and define in the parameters' space the region R as follows:*

$$
\begin{aligned}
R = \Bigg\{ (\upsilon, \sigma) \in \mathbb{R}_- \times \mathbb{R} : 6\,\upsilon^3 + 3\,\sigma\,\upsilon + 18\,\upsilon^2 + 3\,\sigma + 24\,\upsilon + 6 < 0 \quad \textit{and} \\
12\,\upsilon^3 + 6\sigma\,\upsilon + 18\,\upsilon^2 + 3\,\sigma + 48\,\upsilon + 12 \\
- \frac{\left(4\,\upsilon^3 + 2\,\sigma\,\upsilon + 18\,\upsilon^2 + 3\,\sigma + 24\,\upsilon + 6\right)\left(2\,\upsilon^3 + \sigma\,\upsilon + 12\,\upsilon\right)}{6\,\upsilon^3 + 3\,\sigma\,\upsilon + 18\,\upsilon^2 + 3\,\sigma + 24\,\upsilon + 6} < 0 \\
\textit{and} \quad 2\,\upsilon^3 + \sigma\,\upsilon + 12\,\upsilon < 0 \\
\textit{and} \quad 4\,\upsilon^3 + 2\,\sigma\,\upsilon + 18\,\upsilon^2 + 3\,\sigma + 24\,\upsilon + 6 < 0 \Bigg\}.
\end{aligned}
\tag{33}
$$

If $(\upsilon, \sigma) \in R$, *then the parametric polynomial* $q_{\upsilon,\sigma}(t)$ *has a constant, while* $t \in (0, 1)$.

Proof Since $t \in (0, 1)$, then using (31), the constancy sign of $q_{\upsilon,\sigma}$ depends on the constancy sign of $\tilde{q}_{\upsilon,\sigma}$. If the roots $\tilde{q}_{\upsilon,\sigma}$ are with negative real part, then the constancy sign of $\tilde{q}_{\upsilon,\sigma}$ is guaranteed. To do that, let us assume for the moment that t is a complex variable, then, the Routh–Hurwitz criterion ensures that all the roots of the equation $\tilde{q}_{\upsilon,\sigma}(t) = 0$ have negative real part without explicitly computing these roots. The Routh–Hurwitz table gives the following condition:

$$\left\{\upsilon > 0,\ \frac{-2\left(2\upsilon^3 + 9\upsilon^2 + 12\upsilon + 3\right)}{2\upsilon + 3} < \sigma\right\}, \tag{34}$$

which is unfortunately inconsistent since υ is considered to be negative, which means that $\tilde{q}_{\upsilon,\sigma}$ may have roots with a positive or a negative real part.

Let us now investigate the appropriate conditions guaranteeing that the roots of $\tilde{q}_{\upsilon,\sigma}$ which are with real part greater than 0 are in fact necessarily with real part greater than 1. We propose first to do a scaling in the real axis using the change of variable $\hat{q}_{\upsilon,\sigma}(t) = \tilde{q}_{\upsilon,\sigma}(t-1)$, and next we proceed with the symmetric of $\hat{q}_{\upsilon,\sigma}$, i.e., we consider the reflection with respect to the imaginary axis: $\bar{q}_{\upsilon,\sigma}(t) = \hat{q}_{\upsilon,\sigma}(-t)$. The resulting parametric polynomial is defined by

$$\begin{aligned}\bar{q}_{\upsilon,\sigma}(t) = &\left(-4\upsilon^3 - 2\sigma\upsilon - 18\upsilon^2 - 3\sigma - 24\upsilon - 6\right)t^3\\ &+\left(-12\upsilon^3 - 6\sigma\upsilon - 36\upsilon^2 - 6\sigma - 48\upsilon - 12\right)t^2\\ &+\left(-12\upsilon^3 - 6\sigma\upsilon - 18\upsilon^2 - 3\sigma - 48\upsilon - 12\right)t\\ &-4\upsilon^3 - 2\sigma\upsilon - 24\upsilon.\end{aligned} \tag{35}$$

Again, applying the Routh–Hurwitz table of the latter polynomial yields the union of the region R (33) and the following empty region:

$$\begin{aligned}&\Big\{(\upsilon, \sigma) \in \mathbb{R}_- \times \mathbb{R} : 6\upsilon^3 + 3\sigma\upsilon + 18\upsilon^2 + 3\sigma + 24\upsilon + 6 > 0 \quad \text{and}\\ &12\upsilon^3 + 6\sigma\upsilon + 18\upsilon^2 + 3\sigma + 48\upsilon + 12\\ &- \frac{\left(4\upsilon^3 + 2\sigma\upsilon + 18\upsilon^2 + 3\sigma + 24\upsilon + 6\right)\left(2\upsilon^3 + \sigma\upsilon + 12\upsilon\right)}{6\upsilon^3 + 3\sigma\upsilon + 18\upsilon^2 + 3\sigma + 24\upsilon + 6} > 0\\ &\quad \text{and} \quad 2\upsilon^3 + \sigma\upsilon + 12\upsilon > 0\\ &\quad \text{and} \quad 4\upsilon^3 + 2\sigma\upsilon + 18\upsilon^2 + 3\sigma + 24\upsilon + 6 > 0\Big\} = \emptyset,\end{aligned} \tag{36}$$

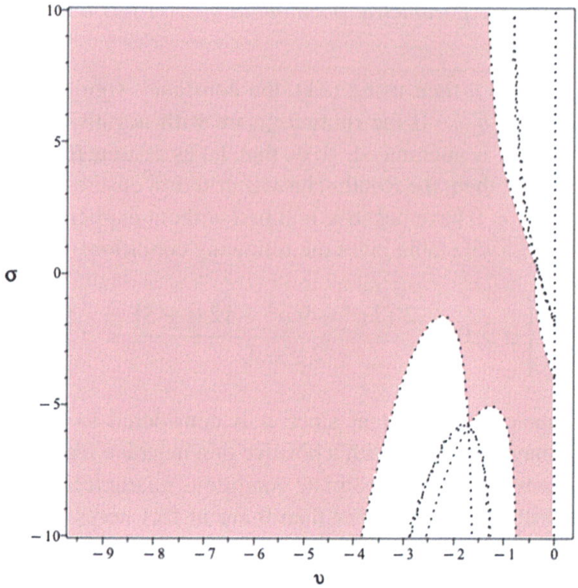

Fig. 4 Plot in the parameters' space of region R given by (33)

thereby terminating the proof.

Figure 4 exhibits a region in the parameters' space (υ, σ), where $\tilde{q}_{\upsilon,\sigma}$ has all its roots with real part greater than 1; this is done thanks to the scaling, the symmetry, and the application of Routh–Hurwitz conditions on the auxiliary polynomial $\bar{q}_{\upsilon,\sigma}$.

Effective Admissible Region

In this section we characterize the region in the original parameters' space (a_2, τ) where one has both properties; the elimination-produced polynomial is real-rooted and Hurwitz, and the integrand $q(t)$ is of a constant sign while $t \in (0, 1)$.

Proposition 4 *Let $\check{\tau}_{k\in\mathbb{N}}$ be the kth real root (ordered increasingly) of*

$$\check{p} = -1296 + 16\left(a_2{}^2 + 4\right)^2 \tau^{12} + \left(-36\, a_2{}^5 + 288\, a_2{}^3 + 1728\, a_2\right)\tau^{10}$$
$$+ \left(-1305\, a_2{}^4 - 1512\, a_2{}^2 - 5904\right)\tau^8$$
$$+ \left(-8352\, a_2{}^3 + 28{,}800\, a_2\right)\tau^6 + \left(-19{,}224\, a_2{}^2 + 120{,}096\right)\tau^4$$
$$- 15{,}552\, \tau^2 a_2.$$

Set $-1000 < a_2 < 2000$. *If one of the following assertions holds:*

1. $0 < \tau < \sqrt{6\,a_2 + 2\,\sqrt{9\,a_2{}^2 + 6}}$
2. $\tau \in (0, \tilde{\tau}_4) \cup (\tilde{\tau}_5, \tilde{\tau}_6)$
3. $\tau \in (0, \check{\tau}_4) \cup (\check{\tau}_5, \check{\tau}_6)$

then $\mathbb{P}$ *is real-rooted and Hurwitz and* $q(t)$ *is of a constant sign for* $t \in (0, 1)$.

Proof As in the proof of Proposition 2, we use the `Maple` routine `Cell Decomposition` with an input consisting of the conditions ensuring the constancy sign of $q_{\upsilon,\sigma}$ (interpreted in terms of s_0, a_2 and τ), the vanishing of the elimination-produced polynomial $\mathbb{P} = 0$, the constraints on the system parameters $-1000 < a_2 < 2000$ and $0 < \tau < 200$, and the negativity of the assigned root $s_0 < 0$. By a posttreatment consisting in merging neighboring cells and exploiting their algebraic characterization, we obtain the announced result.

4.3 Dominancy Sufficient Conditions

Now we are able to conclude our analysis by identifying the region in the parameters' space where the quintuple root at s_0 corresponds to the spectral abscissa of Δ.

Frequency Bound

Lemma 1 *Consider* $\hat{\Delta}$ *the quasipolynomial given in* (28) *with*

$$(\upsilon, \sigma) \in \hat{R} = \Bigg\{ (\upsilon, \sigma) \in \mathbb{R}_- \times \mathbb{R} : -1 < \upsilon < -\frac{3}{4} \quad \text{and} \quad \sigma < \frac{-2\,\left(2\,\upsilon^3 + 9\,\upsilon^2 + 12\,\upsilon + 3\right)}{2\,\upsilon + 3} \quad \text{and} \quad \sigma > \frac{-32\,\upsilon^4 - 96\,\upsilon^3 - 174\,\upsilon^2 + 3\,\sqrt{32\,\upsilon^4 + 56\,\upsilon^3 + 65\,\upsilon^2 + 42\,\upsilon + 9} - 123\,\upsilon - 27}{16\,\upsilon^2 + 24\,\upsilon + 9} \Bigg\}. \tag{37}$$

If $\hat{\Delta}$ *has a root* $s_0 \in \mathbb{R}_+ + i\mathbb{R}_+$, *then* $0 < \Im(s_0) < \pi$.

Proof In order to find an appropriate frequency bound, we first solve the system of conditions defining region R where the constancy sign of $q_{\upsilon,\sigma}$ is ensured, and we obtain many different regions in the parameters' space (υ, σ) $\bigcup_{i=1}^{19} S_i$, where

$$S_1 = \left\{ \upsilon = -\frac{3}{2}, -8 + \frac{\sqrt{29}}{2} < \sigma \right\},$$

$$S_2 = \left\{ \upsilon = -\frac{\sqrt{10}}{2}, -\frac{13{,}453\sqrt{10} - 42{,}473 - 3\sqrt{1{,}559{,}649 - 493{,}188\sqrt{10}}}{947\sqrt{10} - 2986} < \sigma \right\},$$

$$S_3 = \left\{ \upsilon \leq -\frac{\sqrt{10}}{2}, \sigma < -2\,\upsilon^2 - 12 \right\},$$

$$S_4 = \left\{ \upsilon \leq -1 - \frac{\sqrt{2}}{2}, \frac{-2\,(2\,\upsilon^3 + 9\,\upsilon^2 + 12\,\upsilon + 3)}{2\,\upsilon + 3} < \sigma \right\},$$

$$S_5 = \left\{ \upsilon = -1, -14 + 6\sqrt{2} < \sigma < 4 \right\},$$

$$S_6 = \left\{ \upsilon = -\frac{3}{4}, -\frac{43}{4} < \sigma < \frac{19}{8} \right\},$$

$$S_7 = \left\{ \upsilon = -1 + \frac{\sqrt{2}}{2}, -15 + 2\sqrt{2} < \sigma < -\frac{-10{,}640 + 7487\sqrt{2}}{-1042 + 719\sqrt{2}} \right\},$$

$$S_8 = \left\{ \upsilon = \frac{1}{2} - \frac{\sqrt{11}}{2}, -\frac{166{,}520{,}833\sqrt{11} - 552{,}287{,}115}{8{,}535{,}003\sqrt{11} - 28{,}307{,}401} < \sigma < -\frac{28\sqrt{11} - 89}{-4 + \sqrt{11}} \right\},$$

$$S_9 = \{\upsilon = -0.3820880724, -0.33829270 < \sigma < 0.342333120\},$$

$$S_{10} = \{\upsilon = -0.3820880724, -12.29198259 < \sigma < -0.33829270\},$$

$$S_{11} = \left\{ -\frac{\sqrt{10}}{2} < \upsilon < -\frac{3}{2}, \sigma < \frac{-2\,(2\,\upsilon^3 + 9\,\upsilon^2 + 12\,\upsilon + 3)}{2\,\upsilon + 3} \right\},$$

$$S_{12} = \left\{ -\frac{\sqrt{10}}{2} < \upsilon < -\frac{3}{2}, \epsilon < \sigma \right\},$$

$$S_{13} = \Bigg\{ -0.732050808 \leq \upsilon < -0.3820880724, -2\,\upsilon^2 - 12 < \sigma < \frac{-2\,(2\,\upsilon^3 + 9\,\upsilon^2 + 12\,\upsilon + 3)}{2\,\upsilon + 3} \Bigg\},$$

$$S_{14} = \left\{ -1 < \upsilon < -\frac{3}{4}, \epsilon < \sigma < \frac{-2\,(2\,\upsilon^3 + 9\,\upsilon^2 + 12\,\upsilon + 3)}{2\,\upsilon + 3} \right\},$$

$$S_{15} = \left\{ -\frac{3}{2} < \upsilon, \sigma < \frac{-2\,(2\,\upsilon^3 + 9\,\upsilon^2 + 12\,\upsilon + 3)}{2\,\upsilon + 3}, \upsilon < \epsilon < \sigma \right\},$$

$$S_{16} = \left\{ -\frac{3}{4} < \upsilon, \sigma < \frac{-2\,(2\,\upsilon^3 + 9\,\upsilon^2 + 12\,\upsilon + 3)}{2\,\upsilon + 3}, \upsilon < 1 - \sqrt{3}, \epsilon < \sigma \right\},$$

$$S_{17} = \left\{ -0.3820880724 < \upsilon, \sigma < -\epsilon, \upsilon < -0.2928932190, -2.0\,\upsilon^2 - 12.0 < \sigma \right\},$$

$$S_{18} = \left\{ \sigma < \frac{-2\,(2\,\upsilon^3 + 9\,\upsilon^2 + 12\,\upsilon + 3)}{2\,\upsilon + 3}, \upsilon < -1, \epsilon < \sigma, \frac{1}{2} - \frac{\sqrt{11}}{2} < \upsilon \right\},$$

$$S_{19} = \left\{ -0.3820880724 < \upsilon, \sigma < -\frac{-2\left(2\upsilon^3 + 9\upsilon^2 + 12\upsilon + 3\right)}{2\upsilon + 3}, \right.$$

$$\left. \upsilon < -0.2928932190, \epsilon < \sigma \right\},$$

with

$$\epsilon = \frac{-32\upsilon^4 - 96\upsilon^3 - 174\upsilon^2 + 3\sqrt{32\upsilon^4 + 56\upsilon^3 + 65\upsilon^2 + 42\upsilon + 9} - 123\upsilon - 27}{16\upsilon^2 + 24\upsilon + 9}.$$

Let now $s_0 = x_0 + i\omega_0 \in \mathbb{R}_+ + i\,\mathbb{R}_+$ be a root of $\tilde{\Delta}(\lambda) = P_0(\lambda) + P_1(\lambda)\,e^{-\lambda}$ as defined in (28), where $P_0(\lambda) = \sum_{i=0}^{4} \alpha_i \lambda^i$, $P_1(\lambda) = \sum_{i=0}^{3} \beta i\, \lambda^i$, and s_0 satisfy the following equality $|P_0(x_0 + i\omega_0)|^2 e^{2x_0} = |P_1(x_0 + i\omega_0)|^2$. Since $e^{2x} > 1$, the function $F_{\upsilon,\sigma}(x, \omega) = |P_1(x + i\omega)|^2 - |P_0(x + i\omega)|^2$ satisfies $F_{\upsilon,\sigma}(x_0, \omega_0) > 0$. Moreover, the zeros of $F_{\upsilon,\sigma}$ can be characterized by a polynomial of degree 4 in $\Omega = \omega^2$:

$$G_{\upsilon,\sigma}(\Omega, x) = -\Omega^4 + \gamma_{3,\upsilon,\sigma}(x)\,\Omega^3 + \gamma_{2,\upsilon,\sigma}(x)\,\Omega^2 + \gamma_{1,\upsilon,\sigma}(x)\,\Omega + \gamma_{0,\upsilon,\sigma}(x), \tag{38}$$

where $\gamma_{i,\upsilon,\sigma}$, $i = 0, \ldots, 3$, are functions in x parameterized in υ and σ

$$\begin{aligned}
\gamma_{3,\upsilon,\sigma}(x) = & -4x^2 - 8\upsilon x + \frac{4\upsilon^6}{9} + 8\upsilon^5 + \frac{(4\sigma + 468)\upsilon^4}{9} + \frac{(48\sigma + 1344)\upsilon^3}{9} \\
& + \frac{\left(\sigma^2 + 180\sigma + 1692\right)\upsilon^2}{9} \\
& + \frac{\left(6\sigma^2 + 240\sigma + 864\right)\upsilon}{9} + \sigma^2 + 10\sigma + 16,
\end{aligned}$$

$$\begin{aligned}
\gamma_{2,\upsilon,\sigma}(x) = & -6x^4 - 24\upsilon x^3 + \left(\frac{4\upsilon^6}{3} + 24\upsilon^5 - 6\left(-26 - \frac{2\sigma}{9}\right)\upsilon^4 - 6\left(-\frac{224}{3} - \frac{8\sigma}{3}\right)\upsilon^3\right. \\
& \left. - 6\left(-10\sigma - \frac{\sigma^2}{18} - 90\right)\upsilon^2 - 6\left(-\frac{40\sigma}{3} - \frac{\sigma^2}{3} - 48\right)\upsilon + 48 + 26\sigma + 3\sigma^2\right)x^2 \\
& + \left(\frac{16\upsilon^6}{3} + 88\upsilon^5 - 6\left(-\frac{8\sigma}{9} - \frac{260}{3}\right)\upsilon^4\right. \\
& - 6\left(-\frac{88\sigma}{9} - 220\right)\upsilon^3 - 6\left(-\frac{2\sigma^2}{9} - \frac{100\sigma}{3} - 256\right)\upsilon^2 \\
& \left. - 6\left(-112 - \frac{11\sigma^2}{9} - \frac{122\sigma}{3}\right)\upsilon + 96 + 10\sigma^2 + 64\sigma\right)x,
\end{aligned}$$

$$\begin{aligned}
\gamma_{1,\upsilon,\sigma}(x) = & -4x^6 - 24\upsilon x^5 + \left(\frac{4\upsilon^6}{3} + 24\upsilon^5 - 4\left(-39 - \frac{\sigma}{3}\right)\upsilon^4 - 4\left(-112 - 4\sigma\right)\upsilon^3\right. \\
& - 4\left(-129 - \frac{\sigma^2}{12}\right.
\end{aligned}$$

$$- 15\,\sigma)\upsilon^2 - 4\left(-72 - \frac{\sigma^2}{2} - 20\,\sigma\right)\upsilon + 48 + 3\,\sigma^2 + 22\,\sigma\Big)x^4$$

$$+\left(\frac{32\,\upsilon^6}{3} + 176\,\upsilon^5 - 4\left(-260 - \frac{8\,\sigma}{3}\right)\upsilon^4 - 4\left(-652 - \frac{88\,\sigma}{3}\right)\upsilon^3\right.$$

$$- 4\left(-768 - \frac{2\,\sigma^2}{3} - 100\,\sigma\right)\upsilon^2$$

$$\left.- 4\left(-336 - \frac{11\,\sigma^2}{3} - 118\,\sigma\right)\upsilon + 20\,\sigma^2 + 128\,\sigma + 192\right)x^3$$

$$+\Big(32\,\upsilon^6 + 480\,\upsilon^5 - 4\,(-624 - 8\,\sigma)\,\upsilon^4$$

$$- 4\,(-1440 - 80\,\sigma)\,\upsilon^3 - 4\left(-2\,\sigma^2 - 240\,\sigma - 1296\right)\upsilon^2$$

$$- 4\left(-10\,\sigma^2 - 240\,\sigma - 288\right)\upsilon + 48\,\sigma^2 + 96\,\sigma\Big)x^2$$

$$+\Big(32\,\upsilon^6 + 384\,\upsilon^5 - 4\,(-432 - 8\,\sigma)\,\upsilon^4 - 4\,(-576 - 64\,\sigma)\,\upsilon^3$$

$$- 4\left(-2\,\sigma^2 - 144\,\sigma - 144\right)\upsilon^2$$

$$- 4\left(-8\,\sigma^2 - 96\,\sigma\right)\upsilon + 48\,\sigma^2 + 96\,\sigma\Big)x,$$

$$\gamma_{0,\upsilon,\sigma}(x) = - x^8 + \left(- \left(\frac{\sigma}{3} + 16\right)\upsilon - \left(-8 - \frac{\sigma}{3}\right)\upsilon\right)x^7 + \Big(- 12\,\upsilon^2 - (2\,\sigma + 48)\,\upsilon$$

$$- 2\,\sigma - \Big(\frac{2\,\upsilon^3}{3} + 6\,\upsilon^2 + \Big(\frac{\sigma}{3}$$

$$+16)\,\upsilon + \sigma + 4)\left(-\frac{2\,\upsilon^3}{3} - 6\,\upsilon^2 + \left(-8 - \frac{\sigma}{3}\right)\upsilon - 4 - \sigma\right) - (-2\,\sigma - 48)\,\upsilon\Big)x^6$$

$$+\Big(- 8\,\upsilon^3 - (8\,\sigma + 96)\,\upsilon - \left(4\,\upsilon^3 + 36\,\upsilon^2 + (2\,\sigma + 48)\,\upsilon + 6\,\sigma + 12\right)$$

$$\times\Big(-\frac{2\,\upsilon^3}{3} - 6\,\upsilon^2 + \left(-8 - \frac{\sigma}{3}\right)\upsilon$$

$$-4 - \sigma) - \left(\frac{2\,\upsilon^3}{3} + 6\,\upsilon^2 + \left(\frac{\sigma}{3} + 16\right)\upsilon + \sigma + 4\right)$$

$$\times\left(-4\,\upsilon^3 - 24\,\upsilon^2 + (-2\,\sigma - 48)\,\upsilon - 4\,\sigma - 12\right) -$$

$$(-4\,\sigma - 96)\,\upsilon\Big)x^5 + \Big(- \left(16\,\upsilon^3 + 72\,\upsilon^2 + (8\,\sigma + 96)\,\upsilon + 12\,\sigma + 24\right)$$

$$\times\Big(-\frac{2\,\upsilon^3}{3} - 6\,\upsilon^2 + \left(-8 - \frac{\sigma}{3}\right)\upsilon$$

$$-4-\sigma)-\Big(4\,\upsilon^3+36\,\upsilon^2+(2\,\sigma+48)\,\upsilon+6\,\sigma+12\Big)$$
$$\times\Big(-4\,\upsilon^3-24\,\upsilon^2+(-2\,\sigma-48)\,\upsilon-4\,\sigma-12\Big)$$
$$-\left(\frac{2\,\upsilon^3}{3}+6\,\upsilon^2+\left(\frac{\sigma}{3}+16\right)\upsilon+\sigma+4\right)$$
$$\times\Big(-8\,\upsilon^3-72\,\upsilon^2+(-4\,\sigma-96)\,\upsilon-12\,\sigma-24\Big)+32\,\upsilon^3$$
$$+144\,\upsilon^2-(-16\,\sigma-192)\,\upsilon+24\,\sigma+48\Bigg)x^4$$
$$+\Bigg(-\Big(16\,\upsilon^3+72\,\upsilon^2+(8\,\sigma+96)\,\upsilon+12\,\sigma+24\Big)\Big(-4\,\upsilon^3$$
$$-24\,\upsilon^2+(-2\,\sigma-48)\,\upsilon-4\,\sigma-12\Big)$$
$$-\Big(4\,\upsilon^3+36\,\upsilon^2+(2\,\sigma+48)\,\upsilon+6\,\sigma+12\Big)$$
$$\times\Big(-8\,\upsilon^3-72\,\upsilon^2+(-4\,\sigma-96)\,\upsilon-12\,\sigma-24\Big)$$
$$-\left(\frac{2\,\upsilon^3}{3}+6\,\upsilon^2+\left(\frac{\sigma}{3}+16\right)\upsilon+\sigma+4\right)$$
$$\times\Big(-32\,\upsilon^3-144\,\upsilon^2+(-16\,\sigma-192)\,\upsilon-24\,\sigma-48\Big)\Bigg)x^3$$
$$+\Bigg(-\Big(16\,\upsilon^3+72\,\upsilon^2+(8\,\sigma+96)\,\upsilon+12\,\sigma+24\Big)$$
$$\times\Big(-8\,\upsilon^3-72\,\upsilon^2+(-4\,\sigma-96)\,\upsilon-12\,\sigma-24\Big)$$
$$-\Big(4\,\upsilon^3+36\,\upsilon^2+(2\,\sigma+48)\,\upsilon+6\,\sigma+12\Big)$$
$$\times\Big(-32\,\upsilon^3-144\,\upsilon^2+(-16\,\sigma-192)\,\upsilon-24\,\sigma-48\Big)\Bigg)x^2$$
$$-\Big(16\,\upsilon^3+72\,\upsilon^2+(8\,\sigma+96)\,\upsilon+12\,\sigma+24\Big)$$
$$\times\Big(-32\,\upsilon^3-144\,\upsilon^2+(-16\,\sigma-192)\,\upsilon-24\,\sigma-48\Big)x.$$

Let $(\upsilon,\sigma)\in S_{14}$. Since $\upsilon<-3/4$, $\sigma<-2\left(2\,\upsilon^3+9\,\upsilon^2+12\,\upsilon+3\right)/(2\,\upsilon+3)$, and $x>0$, this latter polynomial of degree 4 can be upper bounded with respect to υ and σ by the parameter-free expression

$$G(\Omega,x)=-\,\Omega^4+\frac{1}{9}\left(\frac{1449}{64}+54\,x-36\,x^2\right)\Omega^3$$
$$+\frac{x}{3}\left(\frac{75}{8}-\frac{2967\,x}{64}+54\,x^2-18\,x^3\right)\Omega^2+\frac{x}{3}\left(\frac{1323}{8}-\frac{1575\,x}{8}\right.$$

$$+\frac{579\,x^2}{4}-\frac{7383\,x^3}{64}+54\,x^4-12\,x^5\Big)\,\Omega+\frac{441\,x^3}{8}-\frac{525\,x^4}{8}$$
$$+\frac{361\,x^5}{8}-\frac{1311\,x^6}{64}+6\,x^7-x^8, \tag{39}$$

which admits two real roots, the greatest of which reaches a maximum value at $x^* \approx 0.7627783180$. Thus, $\Omega \approx 5.290134760 < \pi^2$, i.e., $\omega_0 < \pi$. Notice that the region denoted by $\hat{R}$ in the statement of the proposition is equal to S_{14} in the proof.

Dominancy

In principle, one can investigate the dominancy of a given spectral value using the standard Principle argument as recalled in Section 3 or equivalently using the Stepan–Hassard approach. However, here the proof of the dominance is based on a contradiction. Consider $(\upsilon, \sigma) \in \hat{R}$, and assume that there exists $s_0 = x_0 + i\omega_0 \in \mathbb{C}$ a nonzero root of $\hat{\Delta}$ satisfying $\Re(s_0) > 0$. Then, we infer from the imaginary part of (29) that $\int_0^1 q_{\upsilon,\sigma}(t)\sin(t\,\omega_0)\,\mathrm{e}^{-t\,x_0}\,\mathrm{d}t = 0$. Since $\omega_0 < \pi$ from Lemma 1, the function $t \mapsto q_{\upsilon,\sigma}(t)\sin(t\,\omega_0)$ is strictly positive in $(0, 1)$, which contradicts the latter vanishing integral as required to end the dominancy proof.

5 Conclusion

This contribution is intended to be a didactical chapter on the use of the MID-based partial poles placement (PPP) in concrete applications. Namely, it dealt with the modeling of the CNS action on the human balance. Since the PPP paradigm requires computations which often appear to be of complex nature, particularly when some of the system's parameters are left free, we privileged the use of the Computer Algebra System `Maple`. Indeed, `Maple` offers an efficient kernel for effective computations of Gröbner bases. Also, the powerful `CellDecomposition` algorithm in CAD routine is, in our opinion, one of the most efficient symbolic/numeric tools for "solving" semi-algebraic systems depending on parameters. As such, this chapter acts as a bridge between the PPP-based control design methodology, the modeling of sensorial human balance, and the use of symbolic/numeric algorithms provided in the computer algebra system `Maple`.

References

1. A. Akritas, *Elements of Computer Algebra with Applications* (Wiley, Hoboken, 1989)
2. S. Amrane, F. Bedouhene, I. Boussaada, S.-I. Niculescu, On qualitative properties of low-degree quasipolynomials: further remarks on the spectral abscissa and rightmost-roots assignment. Bull. Math. Soc. Sci. Math. Roumanie **61**(109), 361–381 (2018)

3. Y. Asai, Y. Tasaka, K. Nomura, T. Nomura, M. Casadio, P. Morasso, A model of postural control in quiet standing: robust compensation of delay-induced instability using intermittent activation of feedback control. PLoS One **4**, e6169 (2009)
4. T. Balogh, I. Boussaada, T. Insperger, S.-I. Niculescu, Conditions for stabilizability of time-delay systems with real-rooted plant. Int. J. Robust Nonlinear Control **32**, 3206–3224 (2022)
5. M. Bardet, I. Boussaada, A. Chouikha, J.-M. Strelcyn, Isochronicity conditions for some planar polynomial systems {II}. Bulletin des Sciences Mathématiques **135**, 230–249 (2011)
6. F. Bedouhene, I. Boussaada, S.-I. Niculescu, Real spectral values coexistence and their effect on the stability of time-delay systems: Vandermonde matrices and exponential decay. C. R. Math. Acad. Sci. Paris **358**, 1011–1032 (2020)
7. A. Benarab, I. Boussaada, K. Trabelsi, C. Bonnet, Multiplicity-induced-dominancy property for second-order neutral differential equations with application in oscillation damping. Eur. J. Control **69** (2022)
8. A. Benarab, C.A. Molnar, I. Boussaada, K. Trabelsi, T. Insperger, S.-I. Niculescu, Rolling balance board robust stabilization: A MID-based design, in *TDS 2022 - 17th IFAC Workshop on Time Delay Systems*, Montreal (2022)
9. I. Boussaada, S.-I. Niculescu, Characterizing the codimension of zero singularities for time-delay systems. Acta Applicandae Mathematicae **145**, 47–88 (2016)
10. I. Boussaada, S.-I. Niculescu, On the dominancy of multiple spectral values for time-delay systems with applications. IFAC-PapersOnLine **51**, 55–60 (2018)
11. I. Boussaada, A. Chouikha, J.-M. Strelcyn, Isochronicity conditions for some planar polynomial systems. Bulletin des Sciences Mathématiques **135**, 89–112 (2011)
12. I. Boussaada, S. Tliba, S.-I. Niculescu, H. Ünal, T. Vyhlídal, Further remarks on the effect of multiple spectral values on the dynamics of time-delay systems. application to the control of a mechanical system. Linear Algebra Appl. **542**, 589–604 (2018)
13. I. Boussaada, G. Mazanti, S.-I. Niculescu, J. Huynh, F. Sim, M. Thomas, Partial pole placement via delay action: A python software for delayed feedback stabilizing design, in *2020 24th International Conference on System Theory, Control and Computing (ICSTCC)* (2020), pp. 196–201
14. I. Boussaada, S.-I. Niculescu, A. El-Ati, R. Pérez-Ramos, K. Trabelsi, Multiplicity-induced-dominancy in parametric second-order delay differential equations: analysis and application in control design. ESAIM Control Optim. Calc. Var. **26**, Paper No. 57 (2020)
15. I. Boussaada, G. Mazanti, S.-I. Niculescu, A. Leclerc, J. Raj, M. Perraudin, New features of P3delta software: partial pole placement via delay action. IFAC-PapersOnLine **54**, 215–221 (2021). 16th IFAC Workshop on Time Delay Systems TDS 2021
16. I. Boussaada, G. Mazanti, S.-I. Niculescu, The generic multiplicity-induced-dominancy property from retarded to neutral delay-differential equations: when delay-systems characteristics meet the zeros of Kummer functions. C. R. Math. Acad. Sci. Paris **360**, 349–369 (2022)
17. Y.M. Bouzidi, A. Quadrat, F. Rouillier, Certified non-conservative tests for the structural stability of discrete multidimensional systems. Multidimen. Syst. Signal Process. **30**, 31 (2019)
18. G. Buza, J. Milton, L. Bencsik, T. Insperger, Establishing metrics and control laws for the learning process: ball and beam balancing. Biol. Cybern. **114**, 1–11 (2020)
19. J.L. Cabrera, J.G. Milton, On-off intermittency in a human balancing task. Phys. Rev. Lett. **89**, 158702 (2002)
20. J.J. Castillo-Zamora, I. Boussaada, A. Benarab, J. Escareno, Time-delay control of quadrotor unmanned aerial vehicles: a multiplicity-induced-dominancy-based approach. J. Vibr. Control **29**, 2593–2608 (2023)
21. J.R. Chagdes, S. Rietdyk, M.H. Jeffrey, N.Z. Howard, A. Raman, Dynamic stability of a human standing on a balance board. J. Biomech. **46**, 2593–2602 (2013)
22. E. Chumacero-Polanco, J. Yang, Basin of attraction and limit cycle oscillation amplitude of an ankle-hip model of balance on a balance board. J. Biomechan. Eng. **141**, 111007 (2019)
23. D. Cox, J. Little, D. O'Shea, *Ideals, Varieties, and Algorithms. An Introduction to Computational Algebraic Geometry and Commutative Algebra* (Springer, Berlin, 2007)

24. D.R. Cruise, J.R. Chagdes, J.J. Liddy, S. Rietdyk, J.M. Haddad, H.N. Zelaznik, A. Raman, An active balance board system with real-time control of stiffness and time-delay to assess mechanisms of postural stability. J. Biomechan. **60**, 48–56 (2017)
25. C.W. Eurich, J.G. Milton, Noise-induced transitions in human postural sway. Phys. Rev. E **54**, 6681 (1996)
26. J.-C. Faugère, *FGb SALSA Software*
27. P. Gawthrop, I. Loram, H. Gollee, M. Lakie, Intermittent control models of human standing: similarities and differences. Biol. Cybern. **108**, 159–168 (2014)
28. A.D. Goodworth, R. J. Peterka, Influence of stance width on frontal plane postural dynamics and coordination in human balance control. J. Neurophysiol. **104**, 1103–1118 (2010)
29. B. Hassard, Counting roots of the characteristic equation for linear delay-differential systems. J. Differ. Equ. **136**, 222–235 (1997)
30. F.B. Horak, Clinical measurement of postural control in adults. Phys. Therapy **67**, 1881–1885 (1987)
31. T. Insperger, J. Milton, Sensory uncertainty and stick balancing at the fingertip. Biol. Cybern. **108**, 85–101 (2014)
32. T. Kiemel, Y. Zhang, J.J. Jeka, Identification of neural feedback for upright stance in humans: stabilization rather than sway minimization. J. Neurosci. **31**, 15144–15153 (2011)
33. P. Kowalczyk, P. Glendinning, M. Brown, G. Medrano-Cerda, H. Dallali, J. Shapiro, Modelling human balance using switched systems with linear feedback control. J. R. Soc. Inter. **9**, 234–245 (2012)
34. D. Lazard, Computing with parameterized varieties, in *Algebraic Geometry and Geometric Modeling* (Springer, Berlin, 2006), pp. 53–69
35. D. Lazard, F. Rouillier, Solving parametric polynomial systems. J. Symb. Comput. **42**, 636–667 (2007)
36. D. Lehotzky, Numerical methods for the stability and stabilizability analysis of delayed dynamical systems, PhD Thesis, Budapest University of Technology and Economics, 2016
37. D. Lehotzky, I. Tamás, Emberi egyensúlyozás mechanikai modellezése PIDA szabályozó segítségével. Biomechanica Hungarica **7**, 24–33 (2014)
38. I.D. Loram, M. Lakie, Direct measurement of human ankle stiffness during quiet standing: the intrinsic mechanical stiffness is insufficient for stability. J. Physiol. **545**, 1041–1053 (2002)
39. D. Ma, I. Boussaada, J. Chen, C. Bonnet, S.-I. Niculescu, J. Chen, PID control design for first-order delay systems via MID pole placement: Performance vs. robustness. Automatica **137**, 110102 (2022)
40. C. Maurer, R.J. Peterka, A new interpretation of spontaneous sway measures based on a simple model of human postural control. J. Neurophysiol. **93**, 189–200 (2005)
41. G. Mazanti, I. Boussaada, S.-I. Niculescu, T. Vyhlídal, Spectral dominance of complex roots for single-delay linear equations. IFAC-PapersOnLine **53**, 4357–4362 (2020). 21st IFAC World Congress
42. G. Mazanti, I. Boussaada, S.-I. Niculescu, Multiplicity-induced-dominancy for delay-differential equations of retarded type. J. Differ. Equ. **286**, 84–118 (2021)
43. W. Michiels, S.-I. Niculescu, I. Boussaada, A complete characterization of minima of the spectral abscissa and rightmost roots of second-order systems with input delay. IMA J. Math. Control Inf. **40**, 403–428 (2023)
44. J. Milton, J.L. Townsend, M.A. King, T. Ohira, Balancing with positive feedback: the case for discontinuous control. Philosoph. Trans. R. Soc. A Math. Phys. Eng. Sci. **367**, 1181–1193 (2009)
45. J. Milton, R. Meyer, M. Zhvanetsky, S. Ridge, T. Insperger, Control at stability's edge minimizes energetic costs: expert stick balancing. J. R. Soc. Inter. **13**, 20160212 (2016)
46. C.A. Molnar, Human balancing on rolling balance board with adjustable geometry, PhD Thesis, Budapest University of Technology and Economics, 2022
47. C.A. Molnar, T. Balogh, I. Boussaada, T. Insperger, Calculation of the critical delay for the double inverted pendulum. J. Vibr. Control **27**, 356–364 (2021)

48. C.A. Molnar, A. Zelei, T. Insperger, Human balancing on rolling balance board in the frontal plane. IFAC-PapersOnLine **51**, 300–305 (2018)
49. P. Morasso, A. Cherif, J. Zenzeri, Quiet standing: the single inverted pendulum model is not so bad after all. PLOS ONE **14**, 1–20 (2019)
50. G. Moroz, On the real and algebraic decomposition of parametric systems, theses, Université Pierre et Marie Curie - Paris VI, 2008
51. T. Nomura, S. Oshikawa, Y. Suzuki, K. Kiyono, P. Morasso, Modeling human postural sway using an intermittent control and hemodynamic perturbations. Math. Biosci. **245**, 86–95 (2013)
52. J.H. Pasma, T.A. Boonstra, J. van Kordelaar, V.V. Spyropoulou, A.C. Schouten, A sensitivity analysis of an inverted pendulum balance control model. Front. Comput. Neurosci. **11**, 99 (2017)
53. I.J. Pinter, R. van Swigchem, A.J.K. van Soest, L.A. Rozendaal, The dynamics of postural sway cannot be captured using a one-segment inverted pendulum model: a PCA on segment rotations during unperturbed stance. J. Neurophysiol. **100**, 3197–3208 (2008)
54. G. Pólya, G. Szegő, *Problems and Theorems in Analysis, Vol. I: Series, Integral Calculus, Theory of Functions* (Springer, Berlin, 1972)
55. V. Romanovski, D. Shafer, *The Center and Cyclicity Problems: A Computational Algebra Approach* (Birkhäuser Boston, Boston, 2009)
56. F. Rouillier, Solving zero-dimensional systems through the rational univariate representation. Appl. Algebra Eng. Commun. Comput. **9**, 433–461 (1999)
57. T. Schmoderer, I. Boussaada, S.-I. Niculescu, On Boundary Control of the Transport Equation. Assigning Real Spectra and Exponential Decay. IEEE Control Syst. Lett. **7**, 3591–3596 (2023). https://doi.org/10.1109/LCSYS.2023.3337591
58. B. Varszegi, D. Takacs, G. Stepan, S.J. Hogan, Stabilizing skateboard speed-wobble with reflex delay. J. R. Soc. Inter. **13**, 20160345 (2016)
59. R.R. Zana, A. Zelei, Introduction of a complex reaction time tester instrument. Periodica Polytechnica Mechan. Eng. **64**, 20–30 (2020)
60. L. Zhang, G. Stepan, T. Insperger, Saturation limits the contribution of acceleration feedback to balancing against reaction delay. J. R. Soc. Inter. **15**, 20170771 (2018)

Index

K. Ammari et al. (eds.), *Advances in Partial Differential Equations and Control*, Trends in Mathematics, https://doi.org/10.1007/978-3-031-62265-6

The manufacturer's authorised representative in the EU is Springer
Nature Customer Service Centre GmbH, Europaplatz 3, 69115 Heidelberg,
Germany. If you have any concerns regarding our products, please
contact ProductSafety@springernature.com

Printed and bound by CPI Group (UK) Ltd, Croydon, CR0 4YY
01/10/2026
02225863-0004